Serious Fun with Flexagons

SOLID MECHANICS AND ITS APPLICATIONS
Volume 164

Aims and Scope of the Series

The fundamental questions arising in mechanics are: *Why?, How?, and How much?* The aim of this series is to provide lucid accounts written by authoritative researchers giving vision and insight in answering these questions on the subject of mechanics as it relates to solids.

The scope of the series covers the entire spectrum of solid mechanics. Thus it includes the foundation of mechanics; variational formulations; computational mechanics; statics, kinematics and dynamics of rigid and elastic bodies: vibrations of solids and structures; dynamical systems and chaos; the theories of elasticity, plasticity and viscoelasticity; composite materials; rods, beams, shells and membranes; structural control and stability; soils, rocks and geomechanics; fracture; tribology; experimental mechanics; biomechanics and machine design.

The median level of presentation is the first year graduate student. Some texts are monographs defining the current state of the field; others are accessible to final year undergraduates; but essentially the emphasis is on readability and clarity.

For other titles published in this series, go to
www.springer.com/series/6557

Les Pook

Serious Fun with Flexagons

A Compendium and Guide

Springer

Les Pook
Department of Mechanical Engineering
University College London
Torrington Place
London WC1E 7JE
UK
les.pook@tesco.net

ISBN 978-94-007-3680-1 e-ISBN 978-90-481-2503-6
DOI 10.1007/978-90-481-2503-6
Springer Dordrecht Heidelberg London New York

Softcover reprint of the hardcover 1st edition 2009

Printed on acid-free paper

Springer is part of Springer Science+Business Media (www.springer.com)

About the Author

Leslie Philip (Les) Pook was born in Middlesex, England in 1935. He obtained a B.Sc. in metallurgy from the University of London in 1956. He started his career at Hawker Siddeley Aviation Ltd., Coventry in 1956. In 1963 he moved to the National Engineering Laboratory, East Kilbride, Glasgow. In 1969, while at the National Engineering Laboratory, he obtained a Ph.D. in mechanical engineering from the University of Strathclyde. Dr. Pook moved to University College London in 1990. He retired formally in 1998 but remained professionally active in the fields of metal fatigue and fracture mechanics, and is affiliated to University College London as a visiting professor. He now has more tine to pursue long standing interests in recreational mathematics, including flexagons, and in horology, especially synchronous electric clocks. He is a Fellow of the Institution of Mechanical Engineers and a Fellow of the Institute of Materials, Minerals and Mining. Les married his wife Ann in 1960. They have a daughter, Stephanie, and a son, Adrian.

Preface

Flexagons are rings of hinged polygons that have the intriguing property of displaying different pairs of faces when they are flexed. Workable paper models of flexagons are easy to make and entertaining to manipulate. Flexagons have a surprisingly complex mathematical structure, and just how a flexagon works is not obvious on casual examination of a paper model. The aesthetic appeal of flexagons is in their dynamic behaviour rather than the static appeal of, say, polyhedra. One of the attractions of flexagons is that it is possible to explore their dynamic properties experimentally as well as theoretically. Flexagons may be appreciated at three different levels. Firstly as toys or puzzles, secondly as a recreational mathematics topic, and finally as a subject of serious mathematical study.

My book *Flexagons Inside Out* was published in 2003 by Cambridge University Press. Since then there has been an upsurge in interest in flexagons. Enthusiasts can keep in touch through the *Flexagon Lovers Group*, hosted by Yahoo, and moderated with a light touch by Ann Schwartz. Details of some interesting flexagons have been posted by Group members, and I have enjoyed some stimulating exchanges with other members of the Group. The amount of new information available means that *Flexagons Inside Out* is now outdated. Further geometric analysis has also lead to a much better understanding of the behaviour of flexagons, and has in turn led to the discovery of previously unknown flexagons, some of them with entertaining dynamic properties.

Most of the material in the book is new. It is arranged in a logical order appropriate for a textbook on the geometry of flexagons. Extensive cross references are included so that individual chapters do not have to be read in order. Definitions are included in the index so that they can be easily located. It is assumed that the reader already has an interest in flexagons, and has some knowledge of elementary geometry. The book is written so that it can be enjoyed at both the recreational mathematics level, and at the serious mathematics level. In general, detailed proofs are long and tedious, so they are not included. Where there is uncertainty over the accuracy of a conclusion this is made clear in the text. Basic material from *Flexagons Inside Out* is referenced only where needed for clarity but, where appropriate, new material is fully referenced. There are a few errors in *Flexagons Inside Out*, and these are corrected in the present book. In some ways the book is an updated version of the 1962 book length report *Flexagons* by Conrad and Hartline, which is available on the Internet.

A feature of the book is a compendium of over 100 nets for the construction of paper models of some of the more interesting flexagons. These are reproduced at approximately half full size. Many of the nets have not previously been published. The flexagons have been chosen to complement the text, with particular emphasis on demonstrating relationships between different types of flexagon. Three spectacular examples are included. These are the octopus flexagon, the hexa-dodeca-flexagon and the thrice threefold flexagon. Detailed instructions for assembling and manipulating individual flexagons are included for the benefit of those who wish to enjoy flexagons without going into the mathematics. Photographs of some flexagons are included to assist assembly and manipulation.

Most flexigators who move on from making up flexagons from published nets try folding up promising looking nets to see what happens. This bottom up approach has led to the discovery of some interesting flexagons. The top down approach used in this book makes it possible to analyse and understand the dynamic properties of any flexagon. It is also makes it possible to design flexagons having desired properties. Manipulating paper models of the resulting flexagons often reveals unexpected properties that were not predicted theoretically.

January 2009 Les Pook

I do not know how far it is possible to convey to any one who has not experienced it, the peculiar interest, the peculiar satisfaction that lies in a sustained research when one is not hampered by want of money.

H G Wells, *Tono-Bungay*

Contents

1 Introduction 1

1.1 General Features 1
1.2 Terminology 5
1.3 Outline of Book 9
1.4 Making Flexagons 11
1.4.1 General Assembly Instructions 12
1.4.2 The Two Sector First Order Fundamental Square Even Edge Flexagon 13
References 13

2 Polygon Rings 15

2.1 Introduction 15
2.1.1 Multiple Polygons 16
2.1.2 Combinations 17
2.2 Edge Rings of Regular Polygons 17
2.2.1 General Properties 17
2.2.2 Regular Even Edge Rings 20
2.2.3 Regular Odd Edge Rings 21
2.2.4 Compound Edge Rings 23
2.2.5 Irregular Edge Rings 26
2.3 Edge Rings of Irregular Polygons 27
2.3.1 General Properties 27
2.3.2 Even Edge Rings of Silver and Bronze Triangles 28
2.4 Vertex Rings 29
2.4.1 General Properties 29
2.4.2 Vertex Rings of Squares 30
References 31

3 Fundamental Nets 33

3.1 Introduction 33
3.2 First Order Fundamental Edge Nets 34
3.3 Second Order Fundamental Edge Nets 34
3.4 Fundamental Vertex Nets 36
3.5 Fundamental Silver and Bronze Edge Nets 41
References 41

4 Fundamental Edge Flexagons 43

4.1 Introduction 43
4.1.1 Standard Face Numbering Sequences 44
4.1.2 Truncated Flexagons 44
4.2 First Order Fundamental Even Edge Flexagons 45
4.2.1 General Properties 45
4.2.2 Ring Even Edge Flexagons 50
4.2.3 First Order Fundamental Triangle Even Edge Flexagons 51
4.2.4 First Order Fundamental Square Even Edge Flexagons 53
4.2.5 Detailed Analysis of Flexagons 54
4.2.6 First Order Fundamental Pentagon Even Edge Flexagons 58
4.2.7 First Order Fundamental Hexagon Even Edge Flexagons 62
4.2.8 First Order Fundamental Octagon Even Edge Flexagons 64
4.2.9 First Order Fundamental Dodecagon Even Edge Flexagons 66
4.3 Second Order Fundamental Odd Edge Flexagons 68
4.3.1 General Properties 68
4.3.2 Second Order Fundamental Triangle Odd Edge Flexagons 69
4.3.3 A Second Order Fundamental Square Odd Edge Flexagon 72
4.3.4 A Second Order Fundamental 20-gon Odd Edge Flexagon 74
References 75

5 Fundamental Skeletal and Point Flexagons 77

5.1 Introduction 77
5.2 First Order Fundamental Even Skeletal Flexagons 78
5.2.1 General Properties 78
5.2.2 First Order Fundamental Triangle Even Skeletal Flexagons 79
5.2.3 A First Order Fundamental Square Even Skeletal Flexagon 81

5.3 Fundamental Point Flexagons ... 82
5.3.1 General Properties and Unagons ... 82
5.3.2 The Fundamental Triangle Point Flexagon ... 83
5.3.3 The Fundamental Square Point Flexagon ... 84
5.3.4 Fundamental Pentagon Point Flexagons ... 85
5.3.5 The Fundamental Hexagon Point Flexagon ... 86
5.4 Interleaved Fundamental Point Flexagons ... 87
5.4.1 General Properties ... 87
5.4.2 The Interleaved Fundamental Pentagon Point Flexagon ... 89
5.4.3 An Interleaved Fundamental Enneagon Point Flexagon ... 90
5.5 Augmented Fundamental Point Flexagons ... 91
5.5.1 General Properties ... 91
5.5.2 An Augmented Fundamental Triangle Point Flexagon ... 92
5.5.3 An Augmented Fundamental Square Point Flexagon ... 93
5.6 Augmented Interleaved Fundamental Point Flexagons ... 95
5.6.1 General Properties ... 95
5.6.2 Augmented Interleaved Fundamental Triangle Point Flexagons ... 97
5.6.3 An Augmented Interleaved Fundamental Square Point Flexagon ... 100
References ... 101

6 Fundamental Compound Edge Flexagons ... 103

6.1 Introduction ... 103
6.2 General Properties ... 104
6.3 Triangular Fundamental Compound Edge Flexagons ... 107
6.3.1 Some Properties ... 107
6.3.2 A Fundamental Square Compound Edge Flexagon ... 107
6.3.3 A Fundamental Pentagon Compound Edge Flexagon ... 108
6.4 A Square-Like Fundamental Compound Edge Flexagon ... 109
6.4.1 Some Properties ... 109
6.4.2 A Fundamental Hexagon Compound Edge Flexagon ... 109
6.5 Pentagonal Fundamental Compound Edge Flexagons ... 111
6.5.1 Some Properties ... 111
6.5.2 A Fundamental Square Compound Edge Flexagon ... 111
6.5.3 A Fundamental Hexagon Compound Edge Flexagon ... 112
6.6 A Hexagonal Fundamental Compound Edge Flexagon ... 113
6.6.1 Some Properties ... 113
6.6.2 A Fundamental Octagon Compound Edge Flexagon ... 113
6.7 Heptagonal Fundamental Compound Edge Flexagons ... 115
6.7.1 Some Properties ... 115
6.7.2 A Fundamental Hexagon Compound Edge Flexagon ... 115
6.7.3 A Fundamental Decagon Compound Edge Flexagon ... 117
References ... 118

7 Irregular Cycle Flexagons 119
7.1 Introduction 119
7.2 Irregular Cycle Even Edge Flexagons 121
7.2.1 General Properties 121
7.2.2 Derivation of Nets 122
7.2.3 The Irregular Cycle Square Even Edge Flexagon 123
7.2.4 An Irregular Cycle Pentagon Even Edge Flexagon 126
7.2.5 Irregular Cycle Hexagon Even Edge Flexagons 127
7.3 Irregular Cycle Interleaved Point Flexagons 129
7.3.1 General Properties 129
7.3.2 Interleaf Flexes 131
7.3.3 The Irregular Cycle Interleaved Square Point Flexagon 131
7.3.4 Irregular Cycle Interleaved Pentagon Point Flexagons 133
7.3.5 Irregular Cycle Interleaved Hexagon Point Flexagons 135
7.3.6 Augmented Irregular Cycle Interleaved Triangle Point Flexagons 137
7.4 Distinct Face Numbering Sequences 141
7.5 Irregular Cycle Non Interleaved Point Flexagons 142
7.5.1 General Properties 142
7.5.2 The Irregular Cycle Non Interleaved Square Point Flexagon 143
7.5.3 An Irregular Cycle Non Interleaved Pentagon Point Flexagon 143
References 145

8 Degenerate Flexagons 147
8.1 Introduction 147
8.2 Degenerate Even Edge Flexagons 149
8.2.1 General Properties 149
8.2.2 A Degenerate Square Even Edge Flexagon 151
8.2.3 Degenerate Pentagon Even Edge Flexagons 152
8.2.4 Degenerate Hexagon Even Edge Flexagons 154
8.2.5 Degenerate Octagon Even Edge Flexagons 157
8.2.6 A Degenerate Dodecagon Even Edge Flexagon 159
8.3 Degenerate Non Interleaved Point Flexagons 161
8.3.1 General Properties 161
8.3.2 The Degenerate Non Interleaved Square Point Flexagon 163
8.3.3 Degenerate Non Interleaved Pentagon Point Flexagons 164
8.4 Degenerate Irregular Cycle Interleaved Point Flexagons 165
8.4.1 General Properties 165
8.4.2 Degenerate Interleaved Triangle Point Flexagons 166

8.5 Degenerate Compound Edge Flexagons ... 169
8.5.1 General Properties ... 169
8.5.2 The Degenerate Square-Like Hexagon Compound Edge Flexagon ... 171
8.5.3 Degenerate Pentagonal Compound Edge Flexagons ... 172

9 Irregular Ring Even Edge Flexagons ... 175

9.1 Introduction ... 175
9.1.1 Fundamental Irregular Ring Even Edge Flexagons ... 175
9.2 Irregular Ring Triangle Even Edge Flexagons ... 177
9.2.1 General Properties ... 177
9.2.2 An Irregular Ring 12 Triangle Even Edge Flexagon ... 177
9.2.3 An Irregular Ring Eight Triangle Even Edge Flexagon ... 178
9.2.4 An Irregular Ring 16 Triangle Even Edge Flexagon ... 179
9.3 Irregular Ring Square Even Edge Flexagons ... 179
9.3.1 General Properties ... 179
9.3.2 Irregular Ring Six Square Even Edge Flexagons ... 180
9.3.3 Irregular Ring Eight Square Even Edge Flexagons ... 183
9.4 Irregular Ring Pentagon Even Edge Flexagons ... 186
9.4.1 General Properties ... 186
9.4.2 An Irregular Ring Six Pentagon Even Edge Flexagon ... 187
9.4.3 An Irregular Ring Eight Pentagon Even Edge Flexagon ... 190
9.5 Irregular Ring Hexagon Even Edge Flexagons ... 191
9.5.1 General Properties ... 191
9.5.2 An Irregular Ring Six Hexagon Even Edge Flexagon ... 192
9.5.3 Irregular Ring Eight Hexagon Even Edge Flexagons ... 194
9.6 Irregular Ring Dodecagon Even Edge Flexagons ... 197
9.6.1 General Properties ... 197
9.6.2 Irregular Ring Eight Dodecagon Even Edge Flexagons ... 198

10 Irregular Polygon Edge Flexagons ... 201

10.1 Introduction ... 201
10.1.1 Transformation of Polygons ... 202
10.1.2 Stretch Flexagons and Stretch Polygon Rings ... 203
10.2 Irregular Triangle Edge Flexagons ... 203
10.2.1 Irregular Triangles ... 203
10.2.2 Isosceles Triangle Edge Rings ... 204
10.2.3 Bronze Even Edge Rings ... 206
10.2.4 Isosceles Triangle Even Edge Flexagons ... 207
10.2.5 Isosceles Triangle Odd Edge Flexagons ... 214
10.2.6 Scalene Triangle Even Edge Flexagons ... 216
10.2.7 A Partial Overlap Bronze Even Edge Flexagon ... 228
10.2.8 A Scalene Triangle Even Edge Flexagon ... 229

10.3 Irregular Quadrilateral Even Edge Flexagons 231
10.3.1 Irregular Quadrilaterals 231
10.3.2 Irregular Quadrilateral Even Edge Rings 232
10.3.3 A Rectangle Even Edge Flexagon 233
10.3.4 Rhombus Even Edge Flexagons 235
10.3.5 A Trapezium Even Edge Flexagon 239
10.4 Irregular Pentagon Even Edge Flexagons 240
10.4.1 Irregular Pentagons 240
10.4.2 Irregular Pentagon Even Edge Rings 241
10.4.3 An Equiangular Irregular Pentagon Even Edge Flexagon 242
10.4.4 An Irregular Ring Eight Irregular Pentagon Even Edge Flexagon 244
References 244

11 Complex Flexagons 247

11.1 Introduction 247
11.2 Linked Even Edge Flexagons 249
11.2.1 Methods of Linking 249
11.2.2 Linked Hexaflexagons 251
11.2.3 Linked Square Even Edge Flexagons 255
11.2.4 Linked Pentagon Even Edge Flexagons 262
11.2.5 Linked Silver Even Edge Flexagons 265
11.2.6 Linked Bronze Even Edge Flexagons 271
11.3 Linked Point Flexagons 277
11.3.1 Methods of Linking 277
11.3.2 Linked Triangle Point Flexagons 278
11.3.3 Linked Square Point Flexagons 279
11.4 Conjoined Point Flexagons 282
11.4.1 General Properties 282
11.4.2 Conjoined Triangle Point Flexagons 283
11.4.3 A Conjoined Pentagon Point Flexagon 284
11.5 Bundled Odd Edge Flexagons 284
11.5.1 General Properties 284
11.5.2 A Five Sector Bundled Triangle Odd Edge Flexagon 286
11.5.3 The Seven and Six Flexagon 287
11.6 Slipagons 288
11.6.1 General Properties 288
11.6.2 A Trihexaflexagon Slipagon 288
11.6.3 A Partial Overlap Silver Even Edge Slipagon 290
11.7 Coupled Point Flexagons 292
11.7.1 General Properties 292
11.7.2 A Coupled Triangle Point Flexagon 295
11.7.3 A Coupled Square Point Flexagon 296
References 298

12 Miscellaneous Flexagons 299

12.1 Introduction 299
12.2 Three Sector Odd Flexagons 300
12.2.1 General Properties 300
12.2.2 The Three Sector Fundamental Dodecagon Odd Edge Flexagon 300
12.2.3 A Three Sector Isosceles Triangle Odd Edge Flexagon 301
12.2.4 A Three Sector Hexagon Odd Skeletal Flexagon 302
12.3 Degenerate Odd Edge Flexagons 304
12.3.1 General Properties 304
12.3.2 A Degenerate Square Odd Edge Flexagon 304
12.4 Alternating Odd Edge Flexagons 305
12.4.1 General Properties 305
12.4.2 A Square Alternating Odd Edge Flexagon 306
12.5 Flapagons 308
12.5.1 General Properties 308
12.5.2 The Fundamental Square Duplex Flapagon 308
12.5.3 A Square Flapagon–Flexagon Hybrid 309
12.5.4 The Fundamental Isosceles Triangle Triplex Flapagon 309
12.6 Multiplex Edge Flexagons 310
12.6.1 General Properties 310
12.6.2 A Square Duplex Edge Flexagon 311
12.6.3 The Thrice Threefold Flexagon 313
12.7 A Hooke's Joint Flexagon 321
References 323

Index 325

Chapter 1
Introduction

1.1 General Features

Flexagons are a twentieth century discovery (Gardner 1965, 2008; Pook 2003). Arthur H Stone, a postgraduate student at Princeton University in America, discovered them in 1939 while folding strips of paper. Figure 1.1a is a photograph of a trihexaflexagon, which was the first type of flexagon to be discovered. The black and white photographs of flexagons in this book are nearly all of models made either from coloured card, coloured origami paper, or from origami duo paper, which is differently coloured on its two surfaces. The appearance of some flexagons is shown as a line diagram such as the ring of four squares shown in Fig. 1.1b.

Workable paper (or card) models of flexagons are easy to make and entertaining to manipulate. They have the intriguing property of displaying different pairs of faces, sometimes in cyclic order, when they are flexed. Flexagons have a surprisingly complex mathematical structure, and just how a flexagon works is not obvious on casual examination of a paper model. The aesthetic appeal of flexagons is in their dynamic behaviour rather than the static appeal of, say, polyhedra. One of the attractions of flexagons is that it is possible to explore their dynamic properties experimentally as well as theoretically. Manipulation of paper models often reveals configurations that have not been predicted theoretically.

A flexagon is a motion structure that has an infinity of states (positions). An umbrella is an everyday example of a motion structure. An edge flexagon consists of a band of identical polygons hinged at common edges by edge hinges. The individual polygons in a flexagon, called leaves, are usually identical (congruent) and are usually regular convex polygons. However, some flexagons consist of other types of convex polygons, and leaves are not always identical. If one hinge of a band is disconnected the band can be laid flat and used as a *net* to construct a flexagon. Nets are sometimes called templates or friezes. A band of 8 edge hinged squares that has been cut and laid flat as the net for a square even edge flexagon is shown in Fig. 1.2. Assembly and flexing instructions for this flexagon are given in Section 1.4.2. This particular flexagon is a twisted band. This can be seen by disconnecting a hinge

L. Pook, *Serious Fun with Flexagons*, Solid Mechanics and Its Applications 164,
DOI 10.1007/978-90-481-2503-6_1, © Springer Science+Business Media B.V. 2009

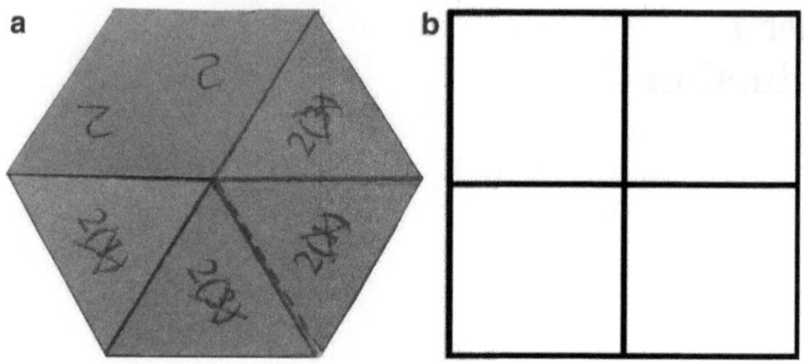

Fig. 1.1 (**a**) A trihexaflexagon as a flat regular even edge ring of six equilateral triangles. (**b**) A flat regular even edge ring of four squares (Les Pook, Flexagons inside out, 2003, © Cambridge University Press 2003, reprinted with permission)

of a paper model, and gently pulling the ends of the band apart. Some edge flexagons are untwisted bands.

A motion structure usually has certain characteristic positions that can conveniently be defined as main positions. For example, a fully opened umbrella is in a main position. A main position of an even edge flexagon is a position that is, in appearance, an untwisted even edge ring of an even number ($2n$) of polygons that are hinged together by edge hinges. Similarly, a main position of an odd edge flexagon is, in appearance, an untwisted odd edge ring of an odd number (n) of polygons that are hinged together. Thus, a main position has two faces. Some flexagons have more than one type of main position. In this book ring refers to the appearance of a flexagon and band to its topological structure. The topological structure of a flexagon is an invariant. This means that the topological structure is always the same no matter what position the flexagon is in.

The trihexaflexagon is an example of an even edge flexagon. A main position is, in appearance, an even edge ring of six equilateral triangles (Fig. 1.1a). The ring is flat and it is a regular edge ring in the sense that all the triangles are the same distance from the centre of the ring. The two sector first order fundamental square even edge flexagon (Fig. 1.2) is, as its name implies, another even edge flexagon. A main position is, in appearance, a flat regular even edge ring of four squares (Fig. 1.1b). The even edge rings shown in Fig. 1.1 are the two possible ways in which an even number of regular convex polygons can be arranged about a point in a plane.

The polygons visible in main positions of a flexagon are called pats. A pat can be either a folded pile of leaves or a single leaf. The pats in a main position of the trihexaflexagon are alternately single leaves and folded piles of two leaves. The pats in a main position of the two sector first order fundamental square even edge flexagon are alternately single leaves and fan folded piles of three leaves. The hinge angle is the angle between the two edge hinges of a polygon, leaf or pat (Fig. 1.3). For example, the hinge angle of the squares shown in Fig. 1.2 is 90°, and the two

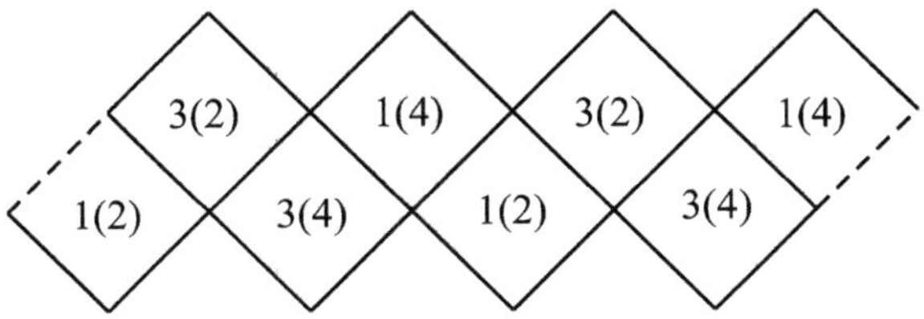

Fig. 1.2 The two sector first order fundamental square even edge flexagon as a band of hinged squares that has been cut and laid flat for use as a net. To assemble the flexagon crease the lines between the squares to form hinges, transfer the number in brackets on the upper face of each square to the reverse face, fold together pairs of squares numbered 3 and 4, and join the ends of the net (dashed lines) using transparent adhesive tape

hinges intersect at a vertex of a square. The two hinges do not always intersect, as shown for a regular hexagon in the figure.

In a flat regular edge ring of regular convex polygons the sum of the hinge angles at the centre of the ring is 360°. Regular edge rings of regular convex polygons, and main positions of flexagons that have the same appearance, are not always flat, and the sum of the hinge angles can be greater or less than 360°. The angle deficit is 360° minus the sum of the hinge angles, and is called the curvature of the ring (Demaine and O'Rourke 2007). In a slant ring the curvature is positive, and in a skew ring it is negative. For example, Fig. 1.4 shows a slant regular odd edge ring of five equilateral triangles. Its curvature is $360° - 5 \times 60° = 60°$. The curvature of the skew regular even edge ring of four regular hexagons shown in Fig. 1.5 is $360° - 4 \times 120° = -120°$.

A vertex flexagon is a band of identical polygons hinged at common vertices by point hinges. Bands can be twisted or untwisted. Point hinges are impossible in a paper model, but short paper strips provide a workable approximation. There are two families of vertex flexagons. These are skeletal flexagons and point flexagons. Skeletal flexagons are not satisfactory as paper models, but are included because of their theoretical interest. Point flexagons are special cases of skeletal flexagons, and are satisfactory as paper models.

A main position of an even skeletal flexagon, is, in appearance, an untwisted even vertex ring of an even number ($2n$) of polygons that are hinged together by point hinges. The point hinges mean that the rings can always be laid flat, and the curvature is indeterminate. A flexagon as a flat regular even vertex ring of four equilateral triangles is shown in Fig. 1.6. Theoretically, vertices of adjacent equilateral triangles coincide, but in the paper model they are separated and connected by narrow strips. This particular ring has an open centre. The two sector first order fundamental even skeletal flexagon is shown collapsed into a twisted band in Fig. 1.7.

A main position of a point flexagon is, in appearance, a polygon vertex pair rather than a vertex ring. The two polygons are connected either by a pair of point hinges or by a single point hinge. This is shown for equilateral triangles in Fig. 1.8.

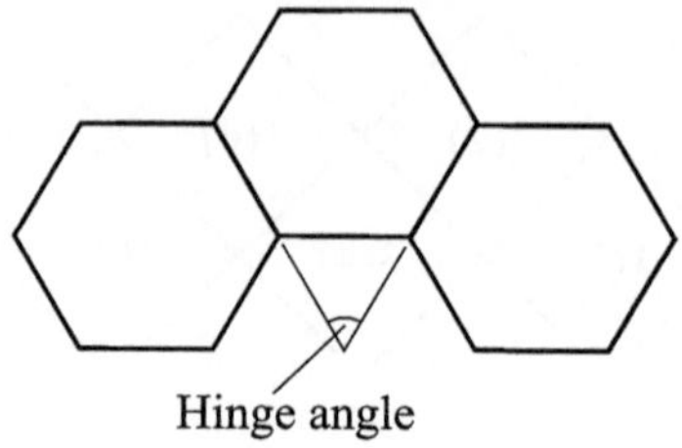

Fig. 1.3 Definition of hinge angle

Fig. 1.4 A flexagon as a slant regular odd edge ring of five equilateral triangles

Fig. 1.5 A flexagon as a skew regular even edge ring of four regular hexagons

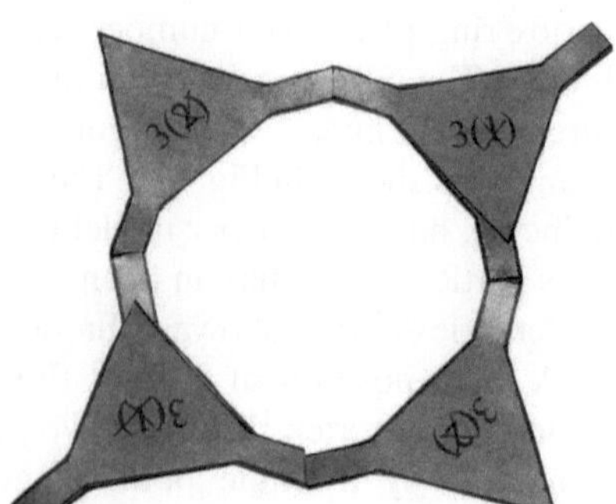

Fig. 1.6 A flexagon as a flat regular even vertex ring of four equilateral triangles. Point hinges approximated by paper strips

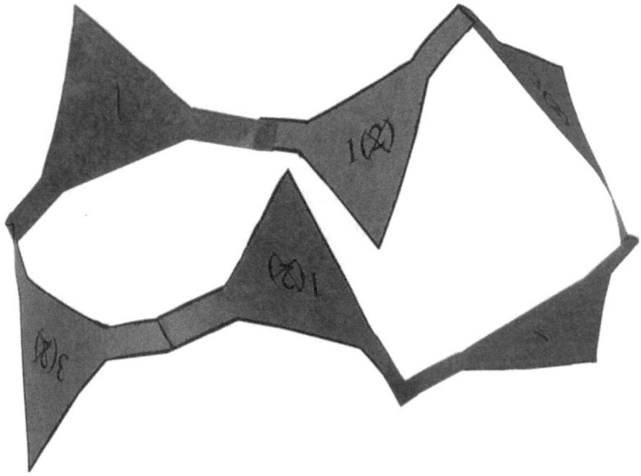

Fig. 1.7 The two sector first order fundamental even skeletal flexagon as an open band

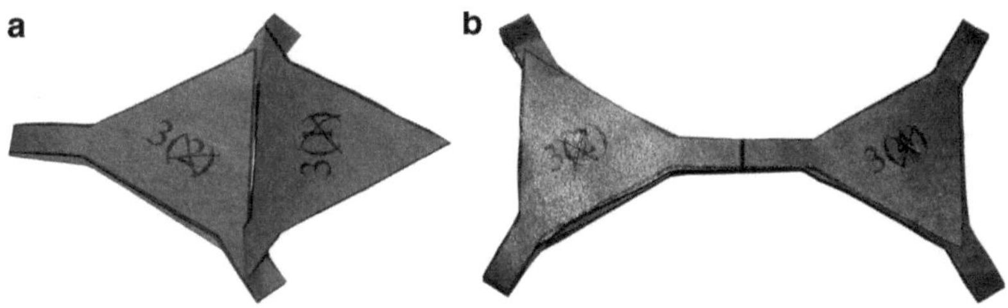

Fig. 1.8 A flexagon as an equilateral triangle vertex pair. Point hinges approximated by paper strips. (**a**) Connected by a pair of point hinges. (**b**) Connected by a single point hinge

Flexagons, in general, exist in infinite series. Usually, only a few members of a series of flexagons are satisfactory as paper models. In this book it is therefore taken as understood that only some early members of a series are being described, for example the first order fundamental even edge flexagons listed in Table 4.1.

1.2 Terminology

Terminology is always a problem in any developing field and appears to be a particular problem with flexagons (Pook 2007). Inevitably, people develop terminology to simplify descriptions of features they are investigating. Equally inevitably, terminologies developed by different people differ, and sometimes conflict. In this book, definitions of descriptive terms and notations are given and indexed when needed, not always when they are first used. Some combinations of descriptive terms are not separately defined or indexed. Terminology has been chosen so as to maintain a balance between clarity and mathematical rigour. As far as is possible usage follows previous practice, but some terms and notations differ from those used in Pook (2003).

The following conventions are used are used when describing flexagons and their characteristics. A family of flexagons is a group of flexagons with some characteristics in common. A characteristic flex for a family of flexagons is a flex that can be used to flex all the members of the family. Most of the descriptions of the dynamic properties of flexagons are based on the use of a characteristic flex. A variety of flexagon is a group of flexagons within a family, all of which have a main position appearance in common. A type of flexagon is a particular flexagon within a variety. All flexagons exist as an enantiomorphic (mirror image) pair. The two members of an enantiomorphic pair are different types of flexagon but are not usually considered to be distinct types of flexagon (Pook 2003). Hence, terminology used does not, in general, distinguish between the two enantiomorphs of a flexagon.

In general, leaves in nets shown are numbered to identify the faces which can be displayed on a flexagon: all the leaves visible on a face of a main position have the same number. Face numbering sequences are arbitrary so different sequences can be applied to the same flexagon. Two face numbering sequences are only regarded as distinct if one cannot be transformed into the other by substitution on a one to one basis. For example, adding 7 to each of the face numbers shown in Fig. 1.2 does not result in a second distinct face numbering sequence. Numbers on nets shown are assigned so as to make descriptions as simple as possible, and also to make it possible to write general assembly instructions (Section 1.4.1). In some special situations faces are also identified using letters, or Roman numerals, or both. Other markings are sometimes added to simplify assembly and flexing.

In geometry, a distinction is often made between an ideal mathematical object and an imperfect physical model of the object. For example, in geometry a line is defined as having zero width, whereas any real line drawn on a piece of paper must have a finite width. Fortunately, paper models of many types of flexagon do approximate closely to a mathematical ideal, and it is not usually necessary to make a distinction between an ideal flexagon and the corresponding paper model.

Mathematically, an idea leaf is a flat polygon that is rigid, of zero thickness, and consists of its one dimensional edges plus its two dimensional interior (Cromwell 1997). Because an ideal leaf is of zero thickness an ideal pat is also of zero thickness. An ideal flexagon consists of a band of ideal leaves that are hinged together by ideal hinges. In an ideal edge hinge the dihedral angle between the two planes containing two hinged leaves may vary between 0°and 360° without constraint. The dihedral angle (Coxeter 1963) is the angle on a section which cuts both planes at 90° (Fig. 1.9). Each leaf is hinged to two other leaves. The angle between the two hinges is the hinge angle (Fig. 1.3). An ideal edge flexagon is an ideal flexagon with ideal leaves which are hinged together by ideal edge hinges. Similarly, an ideal point flexagon is an ideal flexagon with ideal leaves which are hinged together by ideal point hinges.

The concept of an ideal edge flexagon is similar to that of rigid origami in which only a finite number of creases is permitted, between which the paper must stay

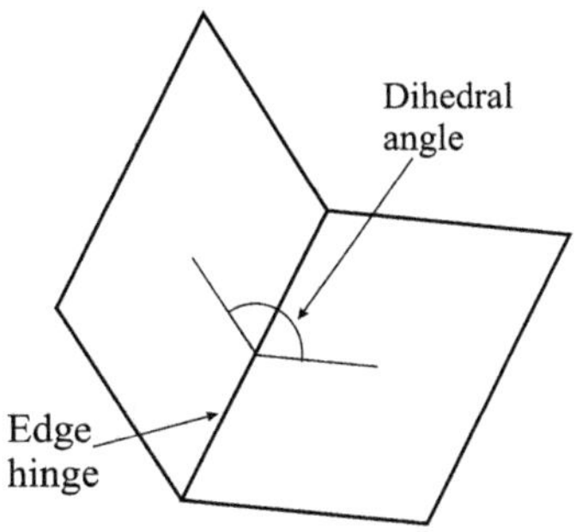

Fig. 1.9 Definition of dihedral angle for an edge hinge

rigid and flat (Demaine and O'Rourke 2007). Leaves are always flat in main positions of edge flexagons, but in some edge flexagons leaves have to be bent in order to flex from one main position to another. In origami terms the leaves are rolled using an infinity of creases. Whether flexes that require leaf bending are legitimate is a matter of taste. A pragmatic approach, used in this book, is that a flex is legitimate provided that it can be carried out in a paper model without too much difficulty. In this approach an ideal flexible leaf is inextensible and of zero thickness, but with some flexibility. Theoretically, the flexibility of paper could be quantified, but in practice this is not helpful.

From a mechanical engineering viewpoint an ideal edge flexagon is a three dimensional linkage (Pook 2003). The formal definition of a linkage (Macmillan 1950) is that it is an assembly of coupled rigid bodies (links) whose freedom of movement is restricted, after the fixture of one link in space, by the constraint imposed by their couplings. Demaine and O'Rourke (2007) give an equivalent definition. The number of degrees of freedom possessed by a linkage is the number of independent parameters needed to completely determine its configuration. For example, two polygons connected by an edge hinge have one degree of freedom in which the dihedral angle changes. In any practical linkage the number of links is finite, so the number of degrees of freedom is also finite. In other words, the links can only follow a finite number of paths relative to each other. In an ideal edge flexagon the links are the rigid leaves, each of which is coupled to two neighbouring leaves by ideal hinges along common edges. An ideal edge flexagon is therefore what is known as a hinged linkage.

Some ideal even edge flexagons, including the hexahexaflexagon (Section 11.2.2) have large numbers of degrees of freedom (Pook 2003), and can therefore by flexed into numerous main positions, in most of which face numbers become mixed up. Possible positions have been investigated in detail for some even edge flexagons, for example by McLean (1979), and by Mitchell (2002). Large numbers of degrees of freedom mean that unwanted positions may occur if a paper model is not flexed correctly. A flexagon that has been accidentally flexed into an unwanted position is said to be muddled. It is easy to get some types badly muddled. It is usually difficult to see how to return an accidentally muddled flexagon to a wanted position.

Avoidance of muddling is the reason why some authors give very detailed instructions on the manipulation of models of some types of even edge flexagon. Such instructions usually include the implicit requirement that rotational symmetry be maintained during flexing. This artificially limits the number of degrees of freedom, and hence unwanted positions are avoided.

Paper models of edge flexagons only approximate to ideal edge flexagons. Paper has finite thickness, and is not rigid. This has two main consequences. Firstly, the finite thickness sometimes makes manipulation of some edge flexagons difficult. Secondly, leaves can be bent during flexing. If this is regarded as permissible then, in some edge flexagons, this extends the range of possible flexes. However, in main positions leaves are always flat. Leaves do not have to be bent while flexing the two sector first order fundamental square even edge flexagon (Fig. 1.2).

Two polygons connected at a common vertex by an ideal point hinge have two degrees of freedom. An ideal point hinge between two polygons is a special case of a Hooke's joint, as used in motor vehicle drivelines (Dunkerley 1910). In an initial position both polygons lie in the same plane, as shown in Fig. 1.10 for two triangles lying in the x-y plane. In one degree of freedom the triangles can be rotated relative to each other about the y-axis. This is equivalent to the degree of freedom of an edge hinge. In the other degree of freedom the triangles can be rotated relative to each other about the y-axis. Possible combinations of the two rotations are restricted by interference between the two triangles. By definition, the triangles cannot be twisted relative to each other. Ideal point flexagons are linkages.

In a compound edge ring alternate polygons are the same distance from the centre of the ring, and alternate hinge angles are the same. By definition, a compound edge ring must be even. A flat compound edge ring of 8 squares is shown in Fig. 1.11. The heavy lines indicate that there is no square in the centre. The curvature of a ring with a hollow centre is calculated in the same way as the curvature of a ring in which the polygons have a common vertex. A more complicated example is the flat compound edge ring of 16 regular octagons shown in Fig. 1.12. The vertices of eight of the hinge angles are on the outside of the ring so these are taken as negative when calculating the curvature.

An irregular edge ring of regular convex polygons is a ring that is neither regular nor compound. For example, Fig. 1.13 shows a flat irregular edge ring of 12 equilateral triangles.

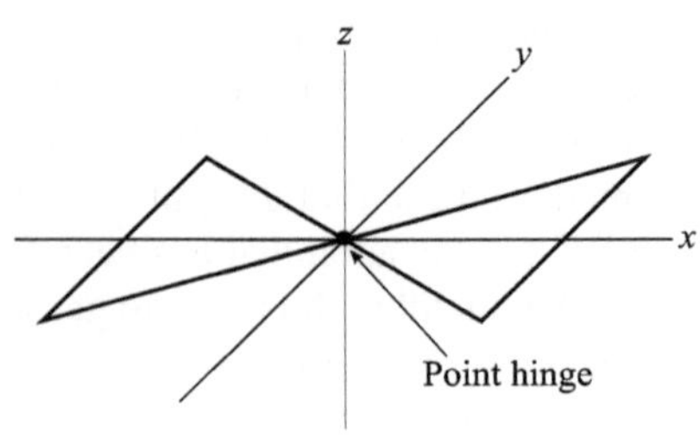

Fig. 1.10 A point hinge connecting two triangles in the x-y plane

Fig. 1.11 A flat compound edge ring of eight squares (Les Pook, Flexagons inside out, 2003, © Cambridge University Press 2003, reprinted with permission)

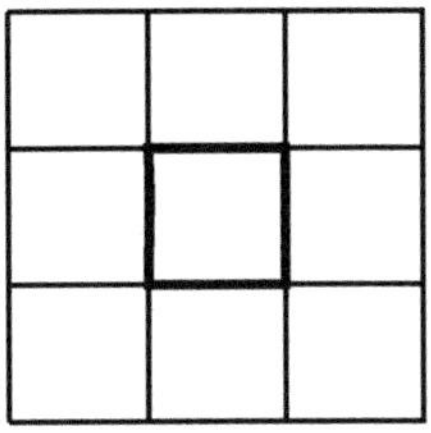

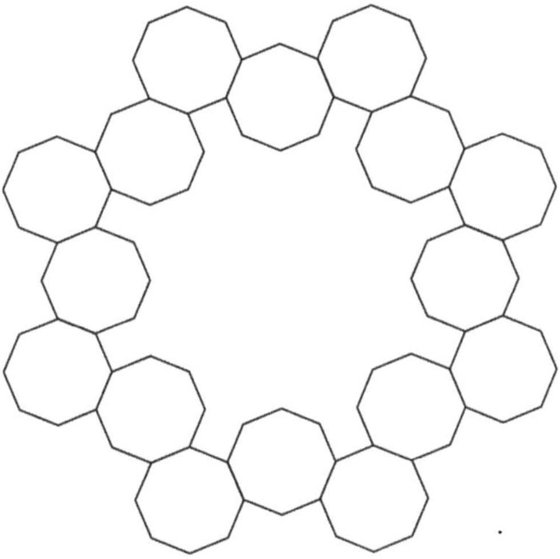

Fig. 1.12 A flat compound edge ring of 16 regular octagons

Fig. 1.13 A flat irregular even edge ring of 12 equilateral triangles

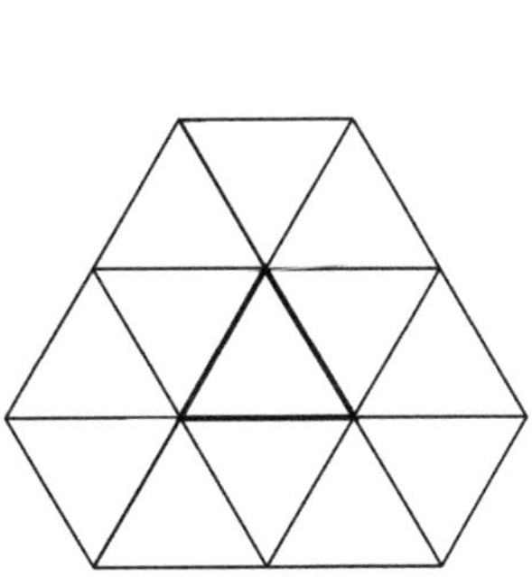

1.3 Outline of Book

There is an infinity of different types of flexagon, so no book on flexagons can be comprehensive. Flexagons whose nets are given in the text were chosen primarily because they are interesting to manipulate. Most are reasonably easy to handle. They have also be chosen to illustrate points made in the text, with particular emphasis on demonstrating relationships between different types of flexagon.

Hinged rings of polygons that have the same appearance as main positions of flexagons are discussed in Chapter 2, including some geometric constraints that restrict permissible rings. Nets used in the construction of flexagons vary widely in appearance, and some are very irregular. However, there are certain fundamental nets, such as the net shown in Fig. 1.2, that have a high degree of symmetry. These fundamental nets are used in the construction of fundamental flexagons, and are described in Chapter 3.

Fundamental edge flexagons, such as the two sector first order fundamental square even edge flexagon (Section 1.4.2) are constructed from fundamental edge nets, and main positions are, in appearance, regular edge rings. Fundamental edge flexagons are described in Chapter 4. Broadly, they are the equivalent of regular polyhedra in that they are constructed from identical regular convex polygons, and have a high degree of symmetry both in structure and in dynamic properties. Fundamental skeletal flexagons and fundamental point flexagons are constructed from fundamental vertex nets, and are described in Chapter 5. Fundamental skeletal flexagons are related to fundamental edge flexagons. Thus, broadly, fundamental skeletal flexagons are also the equivalent of regular polyhedra. Fundamental point flexagons are a special case of fundamental skeletal flexagons. Fundamental compound flexagons are constructed from fundamental edge nets, and are described in Chapter 6. Main positions are, in appearance, compound edge rings, for example the flat compound edge ring of eight squares shown in Fig. 1.11.

In a fundamental flexagon all the main positions that appear as a cycle of main positions is traversed have the same appearance and the same pat structure. However, in an irregular cycle flexagon the pat structure, but not the appearance of main positions, varies as a cycle is traversed. Irregular cycle flexagons are described in Chapter 7.

A precursor flexagon is a flexagon that is modified in some way to form a different type of flexagon. For example, deletion of one or more faces from a precursor flexagon leads to a degenerate flexagon. Most of the flexagons described in Chapters 4–7 can be used as precursors. Degenerate flexagons are described in Chapter 8. A feature of some degenerate flexagons is that they are easier to handle than the precursor flexagons.

Irregular ring even edge flexagons are even edge flexagons with main positions that are, in appearance, irregular even edge rings, for example the flat irregular even edge ring of 12 equilateral triangles shown in Fig. 1.13. Irregular ring fundamental even edge flexagons are made from fundamental edge nets. They are described in Chapter 9, together with degenerate versions.

All the flexagons described in Chapters 4–9 are made from regular convex polygons. However, flexagons can be made from any convex polygon, they are not restricted to regular convex polygons, although only a limited range of irregular shapes results in irregular polygon flexagons whose paper models are reasonably easy to handle. Some of these are described in Chapter 10. Silver flexagons, made from 45°–45°–90° triangles, and bronze flexagons, made from 30°–60°–90° triangles, are of particular interest.

The flexagons described in Chapters 4–10 are all so-called solitary flexagons, for example the two sector first order fundamental square even edge flexagon (Fig. 1.2). Complex flexagons, described in Chapter 11, are made from two or more solitary flexagons. For example, two sector first order fundamental square even edge flexagons can be linked to form a complex flexagon. Its dynamic properties

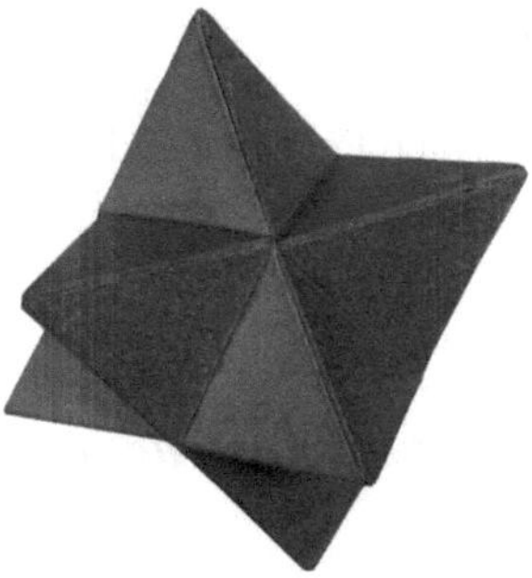

Fig. 1.14 The stella octangula, a compound of two regular tetrahedra

include features of the dynamic properties of the precursor flexagons. Complex flexagons can also incorporate parts of solitary flexagons. Most of the flexagons for which nets have been published are complex flexagons, and include some spectacular examples. For this reason it would have been better to have introduced the concept of a complex flexagon earlier in the book. However, material on solitary flexagons in previous chapters is needed as a preliminary to discussion of complex flexagons. Complex flexagons are broadly equivalent to compound polyhedra, such as the well known stella octangula, which is a compound of two regular tetrahedra (Cromwell 1997), and is shown in Fig. 1.14. Complex flexagons include some of the most interesting types of flexagon.

The miscellaneous flexagons and related structures described in Chapter 12 do not fit conveniently into the classification schemes used in earlier chapters, and include some interesting examples.

1.4 Making Flexagons

Geometric and aesthetic aspects of flexagons can be fully appreciated only by manipulating models. For videos and animations of flexagons being flexed see, for example, Highland Games (2008), Moseley (2008), Sherman (2008a, b), and YouTube (2008).

The nets used to construct flexagons are usually strips. If the polygons are regular, with edge hinges, then nets can be defined as a sequence of hinge angles (Moseley 2008). For example, the hinge angles in the net for the two sector first order fundamental square even edge flexagon (Fig. 1.2) are alternately +90° and −90°. Definition in terms of hinge angles has to be done if nets are generated by computer, but published nets are usually presented as line diagrams, without hinge angle data.

The appearance of flexagon models can be improved by colouring and decorating the faces, or by making them from coloured paper or card. A recent suggestion is to use transparent coloured material for the leaves (Shuttleworth 2006). Numerous decorative schemes have been used on various types of flexagon. Some of the decorative schemes exploit symmetries of flexagons both to create an attractive appearance and to create puzzles, for example Moseley (2008). There are books, for example Mitchell (1999) and Pedersen and Pedersen (1973), that include attractively decorated cut out nets for several types of flexagon. These have the disadvantage that making up the flexagons destroys the book.

1.4.1 General Assembly Instructions

Nets included in Chapters 4–12 are shown at approximately half full size and are satisfactory if 80 g/m^2 printer paper is used. This is a good compromise between rigidity and thickness. When flexing involves bending leaves, the use of origami paper, which is more flexible and creases well, can make flexing easier. A problem with origami paper is that printer ink tends to run and show through. However, origami paper takes pencil well. Some models are neater if made from 160 g/m^2 card with nets enlarged to about three times the size shown in the text. Models of most flexagons can be conveniently kept in transparent A5 (210 × 149 mm) size pockets kept in A5 ring binders. A recommended flexagon for a first attempt is described in the next section.

To assemble nets shown in the book, use the general scheme below. Where needed, different or additional instructions are included in captions for nets. Photographs showing assembly are included for some flexagons. All flexagons exist as enantiomorphic pairs (Section 1.2). The net for one enantiomorph can be converted into the net for the other by interchanging the markings on the faces of each leaf. Sometimes models of flexagons do not flex smoothly. If this is a problem ensure that all the hinges are well creased. If this does not work try using thinner paper or larger leaves. With edge flexagons try trimming a small amount, say 1 mm, from the edges of the net.

1. Make the specified number of copies and cut them out. Cut along any heavy lines.
2. Crease the lines between leaves to form hinges. For point and skeletal flexagons crease the lines across the strips joining leaves. Ensure that adjacent leaves superimpose correctly when folded together.
3. Transfer the numbers, and any other markings, that are in brackets on the upper face of a leaf to the reverse face, and delete from the upper face.
4. If more than one copy is specified join them end to end. For edge hinges join at the dashed lines, using transparent adhesive tape. If the hinges are short tape both sides. For point hinges glue tabs together. When copies have an odd number of leaves turn alternate copies over before joining.
5. Fold leaves with the same number together until only leaves numbered 1 and 2 are visible. An appropriate order is usually obvious, but if in doubt start with the highest number and work downwards. Some point flexagons have to be interleaved during assembly. For instructions see Sections 5.4.2 and 5.6.2.
6. Join the ends of the net to complete assembly.

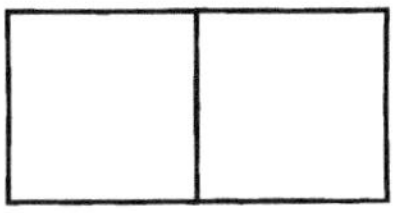

Fig. 1.15 A square edge pair

1.4.2 The Two Sector First Order Fundamental Square Even Edge Flexagon

The two sector first order fundamental square even edge flexagon is recommended as a first attempt at making a flexagon. Its net is shown in Fig. 1.2. The flexagon and its dynamic properties are described in more detail in Sections 4.2.1, 4.2.2, 4.2.4 and 4.2.5. The net is shown half full size. It could either be copied or drawn on squared paper.

As assembled, the flexagon is in a main position with leaves numbered 1 visible on one face and leaves numbered 2 on the other. In appearance, it is a flat regular even edge ring of four squares (Fig. 1.1b). It can be traversed around a cycle of four main positions by using the twofold pinch flex, which is its characteristic flex. Start by folding the flexagon in two, so that only leaves numbered 2 are visible, to reach an intermediate position. This is, in appearance, a square edge pair (Fig. 1.15). There are two ways of folding the flexagon in two so that leaves numbered 2 are visible, only one of which works. Then open the flexagon about the opposite long edge to reach another main position in which leaves numbered 2 and 3 are visible, thus completing a twofold pinch flex. Next, repeat the twofold pinch flex by folding in two so that leaves numbered 3 are visible in an intermediate position, and unfold to reach a main position in which leaves numbered 3 and 4 are visible. Repeat again, folding so that leaves numbered 4 are visible, and unfold so that leaves numbered 1 and 4 are visible. Finally, complete the traverse by folding so that leaves numbered 1 are visible, and unfold to return to the initial main position in which leaves numbered 1 and 2 are visible.

References

Coxeter HSM (1963) Regular Polytopes, 2nd edn. Macmillan, New York

Cromwell PR (1997) Polyhedra. Cambridge University Press, Cambridge

Demaine EJ, O'Rourke J (2007) Geometric Folding Algorithms: Linkages, Origami, Polyhedra. Cambridge University Press, Cambridge

Dunkerley S (1910) Mechanisms, 3 rdth edn. Longmans, Green, London

Gardner M (1965) Mathematical Puzzles and Diversions. Penguin, Harmondsworth

Gardner M (2008) Hexaflexagons, Probability Paradoxes, and the Tower of Hanoi. Cambridge University Press, Cambridge

Highland Games (2008) www.halfpast.demon.co.uk. Accessed 10 December 2008

Macmillan RH (1950) The freedom of linkages. Mathematical Gazette 34:26–37

McLean TB (1979) V-flexing the hexahexaflexagon. American Mathematical Monthly 86:457–466

Mitchell D (1999) The Magic of Flexagons: Manipulative Paper Puzzles to Cut Out and Make. Tarquin, Stradbrooke, Diss, Norfolk

Mitchell, D. (2002). *Silverflexagons 2. The slit-square silverflexagons and related forms*. Water Trade. Distributed by the British Origami Society.
Moseley R (2008) Flexagon portal. www.flexagon.net. Accessed 25 April 2008
Pedersen JJ, Pedersen KA (1973) Geometric Playthings to Color, Cut and Fold. Troubadour, San Francisco, CA
Pook LP (2003) Flexagons Inside Out. Cambridge University Press, Cambridge
Pook LP (2007) Various flexagons. Flexagon Lovers Group posting 330. http://tech.groups.yahoo.com/group/Flexagon_Lovers/. Accessed 22 May 2007
Sherman S (2008a) Dodecaflexagon videos. Flexagon Lovers Group posting 493. http://tech.groups.yahoo.com/group/Flexagon_Lovers. Accessed 25 October 2008
Sherman S (2008b) Hexaflexagon video. Flexagon Lovers Group posting 504. http://tech.groups.yahoo.com/group/Flexagon_Lovers/. Accessed 14 November 2008
Shuttleworth D (2006) Transparent flexagons. Flexagon Lovers Group posting 206. http://tech.groups.yahoo.com/group/Flexagon_Lovers/. Accessed 6 July 2006
YouTube (2008) http://uk.youtube.com. Accessed 1 November 2008

Chapter 2
Polygon Rings

2.1 Introduction

In general, a main position of a flexagon is a position that is, in appearance, a ring of convex polygons (Section 1.1). Consequently, an understanding of the properties of polygon rings is needed for an understanding of some of the properties of flexagons. Polygon rings are clearly defined geometric objects that exist in infinite series. In this chapter it is taken as understood that only the first few members of an infinite series are being described. Polygon rings are described as flat, slant and skew (Section 1.1). These descriptions can also be applied to main positions of flexagons, and they are described as flat main positions, slant main positions and skew main positions.

All the polygon rings described in this chapter, and in other chapters, are hinged. What is meant by an even edge ring, an odd edge ring, and an even vertex ring is defined in Section 1.1. A compound edge ring and an irregular edge ring are defined in Section 1.2. Various aspects of flat edge rings of regular polygons have been discussed by several authors (Conrad and Hartline 1962; Hirst 1995; Dunlap 1997/1998; Griffiths 2001; Pook 2003). Polygon rings illustrated in this chapter have been chosen to complement points made in the text. Polygon rings in subsequent chapters complement descriptions of flexagons.

Most polygon rings are linkages (Section 1.2). A linkage has an infinity of possible positions (states). As a convention, it is assumed that a polygon ring which is a linkage is arranged to be flat, whenever possible, and also that it is arranged to be as symmetrical as possible. This has been done for the edge rings shown in Figs. 1.1, 1.4, 1.5 and 1.11–1.13. The additional degree of freedom in point hinges means that all vertex rings can be laid flat (Fig. 1.6).

Positions that are, in appearance, multiple polygons and combinations appear during the flexing of some flexagons. These are related to polygon rings, and are described in the next two sections.

L. Pook, *Serious Fun with Flexagons,* Solid Mechanics and Its Applications 164,
DOI 10.1007/978-90-481-2503-6_2, © Springer Science+Business Media B.V. 2009

2.1.1 Multiple Polygons

During flexing some flexagons pass through intermediate positions that are, in appearance, a multiple polygon which consists of polygons hinged together, either at a common edge, a common vertex, or at pairs of common vertices. Depending on the number of polygons these multiple polygons are described as polygon pairs, polygon triples, etc. As a convention, it is a assumed that a multiple polygon is arranged to be as symmetrical as possible. Figure 2.1 shows a flexagon as an equilateral triangle edge triple, Fig. 2.2 another flexagon as an equilateral triangle vertex triple connected at a common vertex, Fig. 1.4a flexagon as an equilateral triangle vertex pair connected at a pair of common vertices and at a single common vertex, and Fig. 1.15a square edge pair.

Fig. 2.1 A flexagon as an equilateral triangle edge triple

Fig. 2.2 A flexagon as an equilateral triangle vertex triple connected at a common vertex. Point hinges approximated by paper strips

2.1.2 Combinations

Some edge flexagons have positions that are, in appearance, *combinations* of polygon edge rings and multiple edge polygons. In a combination the components have one or more polygons in common. For example, Fig. 2.3a shows a flexagon as a combination of a regular slant odd edge ring of three equilateral triangles and an equilateral triangle edge triple, with one equilateral triangle in common. Figure 2.3b shows a different version of the combination with two equilateral triangles in common.

2.2 Edge Rings of Regular Polygons

2.2.1 General Properties

From a topological viewpoint three parameters are needed to define an ideal *edge ring* of identical regular convex polygons. Firstly, the type of polygon, secondly, the number of polygons in the edge ring, and thirdly the hinge angle of each polygon (Fig. 1.3). These parameters are the topological invariants for a ring (Gardner 1965, 2008). They are also topological invariants for an ideal edge flexagon. In addition, the curvature is a topological invariant for a ring (Demaine and O'Rourke 2007).

Complete definition of the shape of an ideal edge ring also requires the dihedral angle (Fig. 1.9) between each pair of adjacent polygons. If an edge ring is flat then, by definition, the curvature is zero. The curvature must be zero for it to be possible to lay a ring flat, but this is not a sufficient condition. As a counter example, the irregular edge ring of seven squares shown in Fig. 2.4 has zero curvature but cannot be laid flat.

If the curvature is positive an edge ring is a *slant ring*. A regular slant odd edge ring of three regular polygons is rigid. All slant regular edge rings of four or more

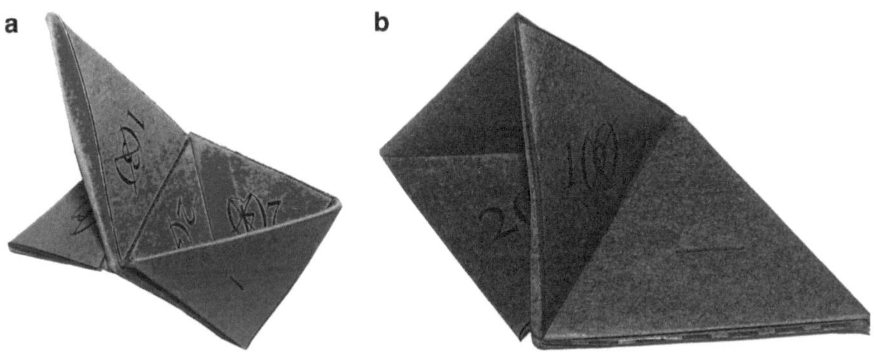

Fig. 2.3 A flexagon as a combination of a regular slant odd edge ring of three equilateral triangles and an equilateral triangle edge triple. (**a**) With one equilateral triangle in common. (**b**) With two equilateral triangles in common

regular convex polygons are linkages. A flexagon as a slant regular odd edge ring of five equilateral triangles, curvature 60°, is shown in Fig. 1.4. This is arranged to be symmetrical in the sense that all the dihedral angles are equal, and it has fivefold rotational symmetry about an axis perpendicular to the ring. A rotational symmetry is a performable operation in which the ring can be turned round so that it coincides with a previous position (Holden 1991).

A slant regular edge ring has two possible configurations, in one the dihedral angles are positive, and in the other they are negative. Available degrees of freedom do not allow the ring to be transformed from one configuration to the other. In practice, paper models of slant edge rings with hollow centres can be transformed if bending of the polygons is allowed, provided that the curvature is not too great. Slant regular even edge rings of regular polygons can be flattened, and are then, in appearance, either an edge pair or an edge strip. Other appearances are sometimes possible. Slant regular odd edge rings of regular polygons cannot be flattened.

In the special case when the curvature is 360° a regular slant edge ring becomes an open ended box edge ring. A box edge ring of four squares is shown in Fig. 2.5, and a flexagon as a box edge ring of five squares in Fig. 2.6. Main positions of flexagons can be called box main positions (or box positions).

If the curvature is negative an edge ring is a skew ring. All skew regular edge rings of regular convex polygons are linkages. A skew regular even edge ring of four regular pentagons is shown in Fig. 2.7, and a flexagon as a skew regular even edge ring of four regular hexagons in Fig. 1.5. These are symmetrical in the sense that all the dihedral angles have the same numerical value, their signs are alternately positive and negative, and a ring has twofold rotational symmetry about an axis perpendicular to the ring. There are two possible configurations, in the second the signs of the dihedral angles are reversed. There is only one degree of freedom so the ring cannot be transformed from one configuration to the other without

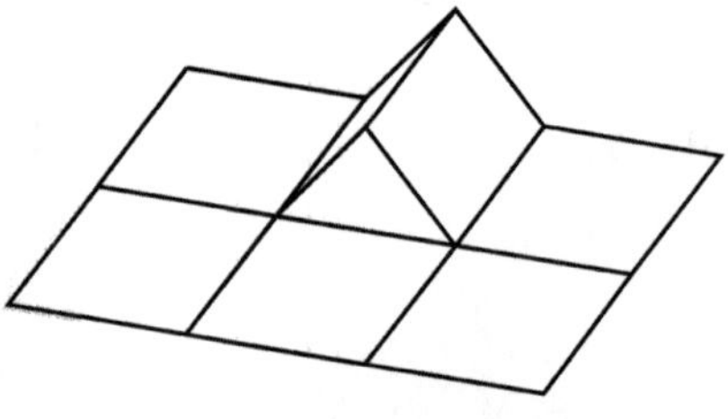

Fig. 2.4 An irregular odd edge ring of seven squares

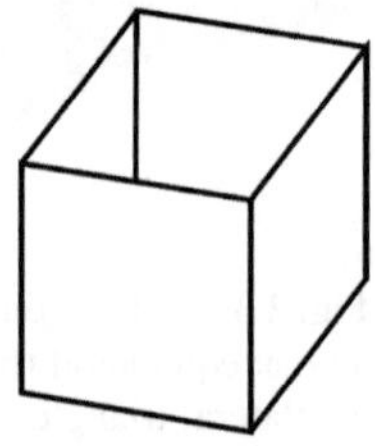

Fig. 2.5 A box edge ring of four squares (Les Pook, Flexagons inside out, 2003, © Cambridge University Press 2003, reprinted with permission)

bending the pentagons. This transformation, using polygon bending, is called a snap flex. In origami terms, mountain folds become valley folds and vice versa. Skew regular even edge rings of six or more regular polygons have sufficient degrees of freedom for the snap flex to be carried out without bending the polygons. When carried out without bending the polygons the snap flex is sometimes called the pass through flex (Hilton and Pedersen 1994). Skew regular even edge rings of regular convex polygons can always be flattened.

Skew regular odd edge rings of regular polygons must contain at least five polygons. If the number of polygons in the ring is a prime number then it cannot be arranged with rotational symmetry. However, it can be arranged so that it has a plane of reflection symmetry. This is a nonperformable operation in which the ring cannot be turned so that it coincides with a previous position. A flexagon as a skew regular odd edge ring of five squares, arranged with a plane of reflection symmetry, is shown in Fig. 2.8.

Fig. 2.6 A flexagon as a box edge ring of five squares

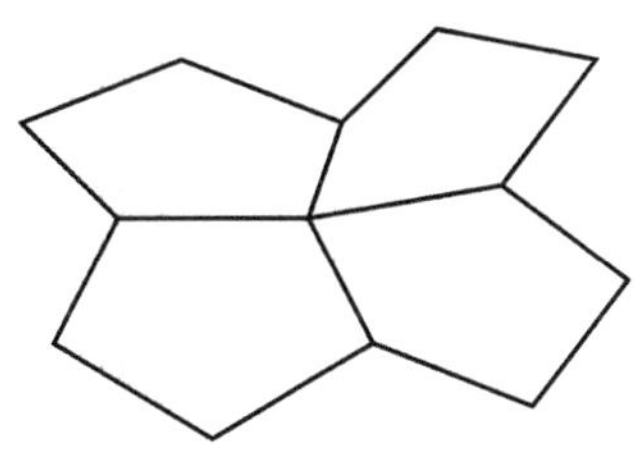

Fig. 2.7 A skew regular even edge ring of four regular pentagons (Les Pook, Flexagons inside out, 2003, © Cambridge University Press 2003, reprinted with permission)

Fig. 2.8 A flexagon as a skew regular odd edge ring of five squares

A composite number is a number that has factors other than 1 and itself. If the number of regular convex polygons in a skew regular odd edge ring is a composite number, then the ring can be arranged with a rotational symmetry that is a factor of the number. For example, a skew regular odd edge ring of nine regular polygons can be arranged with threefold rotational symmetry. Skew regular odd edge rings of regular polygons cannot be flattened if the number of polygons in the ring is a prime number. It is known that some can be flattened if it is a composite number, but it is not known whether this is always possible.

2.2.2 Regular Even Edge Rings

A regular edge ring consists of identical convex polygons. All the hinge angles are the same, as are all the dihedral angles, and all the polygons are the same distance from the centre of the ring. The distance is measured from the centroid (centre of gravity) of a polygon. A slant or flat regular even edge ring consisting of $2n$ polygons has $2n$-fold rotational symmetry about an axis perpendicular to the ring. A skew regular even edge ring has n-fold rotational symmetry about an axis perpendicular to the ring.

All the possible flat regular even edge rings containing up to 12 regular polygons, and with up to 12 edges on a polygon are listed in Table 2.1. In a ring symbol the number outside the brackets is the number of polygons in the ring and the angle inside the brackets is their hinge angle. Thus the ring symbol 6(60°), shows that there are six polygons in the regular edge ring, and that their hinge angle is 60°. A ring symbol contains the information needed to calculate the curvature of the ring.

All the rings in the table are linkages (Section 1.2). The rings of triangles and squares have a common vertex at the centre (Fig. 1.1). The other rings have open centres, for example a ring of eight octagons (Fig. 2.9). Some of the edge rings are related. For example, truncating the squares of a flat regular even edge ring of four squares (Fig. 1.1b) to regular octagons leads to a flat regular even edge ring of four regular octagons.

Table 2.1 Flat regular even edge rings of regular polygons

Polygon type	Number of polygons	Ring symbol
Triangle	6	6(60°)
Square	4	4(90°)
Pentagon	10	10(36°)
Hexagon	6	6(60°)
Octagon	4	4(90°)
Octagon	8	8(45°)
Enneagon	6	6(60°)
Decagon	10	10(36°)
Dodecagon	4	4(90°)
Dodecagon	6	6(60°)
Dodecagon	12	12(30°)

All the possible regular even edge rings of four regular polygons with up to eight edges on the polygons are listed in Table 2.2, including the flat rings listed in Table 2.1. A box edge ring of four squares is shown in Fig. 2.5, a skew regular edge ring of four regular pentagons in Fig. 2.7, a flexagon as a slant regular edge ring of four regular hexagons in Fig. 2.10, and a flexagon as a skew regular even edge ring of four regular hexagons in Fig. 1.5. Further illustrations of regular even edge rings of regular convex polygons are included in Chapters 4 and 8.

2.2.3 *Regular Odd Edge Rings*

A regular odd edge ring consists of an odd number, *n*, of regular polygons. All the hinge angles are the same, and all the polygons are the same distance from the

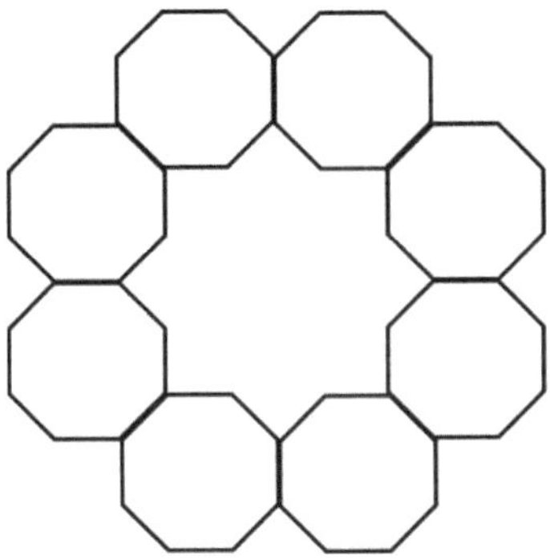

Fig. 2.9 A flat regular even edge ring of eight regular octagons (Les Pook, Flexagons inside out, 2003, © Cambridge University Press 2003, reprinted with permission)

Table 2.2 Regular edge rings of four regular polygons

Polygon type	Number of polygons	Ring symbol	Curvature
Triangle	Slant	4(60°)	120°
Square	Flat	4(90°)	0°
Square	Box	4(0°)	360°
Pentagon	Skew	4(108°)	–72°
Pentagon	Slant	4(36°)	216°
Hexagon	Skew	4(120°)	–120°
Hexagon	Slant	4(60°)	120°
Hexagon	Box	4(0°)	360°
Heptagon	Skew	4(128.6°)	–154.3
Heptagon	Slant	4(77.1°)	51.4°
Heptagon	Slant	4(25.7°)	257.1
Octagon	Skew	4(135°)	–180
Octagon	Flat	4(90°)	0°
Octagon	Slant	4(45°)	180°
Octagon	Box	4(0°)	360°

centre of the ring. A slant or flat regular odd edge ring has n-fold rotational symmetry about an axis perpendicular to the ring. A skew odd edge ring must be of at least five polygons, and can only be arranged with rotational symmetry if n is a composite number (Section 2.2.1).

A selection of flat regular odd edge rings is listed in Table 2.3. The rings of three hexagons and three dodecagons are rigid, the others are linkages. The ring of three hexagons (Fig. 2.11) has a common vertex at the centre. The others have open centres, for example the ring of seven 14-gons (Fig. 2.12).

All the possible regular odd edge rings of five regular polygons with up to eight edges on the polygons are listed in Table 2.4, together with possible rings of decagons. The flat ring of five decagons listed in Table 2.3 is included. All are linkages. The regular slant odd edge ring of five equilateral triangles is shown in Fig. 1.4, the box edge ring of five squares in Fig. 2.6 and the regular skew odd edge ring of five squares in Fig. 2.8. Further illustrations of regular odd edge rings of regular polygons are included in Chapter 4.

Fig. 2.10 A flexagon as a slant regular even edge ring of four regular hexagons

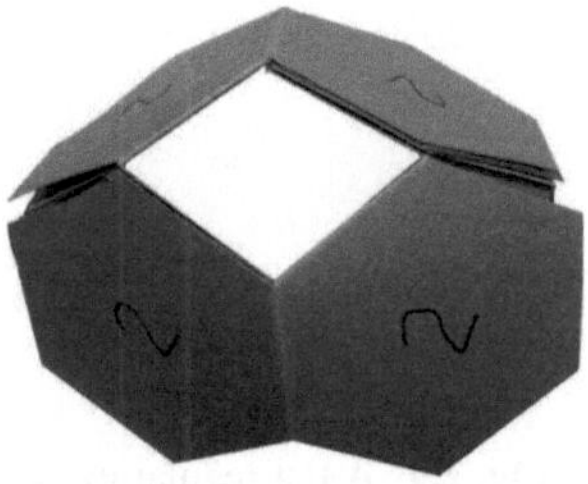

Table 2.3 Flat regular odd edge rings of regular polygons

Polygon type	Number of polygons	Ring symbol
Hexagon	3	3(120°)
Decagon	5	5(72°)
Dodecagon	3	3(120°)
14-gon	7	7(51.4°)
18-gon	9	9(40°)
20-gon	5	5(72°)

Fig. 2.11 A flat regular odd edge ring of three regular hexagons

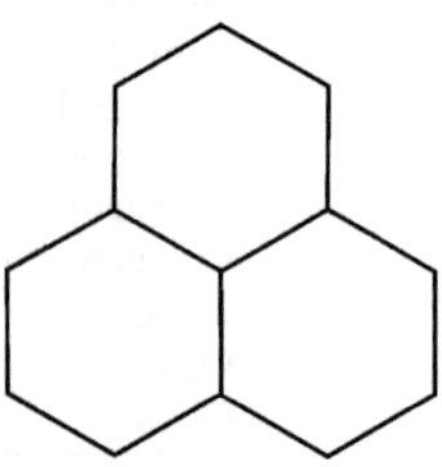

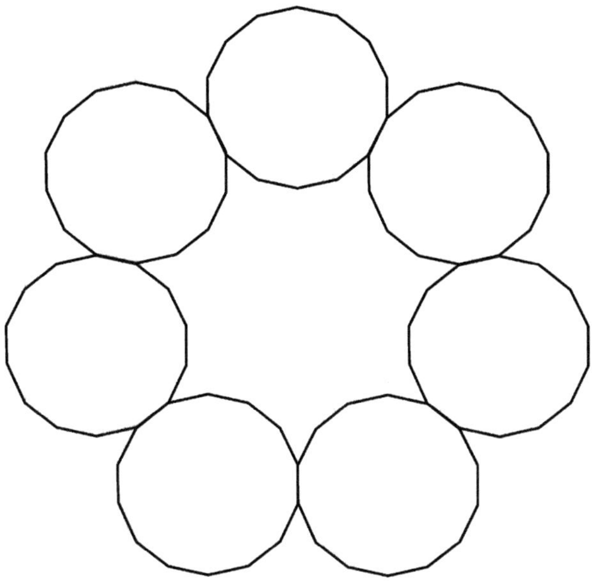

Fig. 2.12 A flat regular odd edge ring of seven regular 14-gons

Table 2.4 Regular odd edge rings of five regular polygons

Polygon type	Number of polygons	Ring symbol	Curvature
Triangle	Slant	5(60°)	60°
Square	Skew	5(90°)	–90°
Square	Box	5(0°)	360°
Pentagon	Skew	5(108°)	–180°
Pentagon	Slant	5(36°)	180°
Hexagon	Skew	5(120°)	–240°
Hexagon	Slant	5(60°)	60°
Hexagon	Box	5(0°)	360°
Heptagon	Skew	5(128.6°)	–282.9°
Heptagon	Skew	5(77.1°)	–25.7°
Heptagon	Slant	5(25.7°)	231.4°
Octagon	Skew	5(135°)	–315°
Octagon	Skew	5(90°)	–90°
Octagon	Slant	5(45°)	135°
Octagon	Box	5(0°)	360°
Decagon	Skew	5(144°)	–360°
Decagon	Skew	5(108°)	–180°
Decagon	Flat	5(72°)	0°
Decagon	Slant	5(36°)	180°
Decagon	Box	5(0°)	360°

2.2.4 Compound Edge Rings

A compound edge ring consists of an even number, $2n$, of identical polygons. Alternate hinge angles are the same and, in general, alternate polygons are the same distance from the centre of the ring. A compound edge ring can be divided into n identical sectors, each containing two polygons. A slant or flat compound edge ring has n-fold rotational symmetry about an axis perpendicular to the ring, and all dihedral angles are the same. In general, slant or flat compound edge rings are rigid. An exception is a flat compound edge ring of eight squares (Fig. 1.11). Skew compound edge rings are, apparently, always linkages. As far as is known, if n is even, a skew compound edge ring is symmetrical in the sense that all the dihedral angles have the same numerical value, their signs are alternately positive and negative, and the ring has n-fold rotational symmetry about an axis perpendicular to the ring. It also appears that, if n is odd, a skew compound edge ring can be arranged with a plane of reflection symmetry. Further, it appears that if n is an odd composite number, then the ring can be arranged with a rotational symmetry that is a factor of the number.

All the possible flat, non overlapping, compound edge rings, containing up to ten regular polygons, with up to 12 edges on the polygons, are listed in Table 2.5. Two flat overlapping compound edge rings are included in the table. In general, compound edge flexagons with main positions that are, in appearance, flat overlapping compound edge rings are not satisfactory as paper models. In a ring symbol for a compound edge ring the number outside the brackets is the number of polygons in the ring and the angles inside the brackets are the hinge angles of alternate polygons. A negative angle means that the vertex of the angle is on the outside of the ring (Section 1.2).

There are no compound edge rings of triangles. There is one possible flat compound edge ring of squares and one of regular pentagons. There are four possible flat compound edge rings of regular hexagons, two of these are included in the table. Possible numbers increase rapidly with the number of edges on constituent

Table 2.5 Flat compound edge rings of regular polygons

Polygon type	Number in ring	Ring symbol
Square	8	4(0°, 90°)
Pentagon	10	5(−36°, 108°)
Hexagon	4	2(60°, 120°)
Hexagon	6	3(0°, 120°)
Octagon	4	2(45°, 135°)[a]
Octagon	8	4(0°, 90°)
Enneagon	6	3(20°, 100°)
Decagon	4	2(36°, 144°)[a]
Decagon	4	2(72°, 108°)
Decagon	10	5(−36°, 108°)
Decagon	10	5(0°, 72°)
Dodecagon	4	2(30°, 150°)[a]
Dodecagon	4	2(60°, 120°)
Dodecagon	6	3(0°, 120°)
Dodecagon	8	4(−30°, 120°)
Dodecagon	8	4(0°, 90°)
Dodecagon	8	4(30°, 60°)

[a]Overlapping ring.

polygons. The maximum possible number of polygons in a flat compound edge ring appears to be twice the number of edges on constituent regular polygon.

In general, alternate polygons are the same distance from the centre of a compound edge ring. However, in the special case of an antibox edge ring all the polygons are the same distance from the centre of the ring. The hinge angles all have the same numerical value, but their signs are alternately positive and negative, so the curvature is 360°, as for box edge rings (Section 2.2.1). Antibox positions, which have the appearance of antibox edge rings, occur during the manipulation of some compound edge flexagons. They are always rigid. An antibox edge ring of six regular hexagons is shown in Fig. 2.13. Its ring symbol is 3(−60°, 60°). An antibox edge ring can be inscribed on an appropriate antiprism, hence its name.

All possible compound edge rings of four regular polygons, with up to eight edges on the polygons, are listed in Table 2.6. A flexagon as a slant compound edge ring of four regular hexagons is shown in Fig. 2.14. Enumeration of possible antibox edge rings is difficult because of the need to check for interference between polygons. For example, the skew edge compound ring of four regular octagons, ring

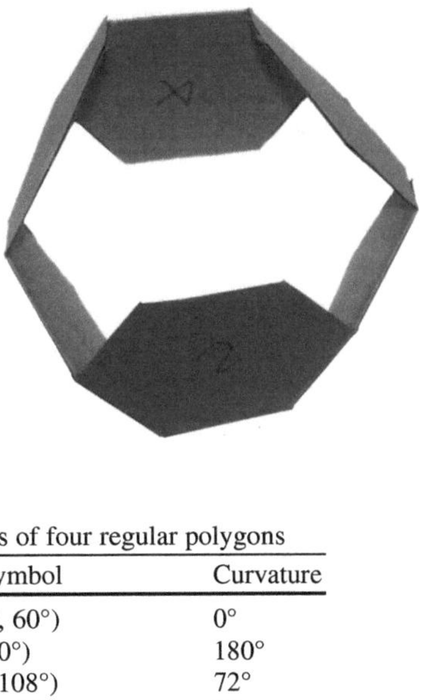

Fig. 2.13 A flexagon as an antibox edge ring of six regular hexagons

Table 2.6 Slant and antibox compound edge rings of four regular polygons

Polygon type	Ring type	Ring symbol	Curvature
Triangle	Antibox	2(−60°, 60°)	0°
Square	Slant	2(0°, 90°)	180°
Pentagon	Slant	2(36°, 108°)	72°
Pentagon	Antibox	2(−36°, 36°),	0°
Pentagon	Antibox	2(−108°, 108°)	0°
Hexagon	Slant	2(0°, 60°)	240°
Hexagon	Antibox	2(−60°, 60°)	0°
Heptagon	Slant	2(−25.7°, 77.1°)	257.1°
Heptagon	Slant	2(25.7°, 77.1°)	154.3°
Heptagon	Antibox	2(−25.7°, 25.7°)	0°
Octagon	Slant	2(0°, 45°)	270°
Octagon	Slant	2(0°, 90°)	180°
Octagon	Slant	2(45°, 90°)	90°
Octagon	Antibox	2(−45°, 45°)	0°

symbol 2(90°, 135°), is impossible because two of the octagons intersect. Further illustrations of compound edge rings are included in Chapters 1 and 6.

2.2.5 *Irregular Edge Rings*

An irregular edge ring is an edge ring of identical regular polygons that is neither regular (Sections 2.2.2 and 2.2.3) nor compound (previous section). There is a wide range of possibilities, with various symmetries. In general, irregular edge rings can be divided into identical sectors, each containing at least three polygons. However, some irregular edge rings cannot be divided into identical sectors and hence have only one sector, which is the complete ring. Irregular edge rings are difficult to classify. The curvature can be calculated but, because there are at least three polygons in a sector, ring symbols are too cumbersome to be useful.

Flat irregular edge rings of equilateral triangles, squares, regular pentagons, regular hexagons and regular heptagons are shown in Figs. 2.15–2.19. Heavy lines show that adjacent polygons are not connected. In each case this is the simplest possible flat irregular edge ring of the particular type of polygon. The flat irregular odd edge ring of five regular hexagons (Fig. 2.18) is a single sector. The flat irregular even edge rings can be divided into two identical sectors. An irregular odd edge ring of seven squares with zero curvature that cannot be laid flat (Section 2.2.1) is

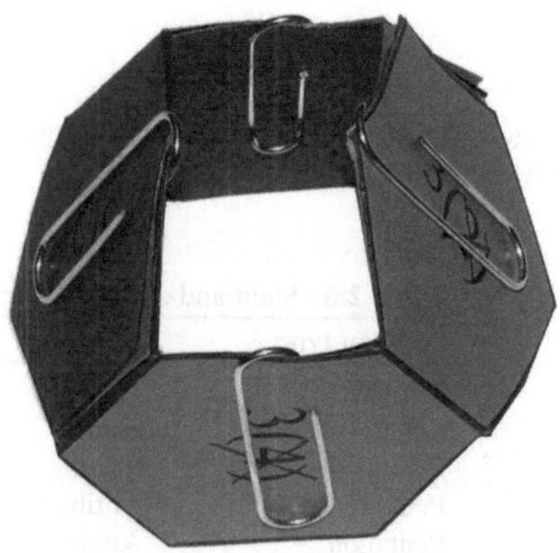

Fig. 2.14 A flexagon as a slant compound edge ring of four regular hexagons

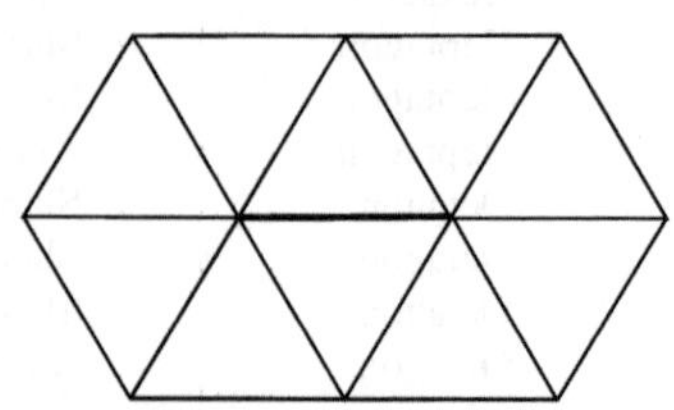

Fig. 2.15 A flat irregular even edge ring of ten equilateral triangles

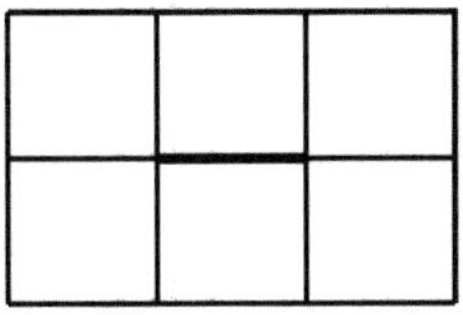

Fig. 2.16 A flat irregular even edge ring of six squares

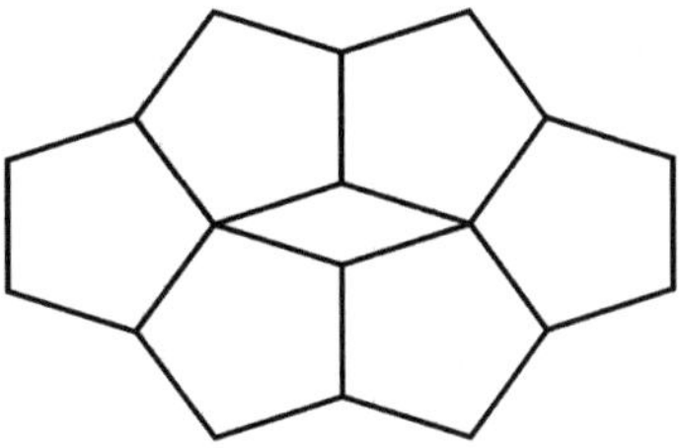

Fig. 2.17 A flat irregular even edge ring of six regular pentagons

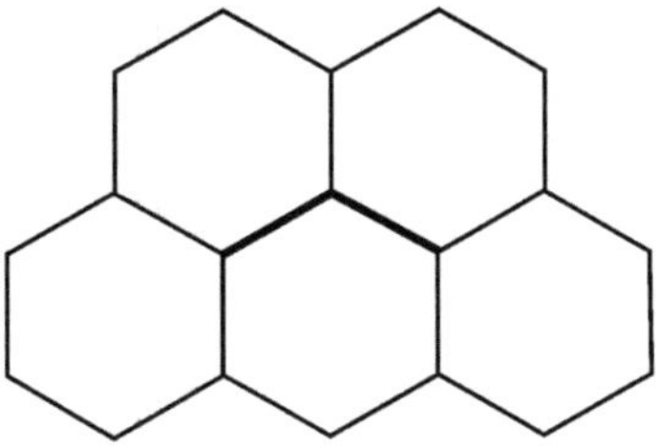

Fig. 2.18 A flat irregular odd edge ring of five regular hexagons

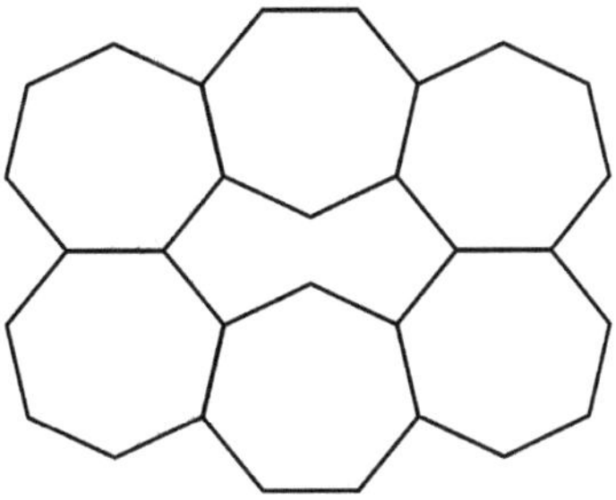

Fig. 2.19 A flat irregular even edge ring of six regular heptagons

shown in Fig. 2.4. Further illustrations of irregular even edge rings of regular polygons are included in Chapter 9.

2.3 Edge Rings of Irregular Polygons

2.3.1 General Properties

An irregular polygon is any convex polygon that is not regular including some, such as rectangles, that have some symmetry. Irregular polygon edge rings can be made from a wide range of irregular polygons. However, only a limited range of irregular polygons results in flexagons whose paper models are satisfactory. An irregular polygon can sometimes be regarded as a partially stellated version of a regular convex polygon. For example, a 45°–45°–90° triangle is a partially stellated regular convex octagon (Fig. 2.20). Similarly, a 30°–60°–90° triangle is a partially stellated regular convex dodecagon. D. Mitchell (2006, personal communication) suggested calling 45°–45°–90° triangles silver triangles and 30°–60°–90° triangles bronze triangles.

Some of the irregular polygons which have been used in published nets for edge flexagons are given by Conrad and Hartline (1962); McIntosh (2000a, b, c, d), Mitchell (1999) and Pook (2003). Even edge rings of silver triangles and bronze triangles are illustrated in the next section. Further illustrations of edge rings of irregular polygons are included in Chapter 10.

The topological invariants for an ideal edge ring of identical irregular polygons are the same as those for an ideal edge ring of identical regular polygons (Section 2.2.1), and are as follows. Firstly, the type of polygon, secondly, the number of polygons in the edge ring, and thirdly the hinge angle of each polygon (Fig. 1.3). Complete definition of the shape of an ideal edge ring also requires the dihedral angle (Fig. 1.9) between each pair of adjacent polygons.

Regular even edge rings of identical irregular convex polygons are always possible. A slant or flat regular even edge ring consisting of $2n$ irregular polygons can have either $2n$-fold or n-fold rotational symmetry, about an axis perpendicular to the ring, depending on both the type of polygon and type of ring. A skew regular even edge ring of irregular polygons always has n-fold rotational symmetry.

Regular odd edge rings of irregular convex polygons are only possible if the polygons are symmetrical about the bisector of the hinge angle (Fig. 1.3). A slant or flat regular odd edge ring consisting of n irregular polygons has n-fold rotational symmetry about an axis perpendicular to the ring. A skew odd edge ring must be of at least five polygons, and can only be arranged with rotational symmetry if n is a composite number (Section 2.2.1).

2.3.2 *Even Edge Rings of Silver and Bronze Triangles*

Even edge flexagons made from silver triangles and bronze triangles are satisfactory as paper models. They are called silver even edge flexagons and bronze even edge flexagons respectively. Possible regular even edge rings of silver and bronze

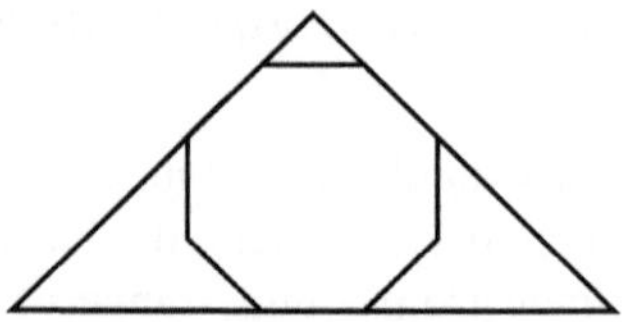

Fig. 2.20 A 45°–45°–90° (silver) triangle as a partially stellated regular convex octagon

triangles are therefore of interest. All the possible flat regular even edge rings of silver and bronze triangles are shown in Figs. 2.21 and 2.22.

2.4 Vertex Rings

2.4.1 General Properties

The first two topological invariants for an ideal vertex ring of regular polygons are the same as those for an ideal edge ring of identical regular polygons (Section 2.2.1), but the third is different. The three are as follows. Firstly, the type of polygon, secondly, the number of polygons in the vertex ring and, thirdly, the relative location of point hinges on each polygon.

The additional degree of freedom in a point hinge (Section 1.2) means that all vertex rings can be laid flat. For example, the regular even vertex ring of four equilateral triangles shown in Fig. 1.6. The additional degree of freedom in a point hinge also means that the curvature is indeterminate (Section 1.1).

A vertex ring corresponding to any edge ring is the dual of the edge ring and can be derived by inscribing polygons with vertices at midpoints of the edges of the polygons of the edge ring. In other words, the polygons are truncated at the vertices so that the vertices become edges and edges become vertices (Taylor 1997). Sherman (2007) assumes that edge hinges become point hinges. That is, he implicitly assumes that an ideal point hinge is the dual of an ideal edge hinge. However, in a rigorous interpretation of duality an ideal point hinge would have one degree of freedom, not two as defined in Section 1.2.

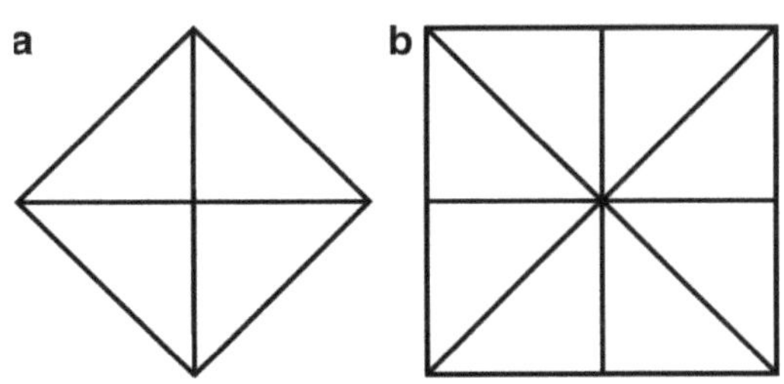

Fig. 2.21 Flat regular even edge rings of 45°–45°–90° (silver) triangles. (**a**) Four triangles. (**b**) Eight triangles (Les Pook, Flexagons inside out, 2003, © Cambridge University Press 2003, reprinted with permission)

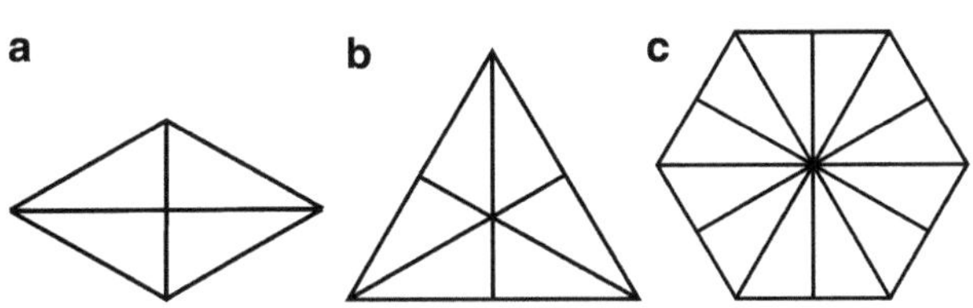

Fig. 2.22 Flat regular even edge rings of 30°–60°–90° (bronze) triangles. (**a**) Four triangles. (**b**) Six triangles. (**c**) Twelve triangles

There are regular even vertex rings corresponding to regular even edge rings (Section 2.2.2), regular odd vertex rings corresponding to regular odd edge rings (Section 2.2.3), compound vertex rings corresponding to compound edge rings (Section 2.2.4), irregular vertex rings corresponding to irregular edge rings (Section 2.2.5), and vertex rings of irregular polygons corresponding to edge rings of irregular polygons (Section 2.3.1). Some even vertex rings of squares are described in the next section.

2.4.2 *Vertex Rings of Squares*

Inscribing squares in the squares of a box edge ring of four squares. (Fig. 2.5) leads to a box vertex ring of four squares with point hinges at diagonally opposite vertices of the squares (Fig. 2.23a). This ring can be flattened, by rotating it as ring, into a flat regular vertex ring of four squares (Fig. 2.23b). In Fig. 2.23 point hinges are indicated by dots, and some squares are slightly separated for clarity.

An edge ring confined to a plane is rigid, but a vertex ring consisting of four or more polygons confined to a plane is usually a linkage, for example the regular vertex ring of four squares, with point hinges at adjacent vertices of the squares, shown in Fig. 2.23c. This was derived by inscribing squares in the squares of a flat regular even edge ring of four squares (Fig. 1.1b). It has fourfold rotational symmetry, and can be manoeuvred into the form shown in Fig. 2.23d, which has twofold rotational symmetry.

In general, vertex rings can be rotated continuously, although rotation is sometimes restricted by interference between polygons. For example, an even vertex ring

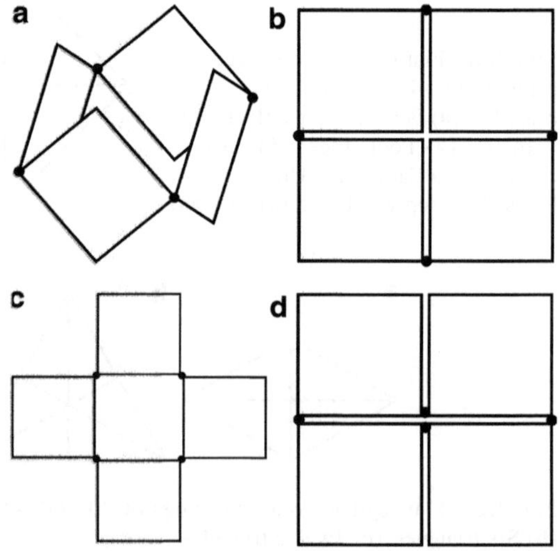

Fig. 2.23 Even vertex rings of four squares. (**a**) Box. (**b**) Flat, hinges diagonally opposite. (**c**) Flat, hinges adjacent (**d**) Flat, hinges adjacent, less symmetrical

of four squares can be rotated continuously so that the configurations shown in Fig. 2.23a and b appear alternately. In this sense vertex rings are analogous to rotating rings consisting of polyhedra hinged at common edges, sometimes called kaleidocycles, as described by Conrad and Hartline (1962); Cundy and Rollett (1981); Engel (1969); Hilton and Pedersen (1994); Schattschneider and Walker (1983).

References

Conrad AS, Hartline DK (1962) *Flexagons*. RIAS Technical Report 62-11. Baltimore, MD: RIAS
Cundy HM, Rollett AR (1981) Mathematical Models, 3 rdth edn. Tarquin, Stradbrooke, Diss, Norfolk, UK
Demaine EJ, O'Rourke J (2007) Geometric Folding Algorithms: Linkages, Origami, Polyhedra. Cambridge University Press, Cambridge
Dunlap RA (1997/1998) Regular polygon rings. Mathematical Spectrum 30:13–15
Engel D (1969) Hybrid flexahedrons. Journal of Recreational Mathematics 2:35–41
Gardner M (1965) Mathematical Puzzles and Diversions. Penguin, Harmondsworth
Gardner M (2008) Hexaflexagons, Probability Paradoxes, and the Tower of Hanoi. Cambridge University Press, Cambridge
Griffiths M (2001) Two proofs concerning 'Octagon loops'. Mathematical Gazette 85:80–84
Hilton P, Pedersen J (1994) Build Your Own Polyhedra. Addison-Wesley, Menlo Park, CA
Hirst A (1995) Can you do it with heptagons? Mathematical Gazette 79:17–29
Holden A (1991) Shapes, Space and Symmetry. Dover, New York
McIntosh HV (2000 a) *Pentagonal Flexagons*. Departamento de Aplicación de Microcomputadoras, Instituto de Ciencias, Universidad Autónoma de Puebla, Puebla, Mexico
McIntosh HV (2000 b) *Tetragonal Flexagons*. Departamento de Aplicación de Microcomputadoras, Instituto de Ciencias, Universidad Autónoma de Puebla, Puebla, Mexico
McIntosh HV (2000 c) *Trigonal Flexagons*. Departamento de Aplicación de Microcomputadoras, Instituto de Ciencias, Universidad Autónoma de Puebla, Puebla, Mexico
McIntosh HV (2000 d) *Heptagonal Flexagons*. Departamento de Aplicación de Microcomputadoras, Instituto de Ciencias, Universidad Autónoma de Puebla, Puebla, Mexico
Mitchell D (1999) The Magic of Flexagons: Manipulative Paper Puzzles to Cut Out and Make. Tarquintions, Stradbrooke, Diss, Norfolk
Pook LP (2003) Flexagons Inside Out. Cambridge University Press, Cambridge
Schattschneider D, Walker W (1983) M C Escher Kaleidocycles. Tarquin, Stradbrooke, Diss, Norfolk
Sherman S (2007) Two hybrid point flexagons. Flexagon Lovers Group posting 376. http://tech.groups.yahoo.com/group/Flexagon_Lovers/. Accessed 27 June 2007
Taylor P (1997) The Complete? Polygon. Nattygrafix, Ipswich

Chapter 3
Fundamental Nets

3.1 Introduction

The nets used to construct flexagons are usually strips of hinged convex polygons. If the polygons are regular and connected by edge hinges, then nets can be completely defined as a sequence of hinge angles. The net used of course determines the appearance and dynamic properties of a flexagon. Some nets can be assembled to produce more than one distinct type of flexagon; this means that to specify a flexagon completely both the net and the method of assembly has to be specified. All flexagons exist as enantiomorphic (mirror image) pairs. The two members of an enantiomorphic pair of flexagons are not usually regarded as distinct types (Section 1.2).

The appearance of nets varies widely, and some are very irregular. Fundamental edge nets have a regular appearance and can be defined as a repeating sequence of hinge angles, without specifying the number of repetitions. In a first order fundamental edge net the hinge angles are all the same, but alternate hinge angles are of opposite sign. The hinge angles of a first order fundamental edge net are the same as the vertex angles of a regular polygon that is associated with the net. Fundamental flexagons also have associated polygons, and they are a key concept in flexagon theory.

A Schläfli symbol completely defines a regular polygon, except for its size. The Schläfli symbol for a regular convex polygon is $\{s\}$, where s is the number of edges (Coxeter 1963; Wenninger 1971). For example, the Schläfli symbol for an equilateral triangle is $\{3\}$ and for a square it is $\{4\}$. The Schläfli symbol for a regular star pentagon (Fig. 3.1) is $\{5/2\}$. The 5 indicates that there are five edges and the 2 indicates that the interior is covered twice. This means that a line drawn from the centre of the regular star pentagon to its exterior crosses two sides. The Schläfli symbol for the associated polygon becomes the net symbol for the corresponding first order fundamental edge net. It completely defines the net, except for its size and the number of repetitions. A net symbol is enclosed in angled brackets. The first or only number in the net symbol is the Schläfli symbol for the constituent polygons. For example, the first order fundamental triangle edge net is made from equilateral triangles, Schläfli symbol $\{3\}$, and the net symbol is $\langle 3 \rangle$.

Second order fundamental edge nets and fundamental vertex nets are modifications of first order fundamental edge nets and the same net symbols can be used.

L. Pook, *Serious Fun with Flexagons*, Solid Mechanics and Its Applications 164,
DOI 10.1007/978-90-481-2503-6_3, © Springer Science+Business Media B.V. 2009

Fig. 3.1 A regular star pentagon (Les Pook, Flexagons inside out, 2003, © Cambridge University Press 2003, reprinted with permission)

The fundamental silver edge net and the fundamental bronze edge net are also modifications, but with different net symbols.

3.2 First Order Fundamental Edge Nets

Short lengths of all the possible first order fundamental edge nets made from regular polygons with up to 12 sides are shown in Figs. 3.2–3.11. Heavy lines show that adjacent polygons are not connected. Hinge angles in a net are all the same, but alternate hinge angles are of opposite sign, and most of the nets have a zigzag appearance. Six of the nets shown in Figs. 3.6–3.11 overlap. Construction of flexagons using these nets is tedious because the nets have to built up from polygon pairs joined together. An alternative is to truncate the polygons to avoid overlapping, and hence make construction easier. For example, Fig. 3.12 shows the first order fundamental octagon edge net ⟨8⟩, with the octagons truncated to irregular hexagons. Truncating the polygons means that individual leaves in pats do not overlap exactly but, in practice, this is not usually a serious drawback.

Suitable lengths of appropriate first order fundamental edge nets are used to construct first order fundamental even edge flexagons (Section 4.2.1), fundamental compound flexagons (Section 6.1) and fundamental irregular ring even edge flexagons (Section 9.1.1). These have main positions that are, in appearance, either regular even edge rings (Section 2.2.2), compound edge rings (Section 2.2.4), or irregular even edge rings (Section 2.2.5).

3.3 Second Order Fundamental Edge Nets

In a second order fundamental edge net the hinge angles are all the same, but alternate pairs of hinge angles are of opposite sign, and the nets have a regular wavelike appearance. Short lengths of all the possible second order fundamental edge nets made from polygons with up to six edges are shown in Figs. 3.13–3.16. A subscript 2 is added to a net symbol in order to distinguish the net from the corresponding first order fundamental edge net with the same associated polygon. In Fig. 3.16 some hexagons should overlap exactly, but vertical hinges have been separated for clarity, as indicated by dashed lines. One of the second order fundamental 20-gon edge nets, $\langle 20/7 \rangle_2$, is shown in Fig. 3.17.

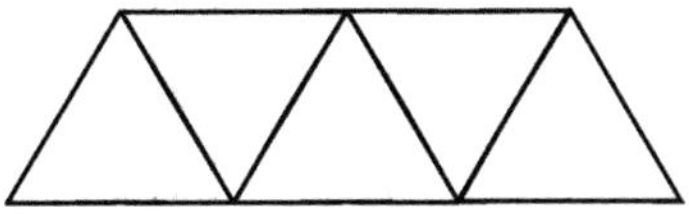

Fig. 3.2 The first order fundamental triangle edge net ⟨3⟩

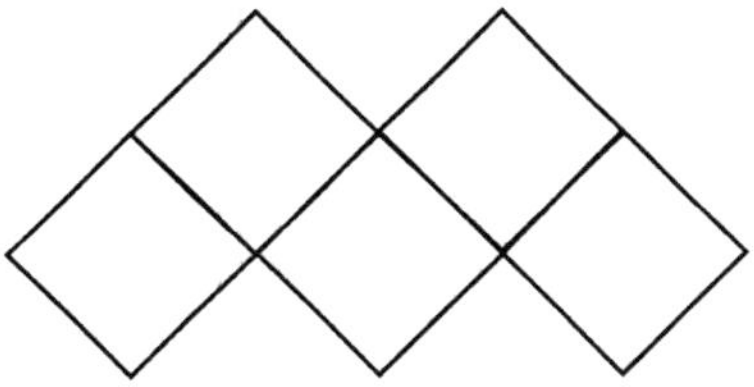

Fig. 3.3 The first order fundamental square edge net ⟨4⟩

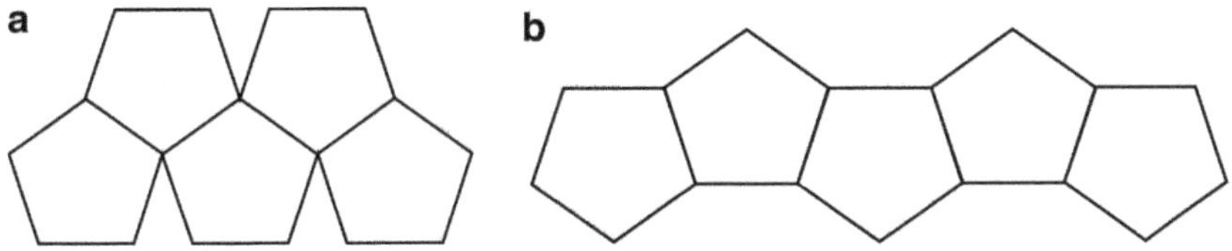

Fig. 3.4 First order fundamental pentagon edge nets. (**a**) ⟨5⟩. (**b**) ⟨5/2⟩

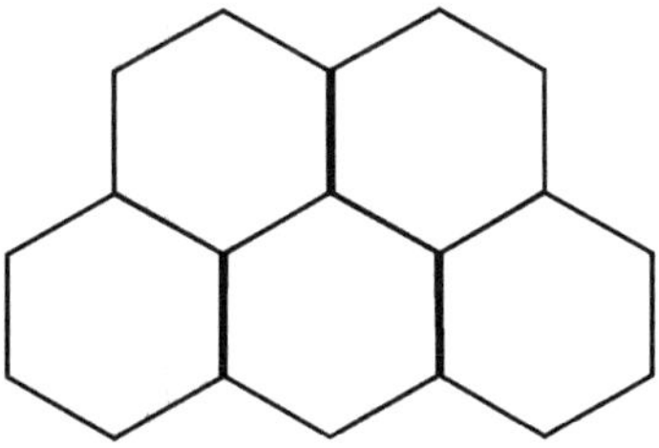

Fig. 3.5 The first order fundamental hexagon edge net ⟨6⟩

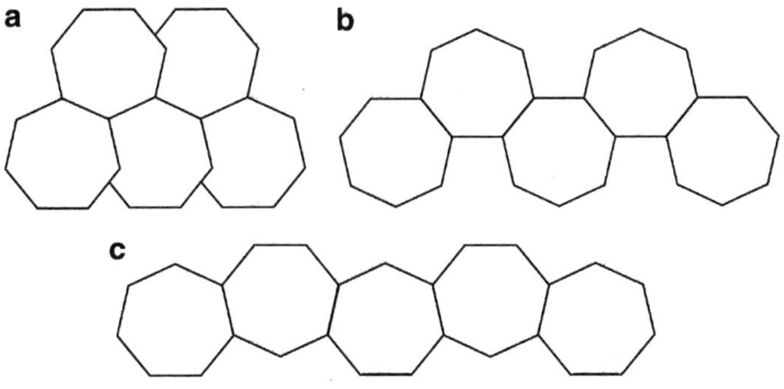

Fig. 3.6 First order fundamental heptagon edge nets. (**a**) ⟨7⟩. (**b**) ⟨7/2⟩. (**c**) ⟨7/3⟩

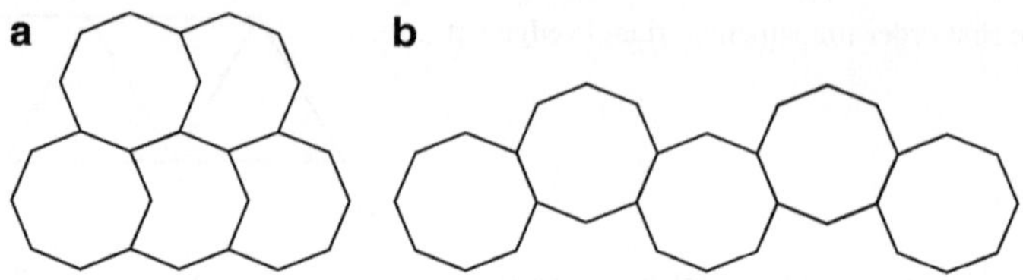

Fig. 3.7 First order fundamental octagon edge nets. (**a**) ⟨8⟩. (**b**) ⟨8/3⟩

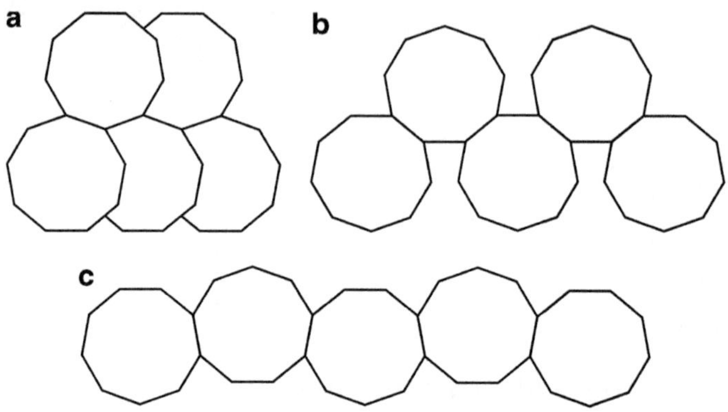

Fig. 3.8 First order fundamental enneagon edge nets. (**a**) ⟨9⟩. (**b**) ⟨9/2⟩. (**c**) ⟨9/4⟩

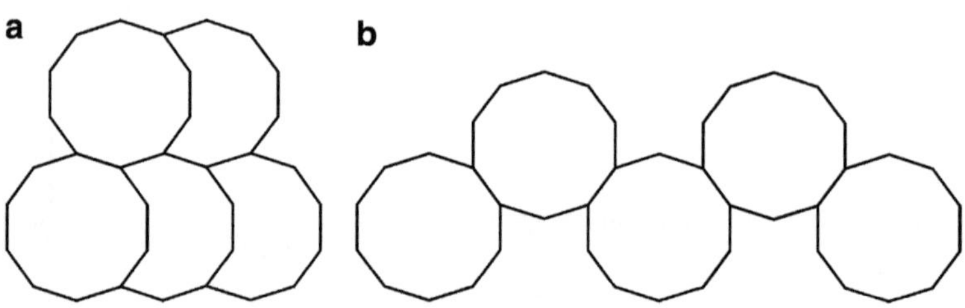

Fig. 3.9 First order fundamental decagon edge nets. (**a**) ⟨10⟩. (**b**) ⟨10/3⟩

Suitable lengths of second order fundamental edge nets are used to construct second order fundamental odd edge flexagons (Section 4.3.1).

3.4 Fundamental Vertex Nets

Any edge ring has a vertex ring equivalent and this is the dual of the edge ring (Section 2.4.1). Similarly, any first order fundamental edge net has a fundamental vertex net equivalent which is the dual of the first order fundamental edge net. The

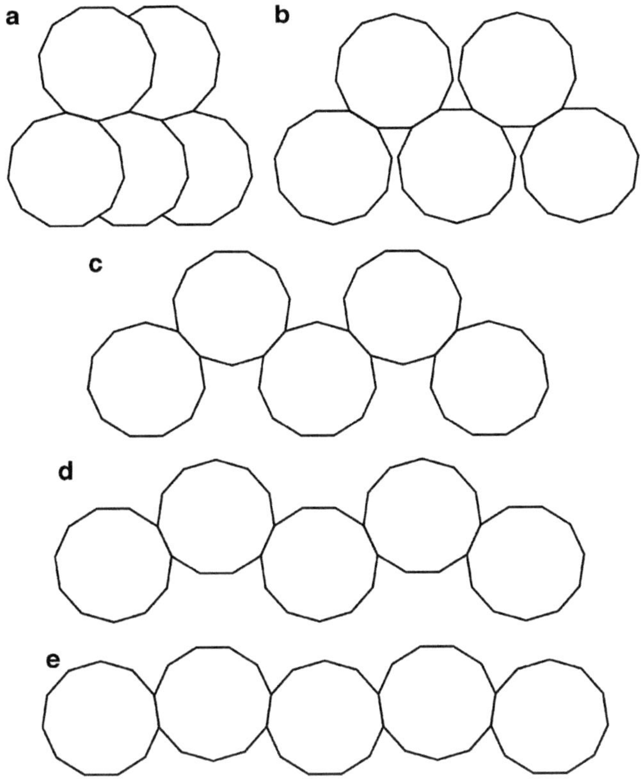

Fig. 3.10 First order fundamental 11-gon edge nets. (**a**) ⟨11⟩. (**b**) ⟨11/2⟩. (**c**) ⟨11/3⟩. (**d**) ⟨11/4⟩. (**e**) ⟨11/5⟩

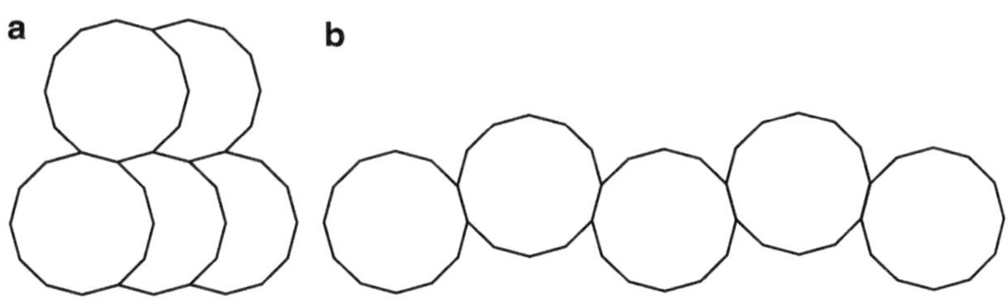

Fig. 3.11 First order fundamental dodecagon edge nets. (**a**) ⟨12⟩. (**b**) ⟨12/5⟩

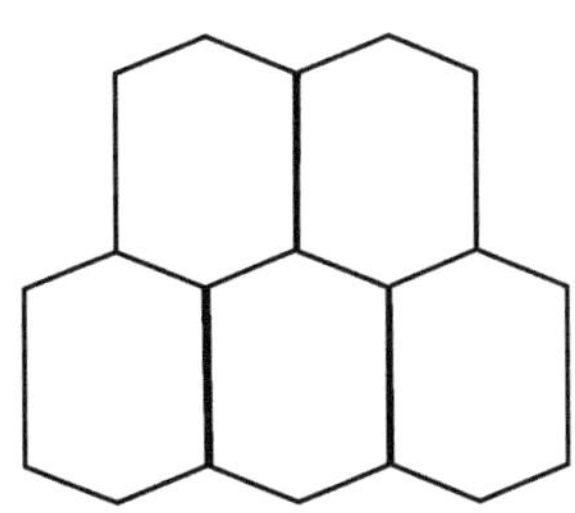

Fig. 3.12 The first order fundamental octagon edge net ⟨8⟩, Octagons truncated to irregular hexagons

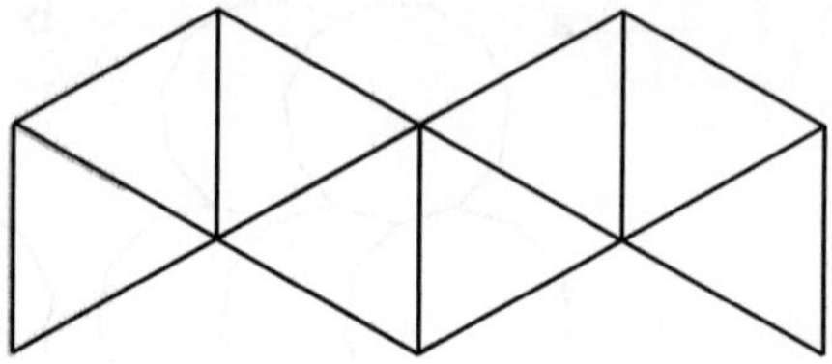

Fig. 3.13 The second order fundamental triangle edge net $\langle 3\rangle_2$

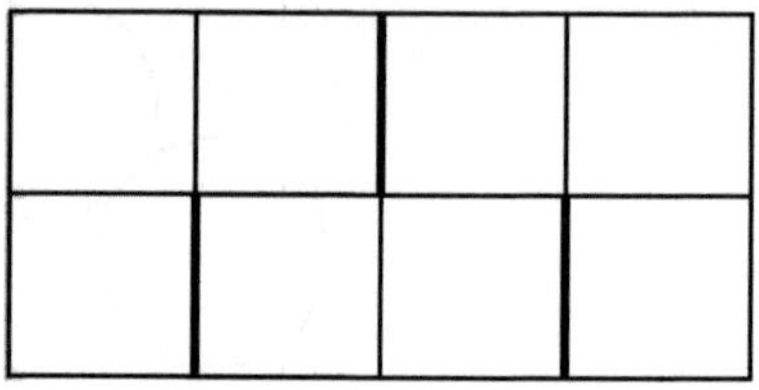

Fig. 3.14 The second order fundamental square edge net $\langle 4\rangle_2$

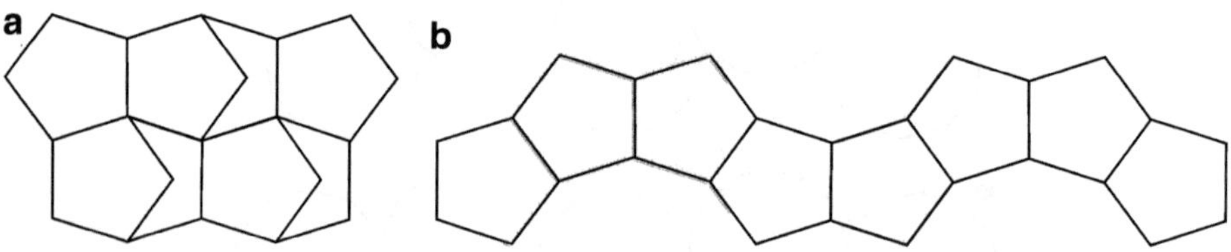

Fig. 3.15 Second order fundamental pentagon edge nets. (**a**) $\langle 5\rangle_2$. (**b**) $\langle 5/2\rangle_2$

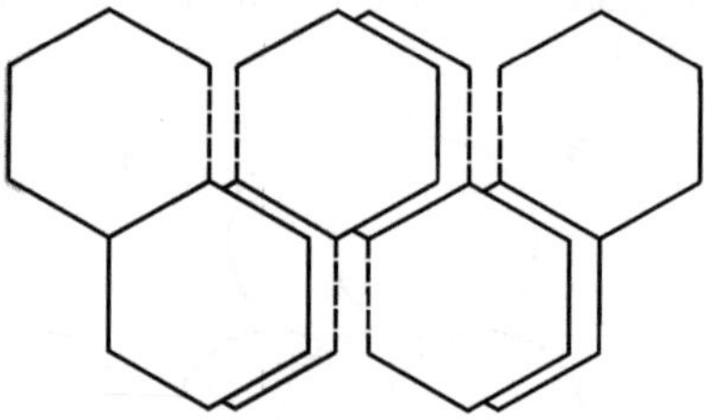

Fig. 3.16 The second order fundamental hexagon edge net $\langle 6\rangle_2$

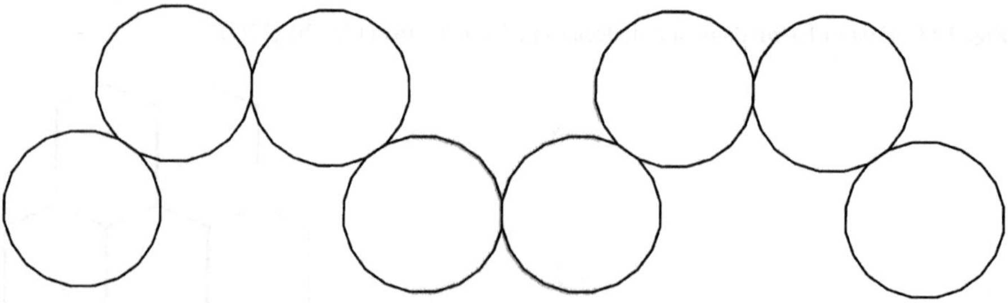

Fig. 3.17 The second order fundamental 20-gon edge net $\langle 20/7\rangle_2$

fundamental vertex net is derived by inscribing polygons in the polygons of the first order fundamental edge net with vertices at midpoints of the edges. Short lengths of fundamental vertex nets, for polygons with up to five edges, derived from first order fundamental edge nets are shown in Figs. 3.18–3.20. Point hinges are indicated by dots. The close relationship makes it convenient to use the same net symbol for both a first order fundamental edge net and the equivalent fundamental vertex net.

Even when confined to a plane, fundamental vertex nets are linkages, so they can be manoeuvred into other forms. For example, a short length of an alternative form of the fundamental square vertex net ⟨4⟩ is shown in Fig. 3.21. For clarity the squares have been slightly separated. This alternative form is not a dual of the corresponding first order fundamental edge net.

Point hinges are impossible in a paper model, but short paper strips provide a workable approximation. Figs. 1.6, 1.8 and 2.2 show paper models of flexagons with equilateral triangles connected by paper strips. Dimensions and orientations of

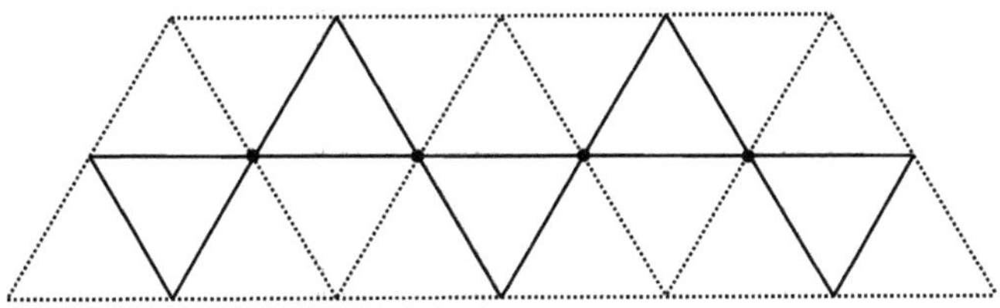

Fig. 3.18 The fundamental triangle vertex net ⟨3⟩ (solid lines) and the first order fundamental triangle edge net ⟨3⟩ (dotted lines)

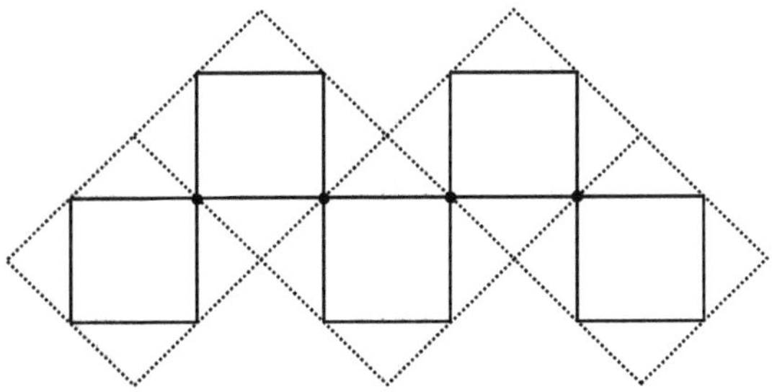

Fig. 3.19 The fundamental square vertex net ⟨4⟩ (solid lines) and the first order fundamental square edge net ⟨4⟩ (dotted lines)

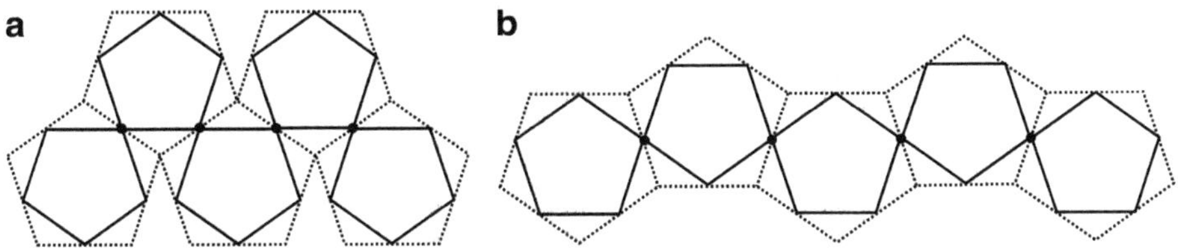

Fig. 3.20 Fundamental pentagon vertex nets (solid lines) and first order fundamental pentagon edge nets (dotted lines). (**a**) ⟨5⟩. (**b**) ⟨5/2⟩

strips are fairly critical for smooth flexing of a paper model. A short length of a practical form of the fundamental pentagon vertex net ⟨5⟩, with the pentagons connected by strips, is shown in Fig. 3.22.

Suitable lengths of practical forms of fundamental vertex nets are used to construct fundamental skeletal flexagons (Section 5.2.1) and fundamental point flexagons (Section 5.3.1).

Paper strips are satisfactory for demonstrating the dynamic properties of point flexagons, but the models are not photogenic. The appearance of point flexagons can be improved by making the leaves from a stiff material, with the same colour on both faces of each leaf, and with point hinges approximated by cord loops through holes near the vertices (Pook 2008). For example, Fig. 3.23 shows a square point flexagon made from Meccano plastic sheets, with point hinges approximated by loops of Meccano cord. It flexes more smoothly than does a point flexagon with paper strip hinges.

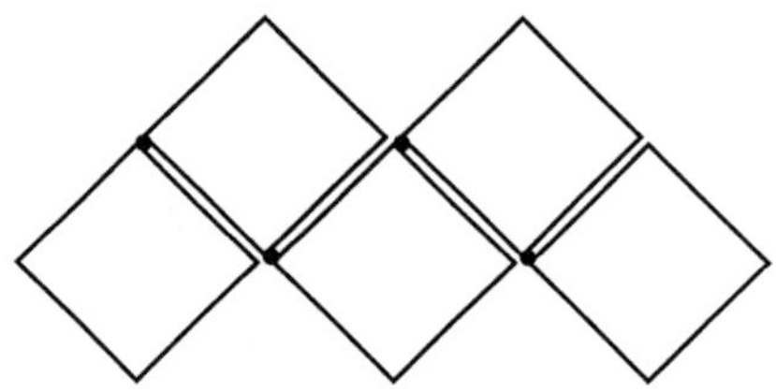

Fig. 3.21 An alternative form of the fundamental square vertex net ⟨4⟩

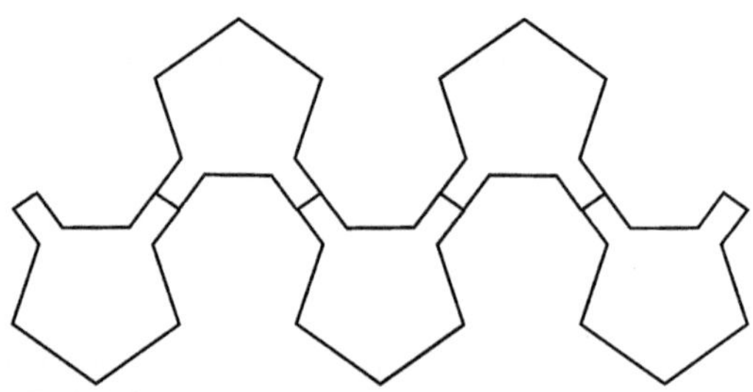

Fig. 3.22 A practical form of the fundamental pentagon vertex net ⟨5⟩

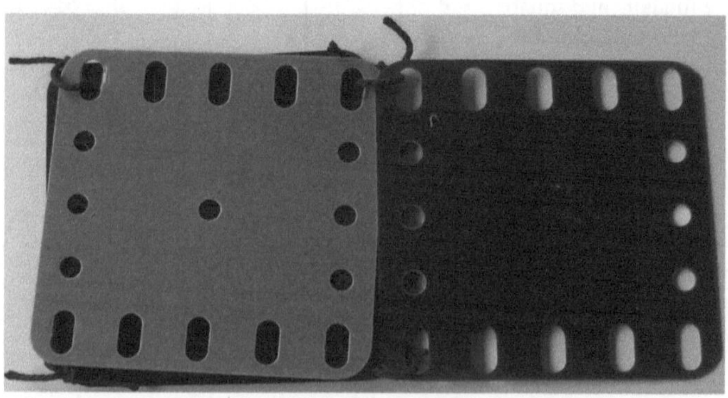

Fig. 3.23 A flexagon as a square vertex pair. Point hinges approximated by cord loops

Fig. 3.24 The fundamental silver edge net ⟨silver⟩

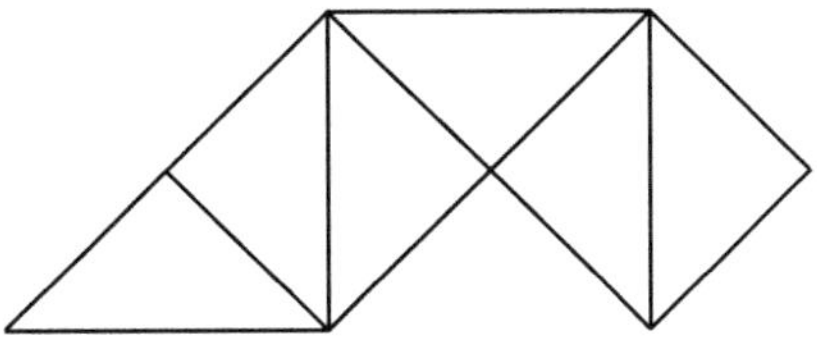

Fig. 3.25 The fundamental bronze edge net ⟨bronze⟩

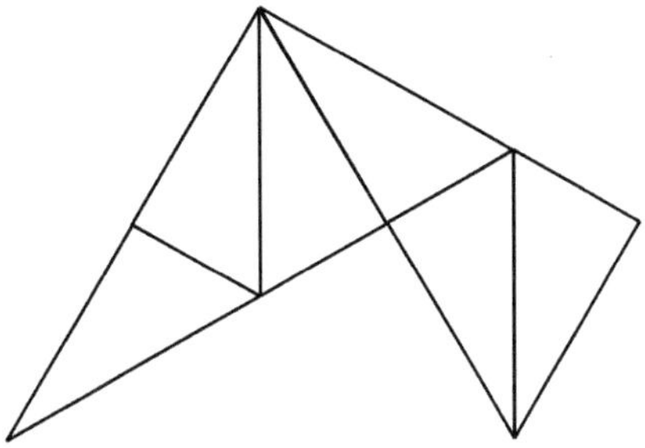

3.5 Fundamental Silver and Bronze Edge Nets

Flexagons made from 45°–45°–90° (silver) triangles and 30°–60°–90° (bronze) triangles(Section 2.3.1) are often regarded as distinct varieties, and there are corresponding fundamental edge nets.

The fundamental silver edge net is derived by using the first order fundamental triangle edge net ⟨3⟩ (Fig. 3.2) as a precursor and replacing the equilateral triangles by silver triangles, with the condition that adjacent silver triangles overlap exactly when folded together. Its net symbol is ⟨silver⟩ and the associated polygon is a silver triangle. The net repeats exactly after six silver triangles, and is shown in Fig. 3.24. Similarly, the fundamental bronze edge net is derived by replacing the equilateral triangles by bronze triangles. Its net symbol is ⟨bronze⟩ and the associated polygon is a bronze triangle. The net repeats exactly after six bronze triangles, and is shown in Fig. 3.25.

Suitable lengths of fundamental silver and bronze edge nets are used to construct fundamental silver even edge flexagons (Section 10.2.4.1) and fundamental bronze even edge flexagons (Section 10.2.6.1).

References

Coxeter HSM (1963) *Regular Polytopes*, 2nd edn. Macmillan, New York

Pook, LP. (2008) A photogenic point flexagon. Flexagon Lovers Group posting 516. http://tech.groups.yahoo.com/group/Flexagon_Lovers/. Accessed 7 December 2008

Wenninger M (1971) *Polyhedron Models*. Cambridge University Press, Cambridge

Chapter 4
Fundamental Edge Flexagons

4.1 Introduction

Fundamental edge flexagons are constructed from fundamental edge nets (Sections 3.2 and 3.3). Broadly, fundamental edge flexagons are the equivalent of regular polyhedra in that they are constructed from identical regular convex polygons, and have a high degree of symmetry in their structure. They also have a high degree of symmetry in their dynamic properties. Pats are, in general, fan folded piles of leaves. Alternate pats can be single leaves. Fundamental edge flexagons are solitary flexagons which means, broadly, that they are equivalent to single polyhedra. Many other types of flexagon can be regarded as modified fundamental edge flexagons, so an understanding of their structure and dynamic properties is an essential prerequisite to the understanding of flexagons in general.

Two families of fundamental edge flexagons are discussed in this chapter. One family is first order fundamental even edge flexagons. In a main position, a first order fundamental even edge flexagon is, in appearance, a regular even edge ring of $2n$ regular convex polygons (Section 2.2.2), where $n \geq 2$. A main position can be divided into S identical sectors containing two adjacent pats, hence $S = n$. The curvature is an important property of main positions, and is calculated in the same way as the curvature of the corresponding ring (Section 1.1). Nets for first order fundamental even edge flexagons are derived by fan folding an appropriate first order fundamental edge net (Section 3.2) onto a regular even edge ring and applying an appropriate face numbering sequence.

The other family is second order fundamental odd edge flexagons. A second order fundamental odd edge flexagon always has one main position which is, in appearance, a regular odd edge ring of n regular convex polygons (Section 2.2.3), where $n \geq 5$. Its curvature is an important property. Each pat in the main position is identical, so it is a sector, and there are n sectors. Nets for second order fundamental odd edge flexagons are derived by fan folding an appropriate second order fundamental edge net (Section 3.3) onto a regular odd edge ring and applying an appropriate face numbering sequence. First order fundamental odd edge flexagons are not possible because the ends of a first order fundamental edge net do not match correctly when fan folded onto an odd edge ring.

L. Pook, *Serious Fun with Flexagons*, Solid Mechanics and Its Applications 164,
DOI 10.1007/978-90-481-2503-6_4, © Springer Science+Business Media B.V. 2009

4.1.1 Standard Face Numbering Sequences

The use of standard face numbering sequences simplifies the description of fundamental edge flexagons, and makes it easier to design their nets. It also avoids ambiguity in deciding which of the cycles that can be traversed in a flexagon is the principal cycle. An appropriate standard face numbering sequence is used for each of the nets given in this chapter, except where noted. For a first order fundamental even edge flexagon, if a sector contains an even number of leaves, then the upper faces of leaves in the first order fundamental edge net are numbered consecutively 1, 3, 3, 5, 5,...1 and the corresponding lower faces 2, 2, 4, 4, 6, 6,.... The standard face numbering sequence is repeated for each sector, and it uniquely identifies each leaf within a sector. A convenient notation for a particular net is 1/2, 3/2. 3/4, 5/4, 5/6,.... where numbers before slashes are for the upper faces, and those after slashes are for the lower faces. For an odd number of leaves in a sector the numbers are 1, 3, 3, 5, 5, 7,... for the upper faces and 2, 2, 4, 4, 6,...1 for the lower faces. In alternate sectors the numbers are applied alternately to the upper and lower faces. For a second order fundamental odd edge flexagons the standard face numbering sequence is 1/3, 2/4, 3/1, 4/3,....

4.1.2 Truncated Flexagons

Truncation means cutting off the vertices of a regular polygon to produce new edges (Taylor 1997). For example, an equilateral triangle can be truncated to a regular hexagon by cutting off all its vertices, and a square can be truncated to a 45°–45°–90° (silver) triangle. Truncating a regular polygon so that the edges of the original regular polygon become points, and the new edges meet these points, produces the dual of the original polygon, with the same number of edges as the original.

A truncated flexagon is an edge flexagon in which the regular convex polygon leaves of a precursor flexagon have been truncated, usually all to the same shape. Hinges may be shortened, but hinge angles are unchanged. Hence, truncation changes the appearance of a flexagon but does not, in general, alter its dynamic properties. However, if the edges of the original edges become points, then the edge hinges become point hinges, and an edge flexagon becomes a skeletal flexagon (Section 5.2.1), with significantly different dynamic properties.

After truncation the leaves of an edge flexagon may or may not overlap exactly. A flexagon in which the leaves do not overlap exactly is a partial overlap flexagon. Truncating equilateral triangles to regular hexagons shortens the hinges of a flexagon and the leaves overlap exactly (Section 8.2.4). After truncating squares to 45°–45°–90° (silver) triangles the hinges of a flexagon are the original length, but the leaves do not overlap exactly (Section 10.2.4.3). Truncation is sometimes used to simplify assembly. For example, after truncating octagons to irregular hexagons (Section 4.2.8) the hinges are the original length, but the leaves do not overlap exactly. In a skeletal flexagon, leaves do overlap exactly.

4.2 First Order Fundamental Even Edge Flexagons

4.2.1 General Properties

First order fundamental even edge flexagons have at least two sectors, and are made from first order fundamental edge nets (Section 3.2). Satisfactory paper models with from two to four sectors can be made from regular convex polygons with up to eight edges on a polygon. Nets for some examples are given below. They are regular cycle flexagons in which all the main positions of a cycle have the same appearance and the same pat structure.

In a main position, a first order fundamental even edge flexagon is, in appearance, a regular even edge ring of $2n$ regular convex polygons (Section 2.2.2) and two faces are visible. If these faces are numbered, say, 1 and 2, then the main position code is 1(2). If the main position is a flat edge ring (Fig. 2.9), or a skew edge ring (Fig. 2.7), then the number outside the brackets, 1, identifies the upper face, and the number inside the brackets, 2, the lower face. The choice of upper face is arbitrary, so codes 1(2) and 2(1) are synonymous. However, if the main position is a slant edge ring (Fig. 2.10), or a box edge ring (Fig. 2.6), then the number outside the brackets identifies the face visible outside the ring, and the two codes identify the two possible configurations of the main position.

In an intermediate position a first order fundamental even edge flexagon is, in appearance, a multiple edge polygon (Section 2.1.1). For example, the equilateral triangle edge triple is shown in Fig. 2.1. Only one face number is visible, and this number is the intermediate position code. In both main and intermediate positions, a first order fundamental even edge flexagon can be divided into S identical sectors; so its net also has S identical sectors. In a main position there are two pats in each sector and $S=n$ (Section 4.1). The number of leaves in a sector is the same as the number of edges on the constituent polygons.

All first order fundamental even edge flexagons are simple twisted bands. The torsion, T, of a flexagon is the number of half twists (180° twists) around the band. If T is odd then a band has only have one side and it is a Möbius band (Gardner 1965, 1978, 2008; Pedersen and Pedersen 1973). If T is even then a band has two sides, and it is not a Möbius band. First order fundamental even edge flexagons exist as enantiomorphic pairs (Section 1.2) so T can be either positive or negative. It is usually taken as positive. The three topological invariants given in Section 2.2.1 are necessary, but not sufficient, for the topological description of first order fundamental even edge flexagons. The torsion is an additional topological invariant that is needed. Once a flexagon is assembled, the torsion cannot be changed. The torsion of a paper model of a flexagon is not usually obvious, but it is nevertheless an important characteristic. In general, flexagons with a high ratio of torsion to number of leaves are more stable and easier to handle. The torsion, T_S, for each of the sectors of a first order fundamental even edge flexagon is given by

$$T_S = s - 2C \tag{4.1}$$

Where $\{s\}$ is the Schläfli symbol for the constituent polygons and C is the denominator of the net symbol. For example, for the net symbol ⟨5⟩ $C=1$ and for the net symbol ⟨5/2⟩ $C=2$. To see the derivation of the equation first note that fan folding a sector into a single pat involves s half twists. Secondly, traversing the inscribed part of a flexagon figure (below) requires C 360° rotations. These rotations introduce $2C$ half twists, but in the opposite direction. Theoretical calculations of the torsion per sector can be checked experimentally by making a paper model of a sector, and counting the number of half twists needed to untwist the sector.

A flexagon symbol $S\langle s, c\rangle$ completely defines a first order fundamental even edge flexagon except for its size and whether the torsion is positive or negative. In a flexagon symbol, S is the number of sectors, $\{s\}$ is the Schläfli symbol for the constituent polygons, and $\langle c\rangle$ is the first order fundamental edge net symbol. The associated polygon for a first order fundamental even edge flexagon is the same as that for its net, so its Schläfli symbol is $\{c\}$ (Section 3.1). Even edge flexagons with the same associated polygon have similar dynamic properties. Any flexagon that has an associated polygon is a solitary flexagon. A characteristic feature of solitary flexagons is that, in a main position, pats are, in general, fan folded piles of leaves. Alternate pats can be single leaves.

A flexagon figure is a convenient graphical representation of a flexagon symbol and it is independent of the number of sectors, S. It may be constructed as described in Section 4.2.5.3. In a flexagon figure the circumscribing polygon (solid lines) is a constituent polygon, $\{s\}$. The inscribed polygon (dotted lines), with its vertices on midpoints of the edges of the circumscribing polygon, is the associated polygon for the flexagon, {c}. A flexagon figure contains most of the information needed to construct a corresponding flexagon (Section 7.2.2). The flexagon figures for all the possible first order fundamental even edge flexagons made from polygons with up to eight edges are shown in Figs. 4.1–4.6.

All first order fundamental even edge flexagons can be flexed by using a pinch flex. This is their characteristic flex. To carry out a pinch flex, starting from a main

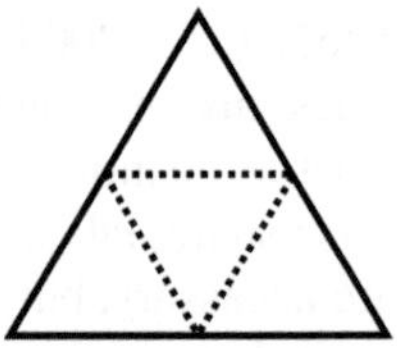

Fig. 4.1 The flexagon figure for the first order fundamental triangle even edge flexagons $S\langle 3, 3\rangle$ (Les Pook, Flexagons inside out, 2003, © Cambridge University Press 2003, reprinted with permission)

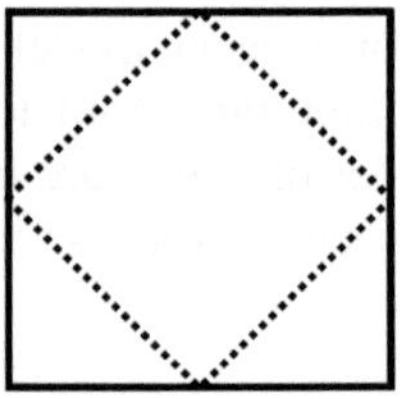

Fig. 4.2 The flexagon figure for the first order fundamental square even edge flexagons $S\langle 4, 4\rangle$ (Les Pook, Flexagons inside out, 2003, © Cambridge University Press 2003, reprinted with permission)

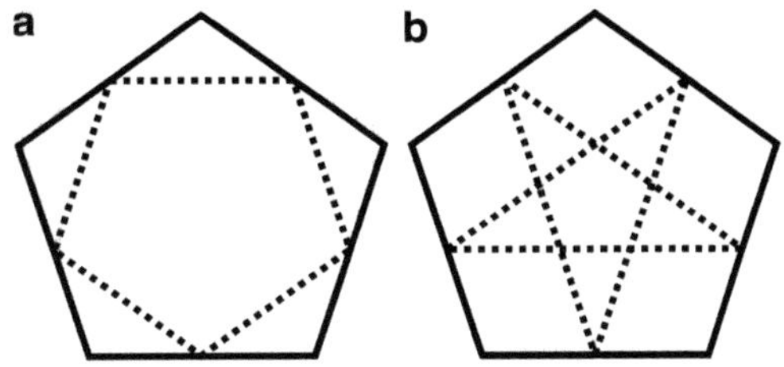

Fig. 4.3 Flexagon figures for first order fundamental pentagon even edge flexagons. (**a**) $S\langle 5, 5\rangle$. (**b**) $S\langle 5, 5/2\rangle$ (Les Pook, Flexagons inside out, 2003, © Cambridge University Press 2003, reprinted with permission)

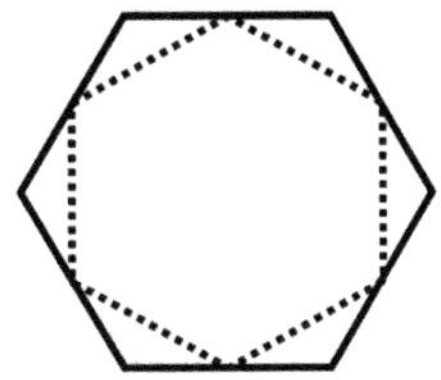

Fig. 4.4 The flexagon figure for the first order fundamental hexagon even edge flexagons $S\langle 6, 6\rangle$ (Les Pook, Flexagons inside out, 2003, © Cambridge University Press 2003, reprinted with permission)

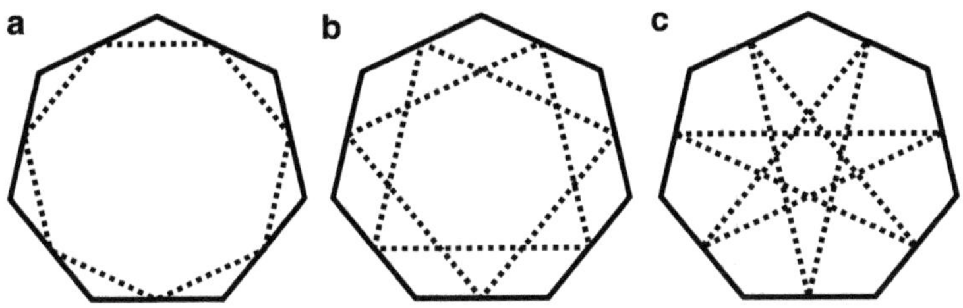

Fig. 4.5 Flexagon figures for first order fundamental heptagon even edge flexagons. (**a**) $S\langle 7, 7\rangle$. (**b**) $S\langle 7, 7/2\rangle$. (**c**) $S\langle 7, 7/3\rangle$

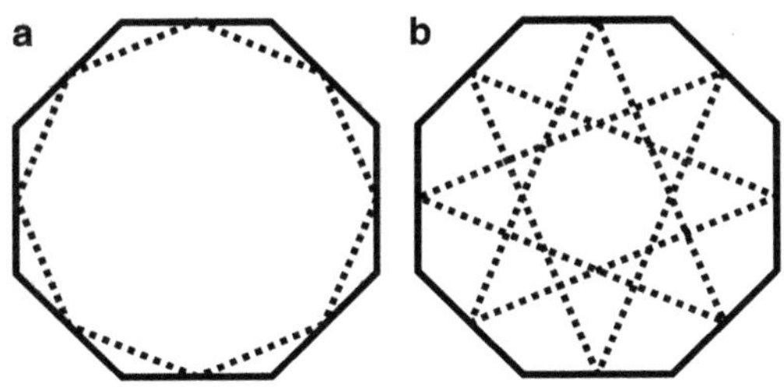

Fig. 4.6 Flexagon figures for first order fundamental octagon even edge flexagons. (**a**) $S\langle 8, 8\rangle$ (Les Pook, Flexagons inside out, 2003, © Cambridge University Press 2003, reprinted with permission) (**b**) $S\langle 8, 8/3\rangle$

position, pinch pairs of pats together to reach an intermediate position. Then unfold the flexagon to reach a new main position. Details of various types of pinch flex are given below in descriptions of flexagons. The dynamic properties of a fundamental flexagon, flexed using the pinch flex, can be characterised by using an intermediate position map. Figures 4.7–4.12 show intermediate position maps for all first order fundamental even edge flexagons consisting of polygons with up to eight edges, with nets numbered using a standard face numbering sequence (Section 4.1.1).

Fig. 4.7 Intermediate position map for the first order fundamental triangle even edge flexagons $S\langle 3, 3\rangle$ (Les Pook, Flexagons inside out, 2003, © Cambridge University Press 2003, reprinted with permission)

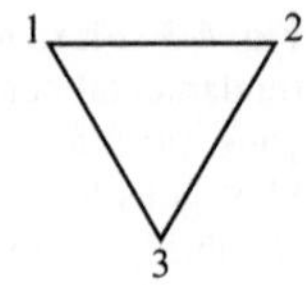

Fig. 4.8 Intermediate position map for the first order fundamental square even edge flexagons $S\langle 4, 4\rangle$ (Les Pook, Flexagons inside out, 2003, © Cambridge University Press 2003, reprinted with permission)

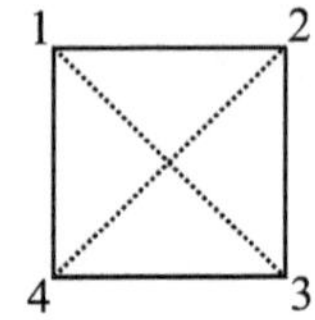

Fig. 4.9 Intermediate position map for the first order fundamental pentagon even edge flexagons $S\langle 5, 5\rangle$ and $S\langle 5, 5/2\rangle$ (Les Pook, Flexagons inside out, 2003, © Cambridge University Press 2003, reprinted with permission)

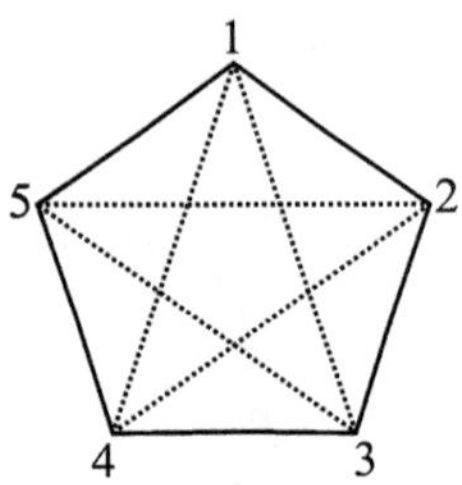

Fig. 4.10 Intermediate position map for the first order fundamental hexagon even edge flexagons $S\langle 6, 6\rangle$ (Les Pook, Flexagons inside out, 2003, © Cambridge University Press 2003, reprinted with permission)

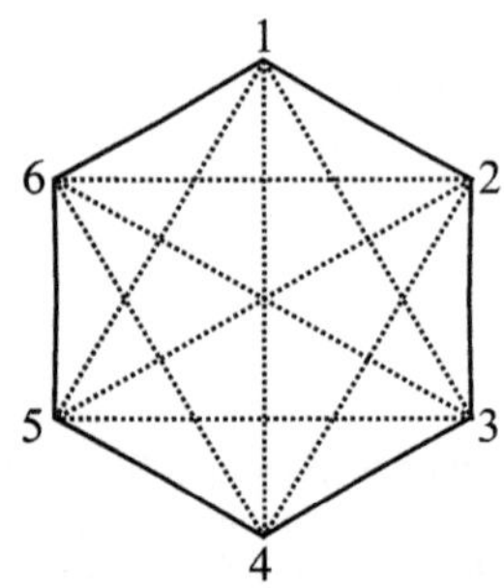

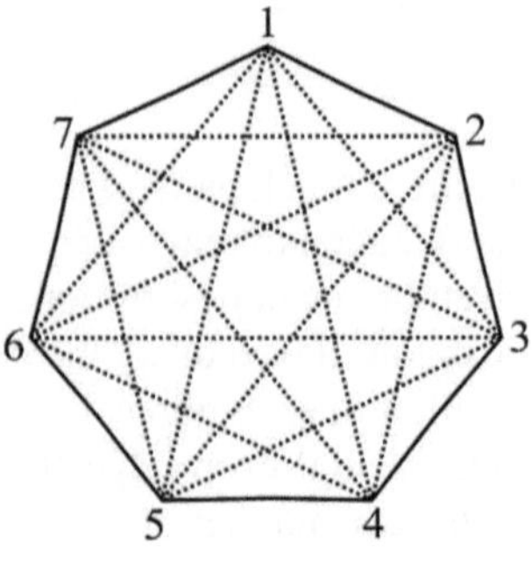

Fig. 4.11 Intermediate position map for first order fundamental heptagon even edge flexagons $S\langle 7, 7\rangle$, $S\langle 7, 7/2\rangle$ and $S\langle 7, 7/3\rangle$

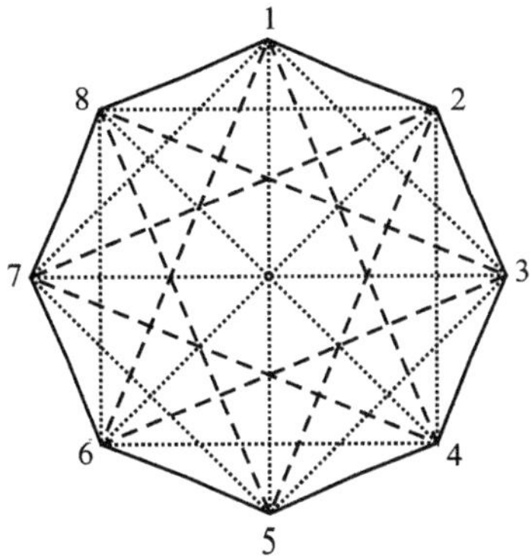

Fig. 4.12 Intermediate position map for the first order fundamental octagon even edge flexagons $S\langle 8, 8\rangle$ and $S\langle 8, 8/3\rangle$

Some of the intermediate position maps also characterise the dynamic properties of other types of flexagon.

In an intermediate position map the sequentially numbered vertices represent intermediate positions. Solid lines joining adjacent intermediate positions represent principal main positions. Thus, the line joining intermediate positions 1 and 2 represents principal main positions 1(2) and 2(1). The convex polygon formed by the solid lines represents the principal cycle. In tables, data for principal cycles are shown in bold. Dotted (or dashed) lines represent subsidiary main positions. Lines of equal length represent main positions of the same appearance and curvature. Lines of different lengths represent main positions with different appearances and curvatures. All the lines that can be drawn linking pairs of intermediate positions represent main positions. Usually, all the main positions in an intermediate position map can be traversed. In a deficient flexagon this cannot be done without disconnecting a hinge, refolding the flexagon, and reconnecting the hinge. Deficient flexagons are usually excluded from lists of possible flexagons. In terms of linkage theory (Demaine and O'Rourke 2007) a deficient flexagon is a locked linkage, so deficient flexagons could also be called locked flexagons.

In principle, it is possible to flex between any two intermediate positions via the main position represented by the line joining them. This is always possible without bending leaves if the main position is a flat main position. If it is skew main position, leaf bending is sometimes required. If the main position is a slant main position or a box position, flexing between the two intermediate positions may be difficult or impossible, even if leaf bending is allowed. In other words, flexing between the two intermediate positions may be difficult or impossible if the curvature of the main position is $>0°$.

In the intermediate position map for a first order fundamental even edge flexagon there is always one principal cycle, and pairs of faces on principal main positions appear in cyclic order. The number of principal main positions in the principal cycle is the same as the number of edges on the constituent polygons. In most first order fundamental even edge flexagons it is also possible to traverse one or more subsidiary cycles. The number of subsidiary main positions in a subsidiary cycle is either the same as the number of edges on the constituent polygons or a factor of the

Table 4.1 General properties of first order fundamental even edge flexagons

Leaf type	Flexagon symbol	Fundamental edge net	Number of cycle types	Torsion per sector
Triangle	*S*⟨3, 3⟩	⟨3⟩	1	1
Square	*S*⟨4, 4⟩	⟨4⟩	1	2
Pentagon	*S*⟨5, 5⟩	⟨5⟩	2	3
Pentagon	*S*⟨5, 5/2⟩	⟨5/2⟩	2	1
Hexagon	*S*⟨6, 6⟩	⟨6⟩	2	4
Heptagon	*S*⟨7, 7⟩	⟨7⟩	3	5
Heptagon	*S*⟨7, 7/2⟩	⟨7/2⟩	3	3
Heptagon	*S*⟨7, 7/3⟩	⟨7/3⟩	3	1
Octagon	*S*⟨8, 8⟩	⟨8⟩	3	6
Octagon	*S*⟨8, 8/3⟩	⟨8/3⟩	3	2
Enneagon	*S*⟨9, 9⟩	⟨9⟩	4	7
Enneagon	*S*⟨9, 9/2⟩	⟨9/2⟩	4	5
Enneagon	*S*⟨9, 9/4⟩	⟨9/4⟩	4	1
Dodecagon	*S*⟨12, 12⟩,	⟨12⟩	4	10
Dodecagon	*S*⟨12, 12/5⟩	⟨12/5⟩	4	2

number of edges. The number of main positions in a cycle is indicated by a prefix number. For example, in a *4-cycle* there are four main positions, all of which have the same appearance.

If a first order fundamental even edge flexagon has two (or more) cycles with the same number of main positions as there are edges on the constituent polygons, then the choice of which cycle is the principal cycle depends on the face numbering sequence used. Use of the standard face numbering sequences described in Section 4.1.1 means that the definition of the principal cycle is unambiguous, and also means that general properties are consistent between different types of first order fundamental even edge flexagons. Some general properties of all possible first order fundamental even edge flexagons, made from polygons with up to nine edges, are given in Table 4.1. Also included are properties of the two possible first order fundamental even edge dodecagon flexagons. The properties included in the table are independent of the number of sectors.

Nets for some first order fundamental even edge flexagons are given below. If desired, any net for a first order fundamental even edge flexagon not described below is easily designed. To do this, start by selecting the appropriate first order fundamental edge net in Table 4.1, and decide on the number of sectors. Next, make the number of leaves in the net, for each of the sectors, equal to the number of edges on the constituent polygons. Finally, apply the appropriate standard face numbering sequence (Section 4.1.1).

4.2.2 Ring Even Edge Flexagons

A ring even edge flexagon is a first order fundamental even edge flexagon that has some main positions that are, in appearance, flat regular even edge rings. All the

Table 4.2 First order fundamental even edge flexagons that are ring even edge flexagons

Leaf type	Number of sectors	Number of polygons in ring	Flexagon symbols
Triangle	3	6	3⟨3, 3⟩
Square	2	4	2⟨4, 4⟩
Pentagon	5	10	5⟨5, 5⟩, 5⟨5, 5/2⟩
Hexagon	3	6	3⟨6, 6⟩
Octagon	2	4	2⟨8, 8⟩, 2⟨8, 8/3⟩
Octagon	4	8	4⟨8, 8⟩, 4⟨8, 8/3⟩
Enneagon	3	6	3⟨9, 9⟩, 3⟨9, 9/2⟩, 3⟨9/4⟩
Dodecagon	2	4	2⟨12, 12⟩, 2⟨12, 12/5⟩

Table 4.3 Properties of first order fundamental triangle even edge flexagons. Principal cycles are in bold

Flexagon symbol	Typical main position	Cycle type	Number of cycles	Main position type	Ring symbol	Sector symbol[a]	Curvature
2⟨3, 3⟩	**1(2)**	**3-cycle**	**1**	**Slant**	**4(60°)**	**⟨3, 3, 2, 1⟩**	**120°**
3⟨3, 3⟩	**1(2)**	**3-cycle**	**1**	**Flat**	**6(60°)**	**⟨3, 3, 2, 1⟩**	**0°**
5⟨3, 3⟩	**1(2)**	**3-cycle**	**1**	**Skew**	**10(60°)**	**⟨3, 3, 2, 1⟩**	**–240°**

[a]See Section 4.2.5.1.

possible ring even edge flexagons with up to nine edges on the constituent polygons and up to ten polygons in the ring are listed in Table 4.2. Also included are two dodecagon ring even edge flexagons.

4.2.3 First Order Fundamental Triangle Even Edge Flexagons

The first order fundamental triangle even edge flexagons described in this section are 2⟨3, 3⟩ and 3⟨3, 3⟩. The latter is the well known trihexaflexagon. It was the first type of flexagon to be discovered (Conrad and Hartline 1962; Pook 2003). Its special name does not fit easily into classification schemes. 'Tri' refers to its three faces, and 'hexa' to the hexagonal appearance of a principal main position. Some of the properties of the two flexagons are given in Table 4.3, also see Tables 4.1 and 4.2. The torsion per sector is 1. The flexagon figure is shown in Fig. 4.1.

Some flexagons exist in different forms with different forms belonging to different families. For convenience, individual forms are sometimes regarded as flexagons in their own right and given separate names. A flex sequence that transforms one form into another is a transformation between flexagons. A transformation between flexagons is possible between the five sector first order fundamental triangle even edge flexagon 5⟨3, 3⟩ and the fundamental irregular ring 12 triangle even edge flexagon type A (Section 9.2.2). Some of the properties of 5⟨3, 3⟩ are given in Table 4.3.

The net for the first order fundamental triangle even edge flexagon 2⟨3, 3⟩ is shown in Fig. 4.13. As assembled, the flexagon is in intermediate position 1. This is, in appearance, an equilateral triangle edge pair. From here it can be opened up into either principal main position 1(2) or principal main position 1(3). These are slant regular even edge rings of four equilateral triangles. It is a deficient flexagon (Section 4.2.1) because the principal 3-cycle shown in the intermediate position map (Fig. 4.7) cannot be traversed without disconnecting a hinge, refolding the flexagon, and reconnecting the hinge. It is included because of its relationship to the trihexaflexagon.

The net for the trihexaflexagon is shown in Fig. 4.14. It is a ring even edge flexagon (previous section). As assembled, it is in principal main position 2(1), which is a flat regular even edge ring of six equilateral triangles (Fig. 1.1a). It can be flexed around the principal 3-cycle shown in the intermediate position map (Fig. 4.7) by using the threefold pinch flex, in which threefold rotational symmetry is maintained during flexing. Starting from principal main position 2(1), pinch pats together in three pairs, with leaves numbered 2 on the outside to reach intermediate position 2, which is an equilateral triangle edge triple (Fig. 2.1). Figure 4.15 shows the flexagon part way between a principal main position and an intermediate position. There are two ways of pinching three pairs of pats together with leaves numbered 2 on the outside, only one of which works. From intermediate position 2, open the trihexaflexagon into principal main position 3(2) and so on round the principal 3-cycle. For an animation of the trihexaflexagon being flexed, see Highland Games (2008). For a truncated flexagon version of the trihexaflexagon, see Section 8.2.4.

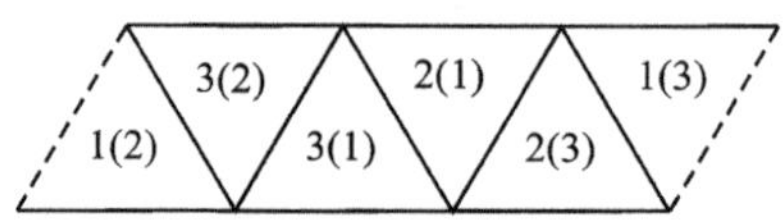

Fig. 4.13 Net for the first order fundamental triangle even edge flexagon 2⟨3, 3⟩. One copy needed. Fold until leaves numbered 1 are visible

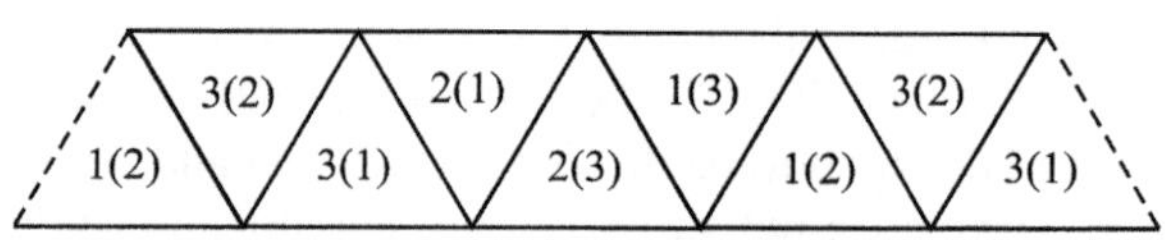

Fig. 4.14 Net for the trihexaflexagon 3⟨3, 3⟩. One copy needed

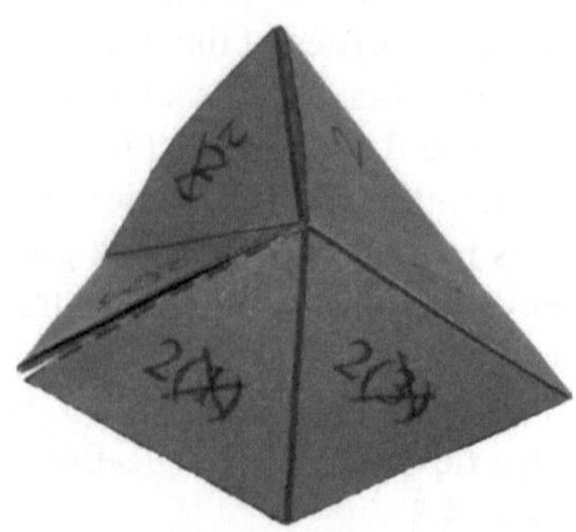

Fig. 4.15 A trihexaflexagon 3⟨3, 3⟩ part way between a principal main position and an intermediate position

4.2.4 First Order Fundamental Square Even Edge Flexagons

Two first order fundamental square even edge flexagons are described in this section. One is the two sector version 2⟨4, 4⟩. This is a ring even edge flexagon (Table 4.2). It is described briefly in Section 1.4.2. For a truncated flexagon version, see Section 10.2.4.3. The other is the four sector version 4⟨4, 4⟩. This four sector version also exists in other forms, some of which are described in Sections 9.3.3 and 12.6.2. Transformations between flexagons with these other forms are possible. The various forms are called, collectively, the octopus flexagon. Some of the properties of 2⟨4, 4⟩ and 4⟨4, 4⟩ are given in Table 4.4, also see Table 4.1. The torsion per sector is 2. The flexagon figure is shown in Fig. 4.2.

A transformation between flexagons is possible between the three sector first order fundamental square even edge flexagon 3⟨4, 4⟩ and the fundamental irregular ring six square even edge flexagon (Section 9.3.2). Some of its properties are given in Table 4.4.

The net for the first order fundamental square even edge flexagon 2⟨4, 4⟩ is shown in Fig. 4.16a. Hinges are identified by adjacent letters. As assembled, the flexagon is in principal main position 2(1). This is, in appearance, a flat regular even edge ring of four squares (Fig. 1.1b). It can be flexed around the principal 4-cycle shown in the intermediate position map (Fig. 4.8) by using the twofold pinch flex. When applied to square flexagons, this is sometimes called the book flex. Starting from principal main position 2(1), fold the flexagon in two about hinges D to reach intermediate position 2, with only leaves numbered 2 visible. This is a square edge pair (Fig. 1.5). Then unfold about hinges B to reach principal

Table 4.4 Properties of first order fundamental square even edge flexagons. Principal cycles are in bold

Flexagon symbol	Typical main position	Cycle type	Number of cycles	Main position type	Ring symbol	Sector symbol[a]	Curvature
2⟨4, 4⟩	**1(2)**	**4-cycle**	**1**	**Flat**	**4(90°)**	**⟨4, 4, 3, 1⟩**	**0°**
2⟨4, 4⟩	1(3)	None	–	Box	4(0°)	⟨4, 4, 2, 2⟩	360°
3⟨4, 4⟩	**1(2)**	**4-cycle**	**1**	**Skew**	**6(90°)**	**⟨4, 4, 3, 1⟩**	**–180°**
3⟨4, 4⟩	1(3)	None	–	Box	6(0°)	⟨4, 4, 2, 2⟩	360°
4⟨4, 4⟩	**1(2)**	**4-cycle**	**1**	**Skew**	**8(90°)**	**⟨4, 4, 3, 1⟩**	**–360°**
4⟨4, 4⟩	1(3)	None	–	Box	8(0°)	⟨4, 4, 2, 2⟩	360°

[a]See Section 4.2.5.1.

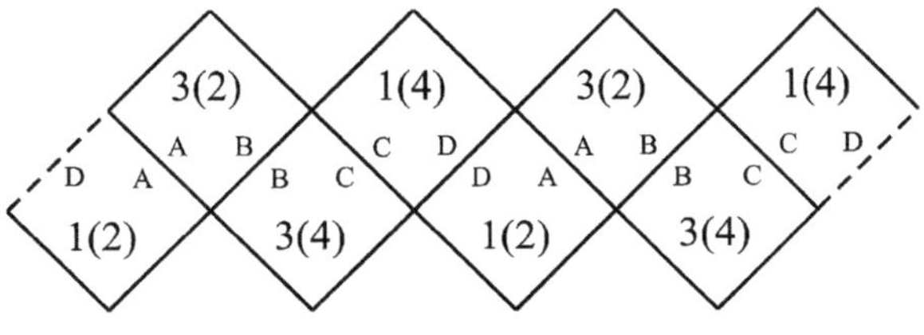

Fig. 4.16 Net for first order fundamental square even edge flexagons. (**a**) 2⟨4, 4⟩. One copy needed. (**b**) 4⟨4, 4⟩. Two copies needed

main position 3(2). By repeating the twofold pinch flex, using appropriate hinges, it is possible to traverse the principal 4-cycle shown in the intermediate position map.

From intermediate position 2, the flexagon can be opened into subsidiary main position 2(4). This is a box edge ring of four squares (Fig. 2.5) with leaves numbered 4 visible inside the box. Such box positions are a special case of subsidiary main positions because they do not form part of a cycle that can be traversed. Lines representing box positions pass through the centre of an intermediate position map. Starting from subsidiary main position 2(4), it is not possible to flex directly to subsidiary main position 4(2), with leaves numbered 2 inside the box. This is because a box position cannot be turned inside out. However, this can be done indirectly by traversing halfway round the principal 4-cycle between intermediate position 2 and intermediate position 4.

The net for the first order fundamental square even edge flexagon 4⟨4, 4⟩ is shown in Fig. 4.16b. As assembled, the flexagon is in principal main position 2(1), which is, in appearance, a skew regular even edge ring of eight squares. It can be flexed around the principal 4-cycle shown in the intermediate position map (Fig. 4.8) by using the fourfold pinch flex, in which fourfold rotational symmetry is maintained during flexing. Starting from principal main position 2(1), pinch pats together in four pairs, with leaves numbered 2 on the outside to reach intermediate position 2, which is a square edge quadruple. From intermediate position 2, open the flexagon into principal main position 3(2) and so on round the principal 4-cycle. From intermediate position 2 the flexagon can be opened into subsidiary main position 2(4). This is a box edge ring of eight squares. Starting from subsidiary main position 2(4) it is not possible to flex directly to subsidiary main position 4(2). However, this can be done indirectly by traversing halfway round the principal 4-cycle between intermediate position 2 and intermediate position 4.

4.2.5 Detailed Analysis of Flexagons

The structure and dynamic properties of some flexagons are very complicated and difficult to describe in detail. Some of the detailed analyses that are possible are illustrated below by detailed analysis of the dynamic properties of the first order fundamental square even edge flexagon 2⟨4, 4⟩ (previous section). These analyses are among those used to derive information on flexagons included in this book.

4.2.5.1 Sector Symbols

Information on the numbers of leaves in individual pats of main positions can be derived from intermediate position maps. For example, consider the intermediate position map for first order fundamental square even edge flexagons $S\langle 4, 4\rangle$, where S is the number of sectors (Fig. 4.8). The solid line joining intermediate positions 1 and 2 represents principal main position 1(2). There are two routes around the principal cycle (solid lines) between these two intermediate positions. These routes pass through, respectively, 1 and 3 principal main positions. This means that alternate pats in a principal main position contain one and three leaves respectively. Similarly, consider intermediate positions 1 and 3. The dotted line joining these positions represents subsidiary main position 1(3), which is a box position. Both routes around the principal cycle between the two intermediate positions pass over two principal main positions, so all four pats contain two leaves. This method can be used to determine numbers of leaves in pats for any main position of any first order fundamental even edge flexagon.

In the flexagon symbol $S\langle s, c\rangle$ (Section 4.2.1) S is the number of sectors, $\{s\}$ is the Schläfli symbol for the constituent polygons and $\langle c\rangle$ is the first order fundamental edge net symbol. A sector symbol, $\langle s, c, m, n\rangle$, defines both the type of flexagon, except for the number of sectors, and the number of leaves in adjacent pats of a main position. In a sector symbol, s and c have the same meanings as in a flexagon symbol, and m and n are the numbers of leaves in the two adjacent pats in a sector of a main position. They are chosen such that $m \geq n$. S is omitted because a sector symbol is independent of the number of sectors. Thus, the sector symbol for the principal main positions of the first order fundamental square even edge flexagon $2\langle 4, 4\rangle$ is $\langle 4, 4, 3, 1\rangle$ (Table 4.4), and for the box positions it is $\langle 4, 4, 2, 2\rangle$. A first order fundamental triangle flexagon $S\langle 3, 3\rangle$ has only one type of main position, and the sector symbol is $\langle 3, 3, 2, 1\rangle$ (Table 4.3).

4.2.5.2 Tuckerman Diagrams

In a Tuckerman diagram (Conrad and Hartline 1962; Gardner 1965, 2008; Hilton and Pedersen 1994; Pook 2003) main position codes for a cycle are shown at the vertices of a polygon, and lines joining the vertices represent intermediate positions. Hence, a Tuckerman diagram is a dual of a cycle in an intermediate position map (Section 4.2.1). In an intermediate position map, vertices represent intermediate positions and the lines joining them represent main positions. The choice of whether to represent cycles by Tuckerman diagrams or intermediate position maps is partly a matter of convenience and partly a matter of taste.

The Tuckerman diagram for the principal cycle of the first order fundamental square even edge flexagon $2\langle 4, 4\rangle$ is shown in Fig. 4.17. Conventionally, in a Tuckerman diagram, main position codes are shown as they appear when a cycle is traversed without turning the flexagon over, and arrows are added to show the direction of traverse. For example, at principal main position 2(1), the upper face is numbered 2 and the lower face is numbered 1. If the flexagon is flexed using the

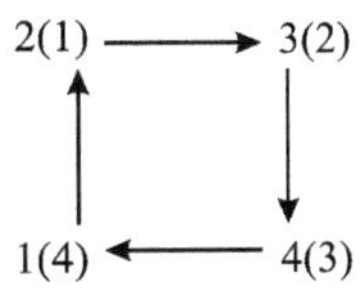

Fig. 4.17 Tuckerman diagram for the principal 4-cycle of the first order fundamental square even edge flexagons $S\langle 4, 4\rangle$ (Les Pook, Flexagons inside out, 2003, © Cambridge University Press 2003, reprinted with permission)

twofold pinch flex, then principal main position 3(2) is reached, as indicated by the arrow. The upper face is numbered 3 and the lower face is numbered 2. If the flexagon is turned over then the direction of traverse is reversed.

4.2.5.3 Full Maps

A full map is more detailed than a Tuckerman diagram, and includes structural details of main positions and intermediate positions (Pook 2003). The full map for the first order fundamental square even edge flexagon 2⟨4, 4⟩ is shown in Fig. 4.18. Its net is shown is shown in Fig. 4.19 with a dotted line joining the centre points of hinges, but without the face markings (cf. Fig. 4.16a). The sketches in the full map are identified by main position codes and intermediate position codes. In the sketches, hinges are identified by the hinge letters shown in Fig. 4.16a, and the dotted line circuits joining the centres of hinges would appear if the flexagon were made from the net shown in Fig. 4.19 using a transparent material, as suggested by

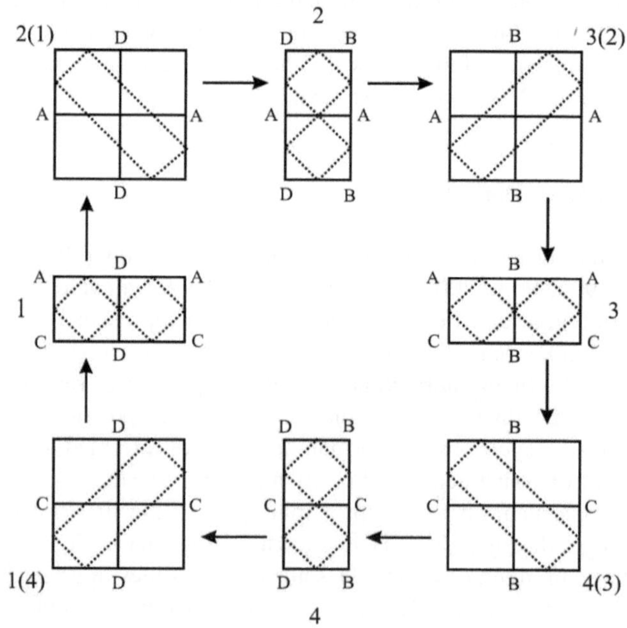

Fig. 4.18 Full map for the first order fundamental square even edge flexagon 2⟨4, 4⟩ (Les Pook, Flexagons inside out, 2003, © Cambridge University Press 2003, reprinted with permission)

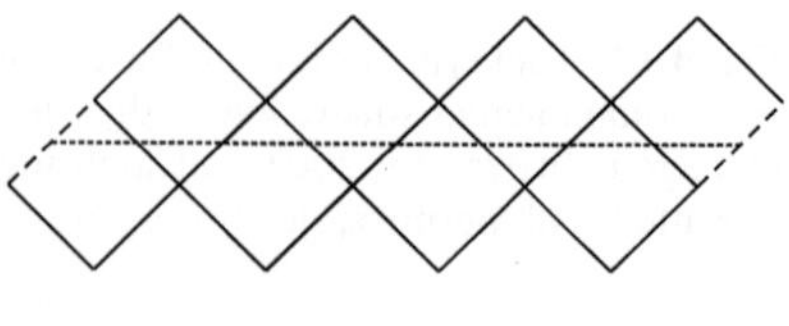

Fig. 4.19 Net for the first order fundamental square even edge flexagon 2⟨4, 4⟩ with a dotted line joining the centre points of hinges

Shuttleworth (2006). The circuits provide a convenient shorthand description of the structure of the pats in various positions, but do not distinguish between the two enantiomorphs of the flexagon. However, it is immediately clear that both principal main positions and intermediate positions have twofold rotational symmetry about a vertical axis through the centre.

Principal main positions can be divided into two sectors, each containing two pats, in two different ways. For example. in principal main position 2(1) (top left diagram) the division into sectors could either be along hinge line A-A or along hinge line D-D. Intermediate positions can also be divided into two identical sectors, each consisting of one pat. The dotted line circuit in each sector of an intermediate position is the inscribed portion of the flexagon figure (Fig. 4.2). In other words, the inscribed polygon of a flexagon figure is the circuit for one pat of an intermediate position. Hence, a flexagon figure can be constructed by inscribing a circuit for an intermediate position at the midpoints of the edges of a constituent polygon. It follows that a flexagon figure can be used to construct the net for a corresponding flexagon (Section 7.2.1).

The full map shows, in more detail than the Tuckerman diagram (Fig. 4.17), what happens when the flexagon is flexed around the principal 4-cycle, in the direction of the arrows, by using the twofold pinch flex. The orientations of the sketches are arbitrary, and have been chosen to assist visualisation of the dynamic properties of the flexagon. As assembled, the flexagon is in principal main position 2(1). To flex to principal main position 3(2), fold the flexagon in two, along hinge line D-D, to reach intermediate position 2, with leaves numbered 1 concealed. Keep the hinge line uppermost, and unfold about hinge line B-B to reach principal main position 3(2). The arrangement of the leaves on face two changes during the flex, in that different vertices appear at the centre of the face. In effect, leaves rotate through 90°, with alternate leaves rotating in opposite directions. This *rotation of leaves* adds to decorative possibilities (Pook 2003). To traverse the complete principal 4-cycle, continue flexing, changing hinges as appropriate at each position.

In a principal main position, there are two degrees of freedom in the sense that rotation can take place about two different hinge lines. In principal main position 2(1) this is either hinge line A-A or hinge line D-D. Four different intermediate positions can be reached from principal main position 2(1). However, two of these intermediate positions are dead ends that do not lead to another principal main position, so they are not shown on the full map. There are four degrees of freedom in the intermediate positions. Thus, at intermediate position 2 the flexagon can be unfolded about hinge lines B-B or D-D, or about hinge line A-A, it can also be opened up into box position 2(4) using hinge lines C-C (not shown) and A-A, and it can be folded in two about hinges A-A.

4.2.5.4 Flexagon Diagrams

In the full map for the first order fundamental square even edge flexagon 2⟨4, 4⟩ (Fig. 4.18), the sketches of all four principal main position diagrams are identical, except for the hinge letters. Removing the hinge letters leads to the flexagon diagram

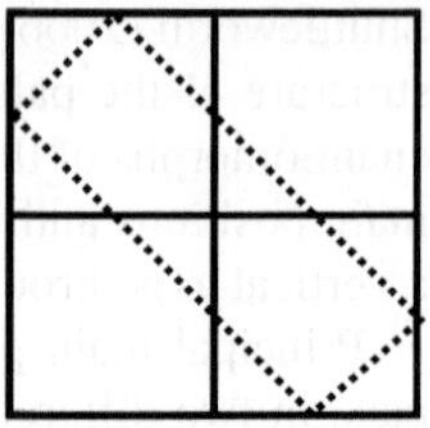

Fig. 4.20 Flexagon diagram for the principal main positions of the first order fundamental square even edge flexagon 2⟨4, 4⟩

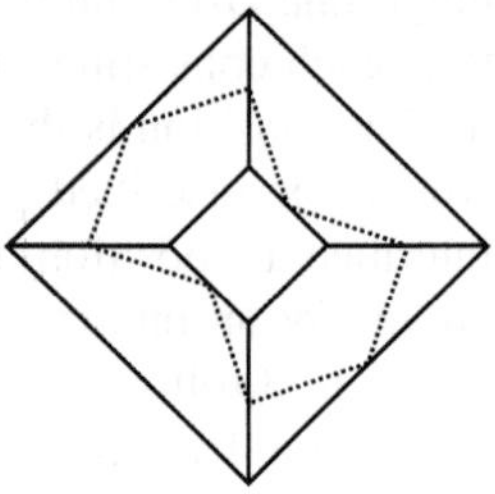

Fig. 4.21 Flexagon diagram, projected onto a plane as a dual map, for the box positions of the first order fundamental square even edge flexagon 2⟨4, 4⟩

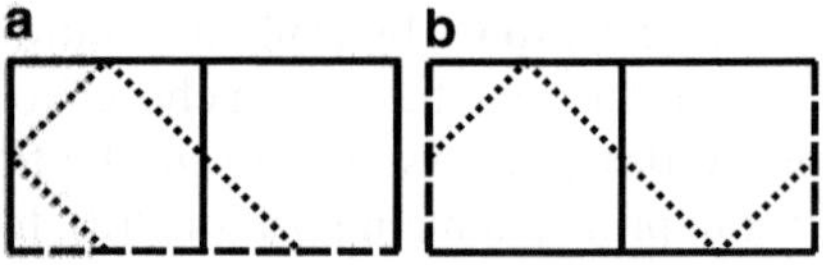

Fig. 4.22 Sector diagrams for the first order fundamental square even edge flexagon 2⟨4, 4⟩. (**a**) Principal main positions. (**b**) Box positions

for the principal main position positions (Fig. 4.20). The flexagon diagram for the box positions is shown in Fig. 4.21. Box positions are not flat, so the diagram is shown projected onto a plane as a dual map (Coxeter 1963). Dividing flexagon diagrams into identical sectors leads to sector diagrams. Division of flexagon diagrams for first order fundamental even edge flexagons into sectors can be done in two different ways. The resultant sector diagrams are an enantiomorphic pair, and are equivalent. Sector diagrams for principal main positions and box positions of the first order fundamental square even edge flexagon 2⟨4, 4⟩ are shown in Fig. 4.22. That for box positions is unfolded onto a plane. Dashed lines show where the division into sectors has been made.

4.2.6 First Order Fundamental Pentagon Even Edge Flexagons

The first order fundamental pentagon even edge flexagons described in this section are 2⟨5, 5⟩, 5⟨5, 5⟩, 2⟨5, 5/2⟩ and 5⟨5, 5/2⟩. The five sector versions, 5⟨5, 5⟩ and 5⟨5, 5/2⟩ are ring even edge flexagons (Section 4.2.2). Some of the properties of the four flexagons are given in Table 4.5, also see Tables 4.1 and 4.2. The torsion per sector is 3 for 2⟨5, 5⟩ and 5⟨5, 5⟩, and 1 for 2⟨5, 5/2⟩ and 5⟨5, 5/2⟩. The flexagon figures are shown in Fig. 4.3.

A transformation between flexagons is possible between the three sector first order fundamental pentagon even edge flexagon 3⟨5, 5⟩ and the fundamental irregular ring six pentagon even edge flexagon type A (Section 9.4.2). Some of the properties of 3⟨5, 5⟩ are given in Table 4.5.

The net for the first order fundamental pentagon even edge flexagon 2⟨5, 5⟩ is shown in Fig. 4.23a. This two sector flexagon is flexed using the twofold pinch flex. As assembled, the flexagon is in intermediate position 1, which is a pentagon edge pair. From here it can be opened into either principal main position 1(2) or principal main position 1(5), which are skew regular even edge rings of four regular pentagons (Fig. 2.7), then closed into a new intermediate position, and so on round the principal 5-cycle shown by the solid lines in the intermediate position map (Fig. 4.9). The continuous path when flexing between adjacent intermediate positions on the principal cycle is aesthetically satisfying.

Intermediate position 1 can also be opened into either subsidiary main position 1(3) or subsidiary main position 1(4), which are slant regular even edge rings of four regular pentagons (Fig. 4.24). The high curvature (216°) means that these cannot be turned inside out, so the subsidiary 5-cycle shown by the dotted lines in the intermediate position map cannot be traversed directly. All the subsidiary main positions can be visited by flexing via principal main positions.

The net for the first order fundamental pentagon even edge flexagon 5⟨5, 5⟩ is shown in Fig. 4.23b. This five sector flexagon is flexed using the fivefold pinch flex. Paper models of the flexagon work reasonably well, but flexing is awkward

Table 4.5 Properties of first order fundamental pentagon even edge flexagons. Principal cycles are in bold

Flexagon symbol	Typical main position	Cycle type	Number of cycles	Main position type	Ring symbol	Sector symbol	Curvature
2⟨5, 5⟩	**1(2)**	**5-cycle**	**1**	**Skew**	**4(108°)**	**⟨5, 5, 4, 1⟩**	**−72°**
2⟨5, 5⟩	1(3)	5-cycle	1	Slant	4(36°)	⟨5, 5, 3, 2⟩	216°
3⟨5, 5⟩	**1(2)**	**5-cycle**	**1**	**Skew**	**6(108°)**	**⟨5, 5, 4, 1⟩**	**−288°**
3⟨5, 5⟩	1(3)	5-cycle	1	Slant	6(36°)	⟨5, 5, 3, 2⟩	144°
5⟨5, 5⟩	**1(2)**	**5-cycle**	**1**	**Skew**	**10(108°)**	**⟨5, 5, 4, 1⟩**	**−720°**
5⟨5, 5⟩	1(3)	5-cycle	1	Flat	10(36°)	⟨5, 5, 3, 2⟩	0°
2⟨5, 5/2⟩	**1(2)**	**5-cycle**	**1**	**Slant**	**4(36°)**	**⟨5, 5/2, 4, 1⟩**	**216°**
2⟨5, 5/2⟩	1(3)	5-cycle	1	Skew	4(108°)	⟨5, 5/2, 3, 2⟩	−72°
5⟨5, 5/2⟩	**1(2)**	**5-cycle**	**1**	**Flat**	**10(36°)**	**⟨5, 5/2, 4, 1⟩**	**0°**
5⟨5, 5/2⟩	1(3)	5-cycle	1	Skew	10(108°)	⟨5, 5/2, 3,2⟩	−729°

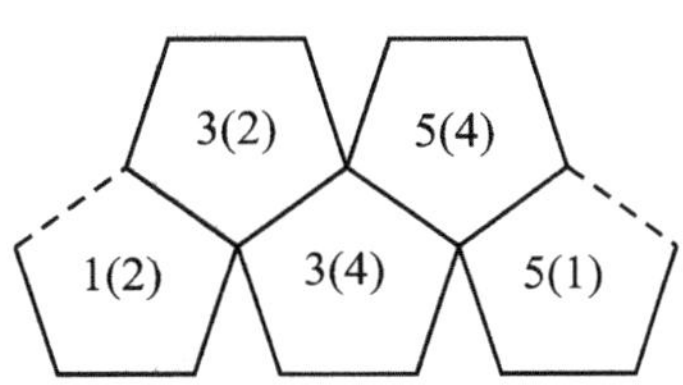

Fig. 4.23 Net for first order fundamental pentagon even edge flexagons and the fundamental pentagon unagon. (**a**) 2⟨5, 5⟩. Two copies needed. Fold until leaves numbered 1 are visible. (**b**) 5⟨5, 5⟩. Five copies needed. Fold until leaves numbered 1 and 3 are visible. (**c**) 1⟨5, 5⟩. One copy needed. Fold until leaves numbered 1 are visible

Fig. 4.24 A flexagon as a slant regular edge ring of four regular pentagons

Fig. 4.25 A flexagon as a pentagon edge quintuple

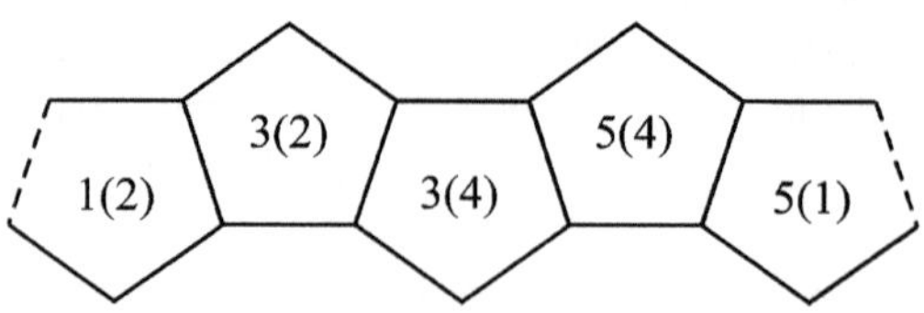

Fig. 4.26 Net for first order fundamental pentagon even edge flexagons. (**a**) 2⟨5, 5/2⟩. Two copies needed. Fold until leaves numbered 1 are visible. (**b**) 5⟨5, 5/2⟩. Five copies needed (Les Pook, Flexagons inside out, 2003, © Cambridge University Press 2003, reprinted with permission)

because of the need to maintain fivefold rotational symmetry with one pair of hands. As assembled, the flexagon is in subsidiary main position 1(3). This is a flat regular even edge ring of ten pentagons. Pinch pats together in pairs to reach an intermediate position, which is a pentagon edge quintuple (Fig. 4.25). Then open it to reach a new subsidiary main position, and so on round the subsidiary 5-cycle shown by the dotted lines in the intermediate position map (Fig. 4.9). Reference to the intermediate position map may be needed to ensure that an intermediate position is opened correctly into the next subsidiary main position. Intermediate positions can also be opened into principal main positions, which are skew regular edge rings of ten regular pentagons, and the principal 5-cycle shown by the solid lines in the intermediate position map can be traversed. Paper clips may be needed to keep the flexagon under control.

The net for the first order fundamental pentagon even edge flexagon 2⟨5, 5/2⟩ is shown in Fig. 4.26a. This two sector flexagon is flexed using the twofold pinch flex. As assembled the flexagon is in intermediate position 1, which is a pentagon edge pair.

From here it can be opened into either subsidiary main position 1(3) or subsidiary main position 1(4), which are skew regular even edge rings of four regular pentagons (Fig. 2.7), then closed into a new intermediate position, and so on round the subsidiary 5-cycle shown by the dotted lines in the intermediate position map (Fig. 4.9).

Intermediate position 1 can also be opened into either principal main position 1(2) or principal main position 1(3), which are slant regular even edge rings of four regular pentagons (Fig. 4.24). The high curvature (216°) means that these cannot be turned inside out, so the principal 5-cycle shown by the solid lines in the intermediate position map cannot be traversed directly. All the principal main positions can be visited by flexing via subsidiary main positions.

In comparison with the first order fundamental pentagon even edge flexagon 2⟨5, 5⟩, the appearances of the principal and subsidiary main positions are interchanged. This interchange can be avoided by using an alternative net for 2⟨5, 5/2⟩ (Fig. 4.27). This has a non standard face numbering sequence. Flexing instructions are identical to those for 2⟨5, 5⟩ (above). In particular, the appearances of principal main positions and subsidiary main positions are the same as those of 2⟨5, 5⟩. In other words, what were principal main positions using the standard face numbering sequence have become subsidiary main positions and vice versa. Hence, the use of a non standard face numbering sequence has lead to ambiguity in deciding which 5-cycle is the principal 5-cycle.

The torsion of the first order fundamental pentagon even edge flexagon 2⟨5, 5/2⟩ is lower than that of 2⟨5, 5⟩, and the flexagon is unstable in the sense that it is easily muddled and also tends to collapse into an open, twisted band. In an open band both faces of all (or most) of the leaves are visible. Figure 4.28 shows 2⟨5, 5/2⟩ as an open, twisted band. The flexagon is easily reassembled from an open band by folding like face numbers together, leaving those of the desired main position visible. Deliberately collapsing a flexagon into an open band, and then reassembling it into another main position, is called a band flex, and is sometimes the easiest way of flexing from one main position to another.

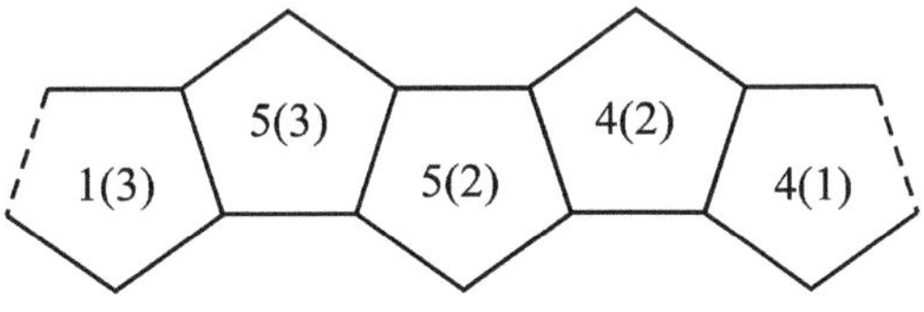

Fig. 4.27 Alternative net for the first order fundamental pentagon even edge flexagon 2⟨5, 5/2⟩. Two copies needed. Fold until leaves numbered 1 are visible

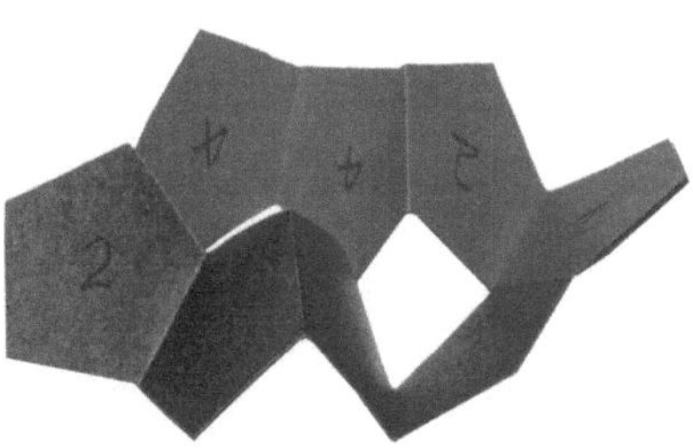

Fig. 4.28 The first order fundamental pent agon even edge flexagon 2⟨5, 5/2⟩ as an open, twisted band

The net for the first order fundamental pentagon even edge flexagon 5⟨5, 5/2⟩ is shown in Fig. 4.26b. This five sector flexagon is very unstable and is unsatisfactory as a paper model.

4.2.7 First Order Fundamental Hexagon Even Edge Flexagons

The first order fundamental hexagon even edge flexagons described in this section are 2⟨6, 6⟩ and 3⟨6, 6⟩. Some of the properties of the two flexagons are given in Table 4.6, also see Tables 4.1 and 4.2. The torsion per sector is 4. The flexagon figure is shown in Fig. 4.4.

The net for the first order fundamental hexagon even edge flexagon 2⟨6, 6⟩ is shown in Fig. 4.29a. This two sector version is reasonably easy to handle, and is flexed using the twofold pinch flex. As assembled, the flexagon is in intermediate position 1. This is, in appearance, a hexagon edge pair. From here it can be opened into either principal main position 1(2) or principal main position 1(6). These are skew regular even edge rings of four regular hexagons (Fig. 1.5). The flexagon can then be closed into a new intermediate position, and so on round the principal 6-cycle shown by the solid lines in the intermediate position map (Fig. 4.10). The

Table 4.6 Properties of first order fundamental hexagon even edge flexagons. Principal cycles are in bold

Flexagon symbol	Typical main position	Cycle type	Number of cycles	Main position type	Ring symbol	Sector symbol	Curvature
2⟨6, 6⟩	**1(2)**	**6-cycle**	**1**	**Skew**	**4(120°)**	**⟨6, 6, 5, 1⟩**	**–120°**
2⟨6, 6⟩	1(3)	3-cycle	2	Slant	4(60°)	⟨6, 6, 4, 2⟩	120°
2⟨6, 6⟩	1(4)	None	–	Box	4(0°)	⟨6, 6, 3, 3⟩	0°
3⟨6, 6⟩	**1(2)**	**6-cycle**	**1**	**Skew**	**6(120°)**	**⟨6, 6, 5, 1⟩**	**–360°**
3⟨6, 6⟩	1(3)	3-cycle	2	Flat	6(60°)	⟨6, 6, 4, 2⟩	0°
3⟨6, 6⟩	1(4)	None	–	Box	6(0°)	⟨6, 6, 3,3⟩	360°

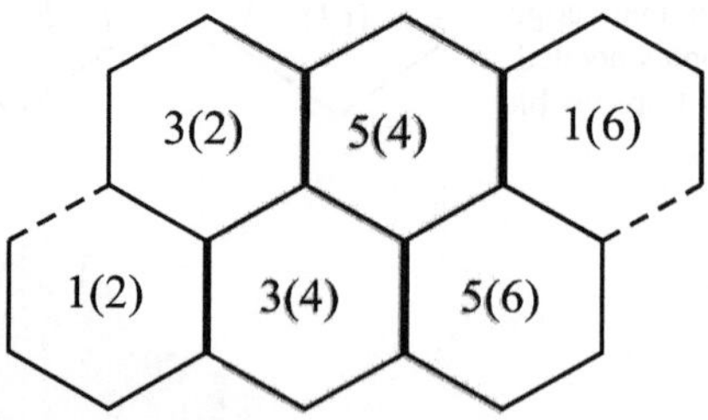

Fig. 4.29 Net for first order fundamental hexagon even edge flexagons. (**a**) 2⟨6, 6⟩. Two copies needed. Fold until leaves numbered 1 are visible. (**b**) 3⟨6, 6⟩. Three copies needed. Fold until leaves numbered 1 and 3 are visible (Les Pook, Flexagons inside out, 2003, © Cambridge University Press 2003, reprinted with permission)

continuous path when flexing between adjacent intermediate positions on the principal cycle is aesthetically satisfying.

Intermediate position 1 can also be opened into either subsidiary main position 1(3) or subsidiary main position 1(5), which are slant regular even edge rings of four regular hexagons. The high curvature (120°) means that these cannot be turned inside out, so the two subsidiary 3-cycles shown by dotted equilateral triangles in the intermediate position map cannot be traversed directly. They can be traversed indirectly via principal main positions. It is also possible to traverse between the two subsidiary main positions via principal main positions. The Tuckerman diagram (Fig. 4.30) is for the subsidiary 3-cycle in which faces with odd numbers are traversed. In the other subsidiary 3-cycle even numbers are traversed. Intermediate position 1 can also be opened into box position 1(4). This is a box edge ring of four regular hexagons (Fig. 4.31a).

The net for the first order fundamental hexagon even edge flexagon 3⟨6, 6⟩ is shown in Fig. 4.29b. This three sector version is a ring even edge flexagon (Section 4.2.2) and it is flexed using the threefold pinch flex. Paper models work reasonably well, provided that care is taken to maintain threefold rotational symmetry during flexing. As assembled, the flexagon is in subsidiary main position 3(1). This is a flat regular even edge ring of six regular hexagons (Fig. 4.31b). Pinch pats together in pairs, with leaves numbered 3 visible, to reach intermediate position 3, which is a hexagon edge triple. Then open the flexagon to reach subsidiary main position 5(3), and so on round the subsidiary 3-cycle shown in the Tuckerman diagram (Fig. 4.30). The principal 6-cycle shown by solid lines in the intermediate position map (Fig. 4.10) can also be traversed. Principal main positions are skew regular even edge rings of six regular hexagons. The intermediate position map shows that there are two subsidiary 3-cycles. To traverse from one subsidiary 4-cycle to the other, flex to an intermediate position, and then to another intermediate

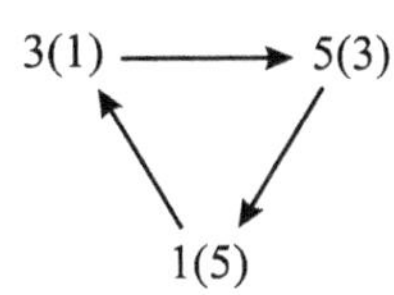

Fig. 4.30 Tuckerman diagram for a subsidiary 3-cycle of the first order fundamental hexagon even edge flexagons 2⟨6, 6⟩ and 3⟨6, 6⟩

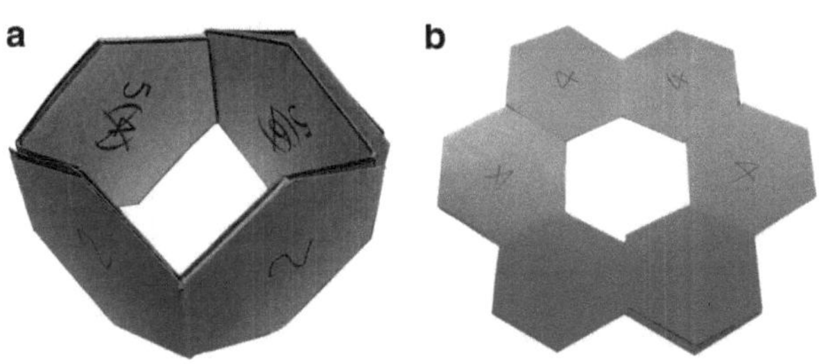

Fig. 4.31 Flexagons as the following. (**a**) A box edge ring of four regular hexagons. (**b**) A flat regular even edge ring of six regular hexagons

position via a principal main position. Intermediate positions can also be opened into box positions, which are box edge rings of six regular hexagons.

4.2.8 First Order Fundamental Octagon Even Edge Flexagons

The first order fundamental octagon even edge flexagons described in this section are 2⟨8, 8⟩ and 2⟨8, 8/3⟩. They are ring even edge flexagons (Section 4.2.2). Some of their properties are given in Table 4.7, also see Tables 4.1 and 4.2. The torsion per sector is 6 for 2⟨8, 8⟩ and two for 2⟨8, 8/3⟩. Flexagon figures are shown in Fig. 4.6.

The net for the first order fundamental octagon even edge flexagon 2⟨8, 8⟩ is shown in Fig. 4.32. The leaves have been truncated to irregular hexagons to make construction easier (Section 3.2), but overlap reasonably well (Fig. 4.33c). The resulting truncated flexagon is a partial overlap flexagon and could also be called an irregular hexagon even edge flexagon. It is an example of a flexagon that belongs to more than one family of flexagons. It is flexed using the twofold pinch flex.

As assembled, the flexagon is in intermediate position 1. This is, in nominal appearance, an octagon edge pair. From here it can be opened into either principal main position 1(2) or principal main position 1(8). These are skew regular even edge ring, of four regular octagons (Fig. 4.33a). The flexagon can then be closed into a new intermediate position, and so on round the principal 8-cycle shown by the solid lines

Table 4.7 Properties of first order fundamental octagon even edge flexagons. Principal cycles are in bold

Flexagon symbol	Typical main position	Cycle type	Number of cycles	Main position type	Ring symbol	Sector symbols	Curvature
2⟨8, 8⟩	**1(2)**	**8-cycle**	**1**	**Skew**	**4(135°)**	**⟨8, 8, 7, 1⟩**	**−180°**
2⟨8, 8⟩	1(3)	4-cycle	2	Flat	4(90°)	⟨8, 8, 6, 2⟩	0°
2⟨8, 8⟩	1(4)	8-cycle	1	Slant	4(45°)	⟨8, 8, 5, 3⟩	180°
2⟨8, 8⟩	1(5)	None	–	Box	4(0°)	⟨8, 8, 4, 4⟩	360°
2⟨8, 8/3⟩	**1(2)**	**8-cycle**	**1**	**Slant**	**4(45°)**	**⟨8, 8/3, 7, 1⟩**	**180°**
2⟨8, 8/3⟩	1(3)	4-cycle	2	Flat	4(90°)	⟨8, 8/3, 6, 2⟩	0°
2⟨8, 8/3⟩	1(4)	8-cycle	1	Skew	4(135°)	⟨8, 8/3, 5, 3⟩	−180°
2⟨8, 8/3⟩	1(5)	None	–	Box	4(0°)	⟨8, 8/3, 4, 4⟩	360°

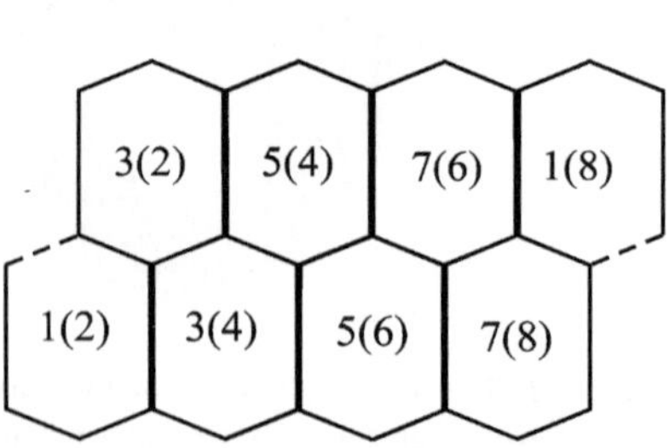

Fig. 4.32 Net for the first order fundamental octagon (irregular hexagon) even edge flexagon 2⟨8, 8⟩. Two copies needed. Fold until leaves numbered 1 are visible

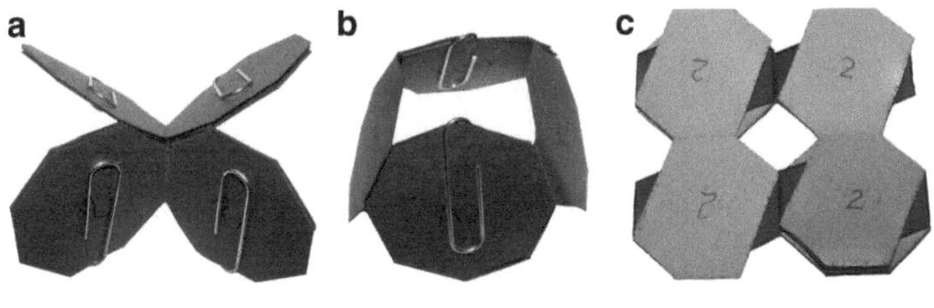

Fig. 4.33 Flexagons as regular even edge rings of four regular octagons. (**a**) Skew. (**b**) Slant. (**c**) Flat, truncated to irregular hexagons

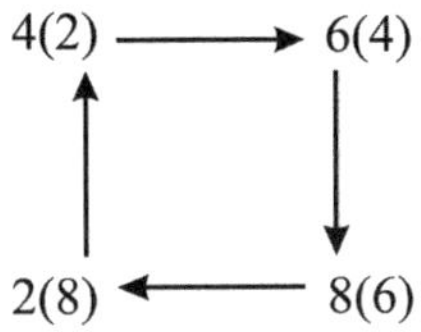

Fig. 4.34 Tuckerman diagram for a subsidiary 4-cycle of the first order fundamental octagon even edge flexagons 2⟨8, 8⟩ and 2⟨8, 8/3⟩

in the intermediate position map (Fig. 4.12). Intermediate positions can be opened into box positions, which are box edge rings of four regular octagons.

A subsidiary 8-cycle, in which the subsidiary main positions are slant regular even edge rings of four regular octagons (Fig. 4.33b), can be traversed, as shown by the dashed lines in the intermediate position map. Start by opening intermediate position 1 into either subsidiary main position 1(4) or subsidiary main position 1(6). The high curvature (180°) means that these cannot be turned inside out, so the subsidiary 8-cycle cannot be traversed directly, but it can be traversed indirectly via principal main positions.

The two subsidiary 4-cycles, shown by dotted squares in the intermediate position map, can be traversed by using the twofold pinch flex. The subsidiary main positions are flat regular even edge rings of four regular octagons (Fig. 4.33c). The Tuckerman diagram (Fig. 4.34) is for the subsidiary 4-cycle, in which faces with even numbers are traversed. In the other subsidiary 4-cycle, odd numbers are traversed. To traverse from one subsidiary 4-cycle to the other, flex to an intermediate position, and then to another intermediate position via a principal main position.

The net for the first order fundamental octagon even edge flexagon 2⟨8, 8/3⟩ is shown in Fig. 4.35. Truncation of leaves is not needed. As assembled the flexagon is in intermediate position 1. This is an octagon edge pair. From here it can be opened into either subsidiary main position 1(4) or subsidiary main position 1(6). These are skew regular even edge rings of four regular octagons (Fig. 4.33a). The flexagon can then be closed into a new intermediate position, and so on round the subsidiary 8-cycle shown by the dashed lines in the intermediate position map (Fig. 4.12). Intermediate positions can be opened into box positions, which are box rings of four regular octagons.

The principal 8-cycle, in which the principal main positions are slant regular even edge rings of four regular octagons (Fig. 4.33b), can be traversed, as shown

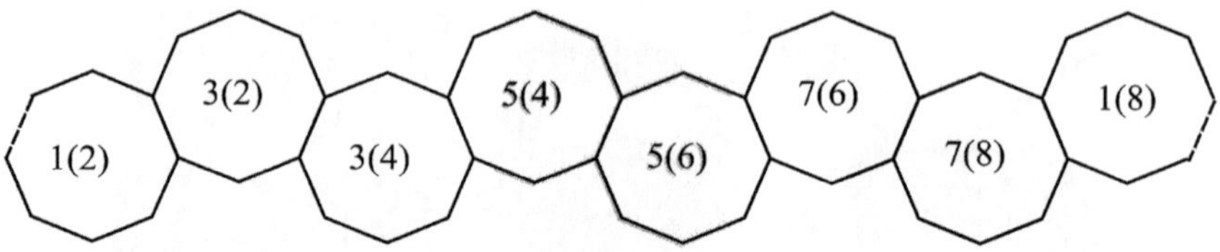

Fig. 4.35 Net for the first order fundamental octagon even edge flexagon 2⟨8, 8/3⟩. Two copies needed. Fold until leaves numbered 1 are visible

by the solid lines in the intermediate position map. Start by opening intermediate position 1 into either principal main position 1(2) or principal main position 1(8). The high curvature (180°) means that these cannot be turned inside out, so the principal 8-cycle cannot be traversed directly, but it can be traversed indirectly via subsidiary main positions. The torsion is lower than that of 2⟨8, 8⟩ and the flexagon is unstable. Thus, the principal 8-cycle can also be traversed by using the band flex.

The two subsidiary 4-cycles, shown by dotted lines in the intermediate position map, can be traversed by using the twofold pinch flex. The subsidiary main positions are flat regular even edge rings of four regular octagons (Fig. 4.33c). The Tuckerman diagram (Fig. 4.34) is for the subsidiary 4-cycle in which faces with even numbers are traversed. In the other subsidiary 4-cycle odd numbers are traversed. To traverse from one subsidiary 4-cycle to the other, flex to an intermediate position, and then to another intermediate position via a subsidiary main position.

4.2.9 First Order Fundamental Dodecagon Even Edge Flexagons

The first order fundamental dodecagon even edge flexagons 2⟨12, 12⟩ and 2⟨12, 12/5⟩ are described in this section. They are ring even edge flexagons (Section 4.2.2). Some of their properties are given in Table 4.8, also see Tables 4.1 and 4.2. The torsion per sector is 10 for 2⟨12, 12⟩ and two for 2⟨12, 12/5⟩. Flexagon figures are shown in Fig. 4.36.

The nets for the first order fundamental dodecagon even edge flexagons 2⟨12, 12⟩ 2⟨12, 12/5⟩ are shown in Figs. 4.37 and 4.38. The leaves of 2⟨12, 12⟩. have been truncated to irregular hexagons to make construction easier (Section 3.2). The resulting truncated flexagon is a partial overlap flexagon and could also be called an irregular hexagon even edge flexagon. It is an example of a flexagon that belongs to more than one family of flexagons. Both flexagons are very unstable and are unsatisfactory as paper models, but they are included because they are related to bronze flexagons (Section 10.2.6.1).

The dynamic properties are very complicated. This arises because 12 has factors of 2, 3, 4 and 6. Smaller numbers have, at most, two different factors. The table shows that, theoretically, each flexagon can be flexed around 11 different cycles,

Table 4.8 Properties of first order fundamental dodecagon even edge flexagons 2⟨12, 12⟩ and 2⟨12, 12/5⟩. Principal cycles are in bold

Flexagon symbol	Typical main position	Cycle type	Number of cycles	Main position type	Ring symbol	Sector symbols	Curvature
2⟨12, 12⟩	**1(2)**	**12-cycle**	**1**	**Skew**	**4(150°)**	**⟨12, 12, 11, 1⟩**	**–240°**
2⟨12, 12⟩	1(3)	6-cycle	2	Skew	4(120°)	⟨12, 12, 10, 2⟩	–120°
2⟨12, 12⟩	1(4)	4-cycle	3	Flat	4(90°)	⟨12, 12, 9, 3⟩	0°
2⟨12, 12⟩	1(5)	3-cycle	4	Slant	4(60°)	⟨12, 12, 8, 4⟩	120°
2⟨12, 12⟩	1(6)	12-cycle	1	Slant	4(30°)	⟨12, 12, 7, 5⟩	240°
2⟨12, 12⟩	1(7)	None	–	Box	4(0°)	⟨12, 12, 6, 6⟩	360°
2⟨12, 12/5⟩	**1(2)**	**12-cycle**	**1**	**Slant**	**4(30°)**	**⟨12, 12/5, 11, 1⟩**	**240°**
2⟨12, 12/5⟩	1(3)	6-cycle	2	Skew	4(120°)	⟨12, 12/5, 10, 2⟩	–120°
2⟨12, 12/5⟩	1(4)	4-cycle	3	Flat	4(90°)	⟨12, 12/5, 9, 3⟩	0°
2⟨12, 12/5⟩	1(5)	3-cycle	4	Slant	4(60°)	⟨12, 12/5, 8, 4⟩	120°
2⟨12, 12/5⟩	1(6)	12-cycle	1	Skew	4(150°)	⟨12, 12/5, 7, 5⟩	–240°
2⟨12, 12/5⟩	1(7)	None	–	Box	4(0°)	⟨12, 12/5, 6, 6⟩	360°

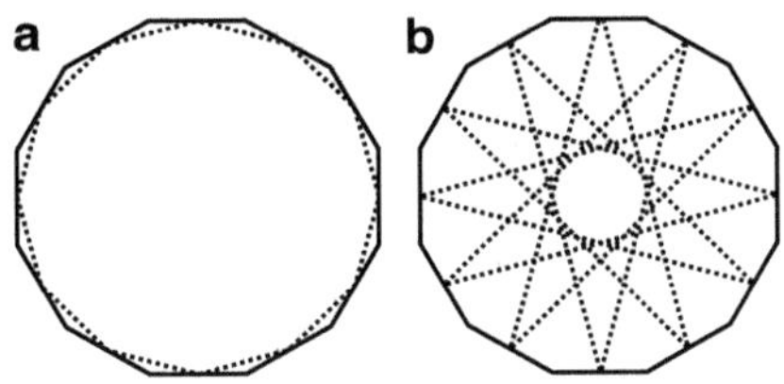

Fig. 4.36 Flexagon figures for first order fundamental dodecagon even edge flexagons. (**a**) *S*⟨12, 12⟩. (**b**) *S*⟨12, 12/5⟩

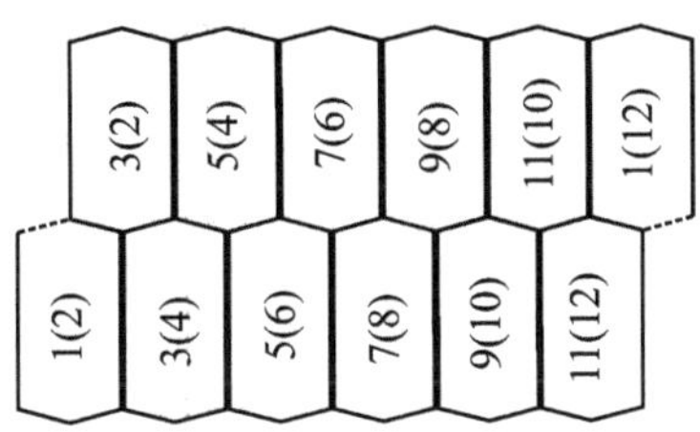

Fig. 4.37 Net for the first order fundamental dodecagon (irregular hexagon) even edge flexagon 2⟨12, 12⟩. Join the two parts of the net at A-A. Two copies needed. Fold until leaves numbered 1 and 3 are visible

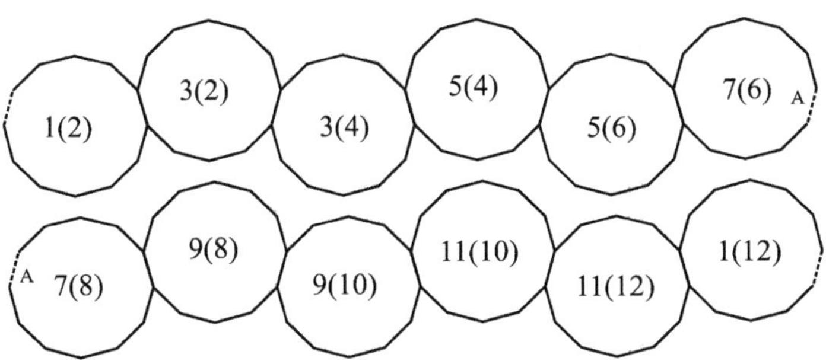

Fig. 4.38 Net for the first order fundamental dodecagon even edge flexagon 2⟨12, 12/5⟩. Join the two parts of the net at A-A. Two copies needed. Fold until leaves numbered 1 and 4 are visible

with five different types of main position appearance. This can be done by using the band flex. However, this is tedious and paper clips are needed to keep the flexagon under control.

4.3 Second Order Fundamental Odd Edge Flexagons

4.3.1 General Properties

Second order fundamental odd edge flexagons have at least five sectors, and are made from second order fundamental edge nets (Section 3.3). Satisfactory paper models with five or seven sectors can be made from regular convex polygons with up to 20 edges on a polygon. Nets for some examples are given below. Flexing options differ from those of first order fundamental even edge flexagons (Section 4.2.1).

In a principal main position, a second order fundamental odd edge flexagon is, in appearance, a regular odd edge ring of n regular convex polygons (Section 2.2.3), where $n \geq 5$. All pats are identical, so each pat is a sector, and there are $S = n$ sectors. Each pat in a principal main position consists of a folded pair of leaves. Main position codes are used to identify principal main positions. If a principal main position is flat, slant or box, then it can be arranged with n-fold rotational symmetry, but this is not possible for skew principal main positions. Principal main positions of some second order fundamental odd edge flexagons that are slant, box or skew cannot be arranged to lie flat. A subscript 2 is added to flexagon symbols to indicate second order flexagon. Flexagon figures and associated polygons are not appropriate.

Sector diagrams have a similar appearance to flexagon figures for first order fundamental even edge flexagons, and always have a plane of reflection symmetry. For example, the sector diagram for second order fundamental triangle odd edge flexagons $S\langle 3, 3\rangle_2$ is shown in Fig. 4.39. A sector diagram contains most of the information needed to construct the corresponding flexagon (Section 7.2.2).

Second order fundamental odd edge flexagons are solitary flexagons. They are twisted bands, and the torsion per sector is 1. The total torsion is equal to the number of sectors and is always odd, so the flexagons are Möbius bands. If a standard face numbering sequence (Section 4.1.1) is used, then second order fundamental odd edge flexagons can be assembled either in principal main position 1(2) or principal main position 3(4). The torsions of these principal main positions are of opposite signs, so it is never possible to flex between them.

All second order fundamental odd edge flexagons can be flexed by using pocket flexes. (Sherman 2007; Pook 2007). This is their characteristic flex. These are used

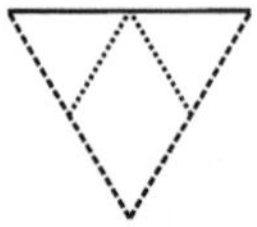

Fig. 4.39 The sector diagram for second order fundamental triangle odd edge flexagons $S\langle 3, 3\rangle_2$

to flex to minor main positions. A pocket flex is a local flex because some pats are left unchanged. Some can also be flexed by using asymmetric pinch flexes, which do not have rotational symmetry. Details of these flexes are given below in descriptions of flexagons. Further, some can be flexed by using pinch flexes similar to those used for first order fundamental even edge flexagons. In these, the rotational symmetry of the pinch flex is a factor of the number of sectors in a principal main position. For example, the second order fundamental odd edge flexagon $9\langle 3, 3\rangle_2$, which has nine sectors, can be flexed by using a threefold pinch flex.

Second order fundamental even edge flexagons are possible, but they are differently numbered versions of other types of flexagon and transformations between flexagons are possible. For example, the second order fundamental triangle even edge flexagon $6\langle 3, 3\rangle_2$ is a differently numbered tetrahexaflexagon (Section 11.2.2).

4.3.2 Second Order Fundamental Triangle Odd Edge Flexagons

The second order fundamental triangle odd edge flexagons $5\langle 3, 3\rangle_2$, $7\langle 3, 3\rangle_2$ and $9\langle 3, 3\rangle_2$ are described in this section. The second order fundamental triangle edge net $\langle 3\rangle_2$ (Fig. 3.13) is used. The sector diagram is shown in Fig. 4.39. The net for $5\langle 3, 3\rangle_2$ is shown in Fig. 4.40. As assembled, the flexagon is either in principal main position 1(2) or in principal main position 2(1). These are slant regular odd edge rings of five equilateral triangles (Fig. 1.2). In the main position codes, the number outside the brackets indicates the face number visible on the outside of the ring. The positive curvature (60°) means that it is not possible to flex between the two forms of the principal main position.

To flex to a minor main position using a pocket flex, start by pinching two pats together to reach an intermediate position (Fig. 4.41a). With two pats in line, open the flexagon to reach a double pyramid position (Fig. 4.41b). This is a combination of a regular slant even edge ring of four equilateral triangles and a regular slant odd edge ring of three equilateral triangles with one equilateral triangle in common. Close the square pyramid to reach a minor main position. This is a combination of a regular slant odd edge ring of three equilateral triangles and an equilateral triangle edge pair with one equilateral triangle in common (Fig. 4.41c). Starting from either principal main position 1(2), or from principal main position 2(1), the pocket flex can, in each case, be carried in five different ways. All ten subsidiary main positions have the same appearance. Further flexing is not possible. A pocket flex always reduces the number of pats left in an edge ring by two.

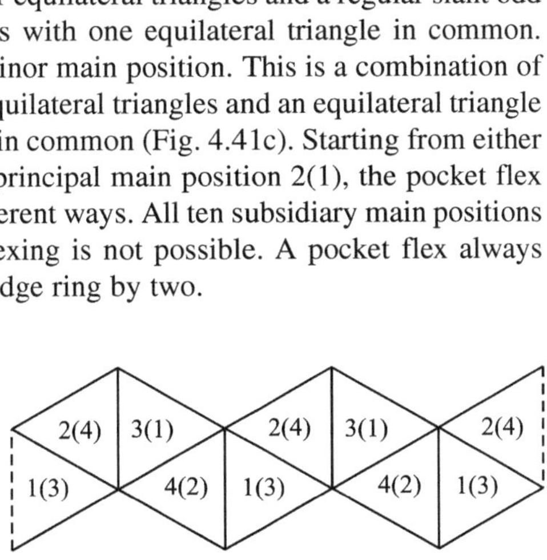

Fig. 4.40 Net for the second order fundamental triangle odd edge flexagon $5\langle 3, 3\rangle_2$. One copy needed. Starting from one end of the net, fold each leaf numbered 3 onto a leaf numbered 4

A principal main position can be turned inside out, so it is possible to flex between the two forms of a principal main position. Starting from principal main position 1(2), first flex to a first intermediate position (Fig. 4.49a). Then fold together the two leaves numbered 1 at top centre and top left of Fig. 4.49a to reach a second intermediate position (Fig. 4.49c). Next, pinch together the two pats at bottom left of the photograph to reach another first intermediate position. Finally, fold under the lower pair of pats (Fig. 4.49a) to reach principal main position 2(1). Turning the principal main position inside out makes another five minor main positions available, making a total of 10.

4.3.4 A Second Order Fundamental 20-gon Odd Edge Flexagon

Possible second order fundamental 20-gon odd edge flexagons have flexagon symbols of the forms $S\langle 20, 20\rangle_2$, $S\langle 20, 20/3\rangle_2$, $S\langle 20, 20/7\rangle_2$ and $S\langle 20, 20/9\rangle_2$. The flexagon described in this section is $5\langle 20, 20/7\rangle_2$. The second order fundamental 20-gon edge net $\langle 20/7\rangle_2$ Fig. (3.17) is used. The sector diagram is shown in Fig. 4.50. Its net is shown in Fig. 4.51.

As assembled, the flexagon is in principal main position 1(2). This is, in appearance, a flat regular odd edge ring of five regular 20-gons (Fig. 4.52). Flat principal

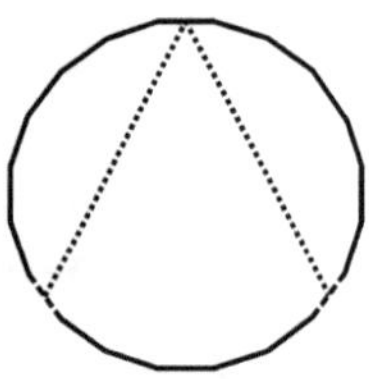

Fig. 4.50 Sector diagram for second order fundamental 20-gon odd edge flexagons $S\langle 20, 20/7\rangle_2$

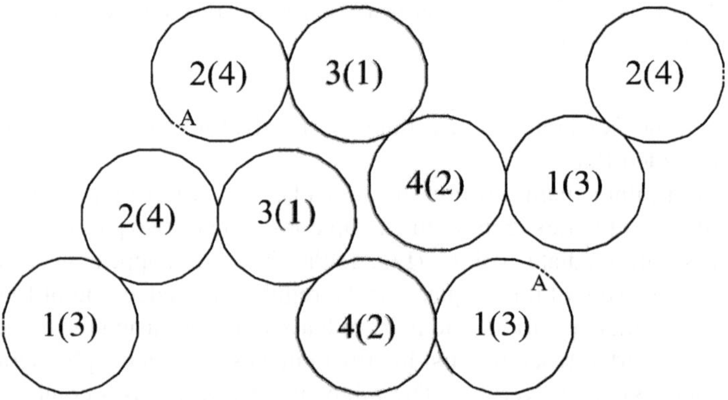

Fig. 4.51 Net for the second order fundamental 20-gon odd edge flexagon $5\langle 20, 20/7\rangle_2$. One copy needed. Join the two parts of the net at A-A. Starting from one end off the net, fold each leaf numbered 3 onto a leaf numbered 4

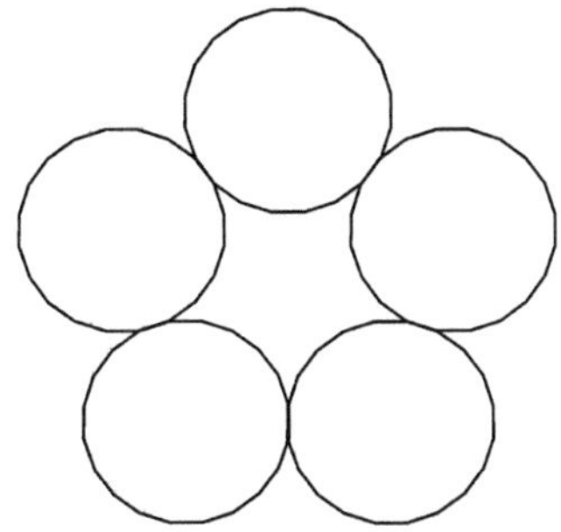

Fig. 4.52 A flat regular even edge ring of five 20-gons

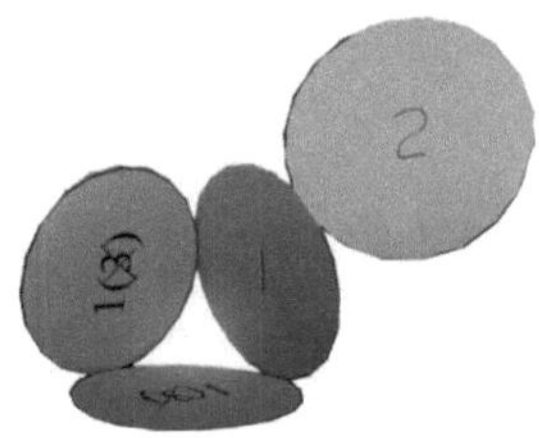

Fig. 4.53 A flexagon as a combination of a slant regular edge ring of three regular 20-gons and an edge pair of regular 20-gons

main positions are not possible for second order fundamental odd edge flexagons made from regular polygons with a smaller number of edges. Despite the short hinges, the flexagon is easy to handle.

A pocket flex can be used to reach a minor main position (Fig. 4.53). This is a combination of a slant regular edge ring of three regular 20-gons and an edge pair of regular 20-gons. Starting from one principal main position 1(2), a pocket flex can be carried in ten different ways, so there are ten minor main positions that all have the same appearance.

References

Conrad AS, Hartline DK (1962) *Flexagons*. RIAS Technical Report 62-11. RIAS, Baltimore, MD

Coxeter HSM (1963) Regular Polytopes, 2nd edn. Macmillan, New York

Demaine EJ, O'Rourke J (2007) Geometric Folding Algorithms: Linkages, Origami, Polyhedra. Cambridge University Press, Cambridge

Gardner M (1965) Mathematical Puzzles and Diversions. Penguin, Harmondsworth

Gardner M (1978) Mathematical games. *Scientific American* 239:18–19, 22, 24–25

Gardner M (2008) Hexaflexagons, Probability Paradoxes, and the Tower of Hanoi. Cambridge University Press, Cambridge

Highland Games (2008) www.halfpast.demon.co.uk. Accessed 10 December 2008

Hilton P, Pedersen J (1994) Build Your Own Polyhedra. Addison-Wesley, Menlo Park, CA

Pedersen JJ, Pedersen KA (1973) Geometric Playthings to Color Cut and Fold . Troubadour, San Francisco, CA

Pook LP (2003) Flexagons Inside Out. Cambridge University Press, Cambridge

Pook LP (2007) Results of pocket flexes. Flexagon Lovers Group posting 430. http://tech.groups.yahoo.com/group/Flexagon_Lovers/. Accessed 3 December 2007

Sherman S (2007) Triangle flexagon bestiary. http://loki3.com/flex/triangles.html. Accessed 3 June 2007
Shuttleworth D (2006) Transparent flexagons. Flexagon Lovers Group posting 206. http://tech.groups.yahoo.com/group/Flexagon_Lovers/. Accessed 6 July 2006
Taylor P (1997) The Complete?. Nattygrafix, Polygon, Ipswich

Chapter 5
Fundamental Skeletal and Point Flexagons

5.1 Introduction

Fundamental skeletal flexagons are constructed from fundamental vertex nets (Section 3.4). Main positions are, in appearance, regular vertex rings (Section 2.4.1). Pats are either single leaves or fan folded piles of leaves. Fundamental skeletal flexagons are related to fundamental edge flexagons. Thus, broadly, fundamental skeletal flexagons are also the equivalent of regular polyhedra. They are solitary flexagons (Sections 4.1 and 4.2.1). There is a fundamental skeletal flexagon corresponding to every fundamental edge flexagon (Section 4.1). In particular, they have the same associated polygon.

Point flexagons are a recent development, first described by Sherman (2007a). Main positions are, in appearance, polygon vertex pairs (Section 2.1.1) and intermediate positions are, in appearance, single polygons. A point flexagon has one sector. Leaves can be regarded as rigid because, unlike edge flexagons, there are never flexes in which leaf bending is needed. In this sense, point flexagons are ideal flexagons (Section 1.2).

Fundamental point flexagons are constructed from fundamental vertex nets. They can be regarded as variants of first order fundamental skeletal flexagons, in which the number of sectors is reduced to one. Hence, they have first order fundamental edge flexagon equivalents, and the same associated polygons. The number of main positions in a cycle is equal to the number of edges on the constituent polygons. Edge flexagons are usually assembled by folding together pairs of leaves that have the same face number. Fundamental point flexagons are assembled in this way, but interleaved fundamental point flexagons (and some other types of point flexagon) are interleaved during assembly (Pook 2007a; Sherman 2007b, c). Unlike most flexagons, which are un-knotted bands, some interleaved point flexagons can be assembled either as knotted bands or as un-knotted bands.

In an augmented fundamental point flexagon, the number of main positions in a cycle is, in general, a multiple of the number of edges on the constituent polygons, and is the same as the number of edges on an associated polygon. Similarly, there are augmented interleaved fundamental point flexagons.

L. Pook, *Serious Fun with Flexagons,* Solid Mechanics and Its Applications 164,
DOI 10.1007/978-90-481-2503-6_5, © Springer Science+Business Media B.V. 2009

The three topological invariants given in Section 2.4.1 are necessary, but not sufficient, for the topological description of skeletal flexagons and point flexagons. The torsion is an additional topological invariant that is needed (Section 4.2.1), and is sufficient for point flexagons that are simple bands. If a point flexagon is a knotted band, a figure-of-eight band, or a more complicated band, then an additional topological invariant describing the type of band is needed. The curvature of skeletal flexagons in main positions is indeterminate (Section 1.1). Curvature cannot be defined for a main position of a point flexagon.

First order fundamental even skeletal flexagons, fundamental point flexagons, interleaved fundamental point flexagons, associated fundamental point flexagons, and associated interleaved fundamental point flexagons are discussed in this chapter. They are all regular cycle flexagons in which all the main positions of a cycle have the same appearance and the same pat structure. Point hinges, used in nets used to construct skeletal and point flexagons, are impossible in a paper model, but short paper strips provide a workable approximation (Section 3.4), and are used in the nets included in this chapter.

5.2 First Order Fundamental Even Skeletal Flexagons

5.2.1 General Properties

Fundamental skeletal flexagons are made from fundamental vertex nets (Section 3.4), and a standard face numbering sequence (Section 4.1.1) is used. They are very unstable (Pook 2007b; Sherman 2007b). In general, paper clips are needed to hold the pats together, rearranging the clips as needed to allow each flex. Skeletal flexagons are a special case of truncated flexagons (Section 4.1.2). All first order fundamental even skeletal flexagons can be flexed by using pinch flexes. These are their characteristic flexes.

Every fundamental edge flexagon (Section 4.1) has a dual that is a fundamental skeletal flexagon. This is derived by inscribing polygons with vertices at midpoints of the edges of the polygons of the fundamental edge flexagon. Thus, the edge hinges along edges of the original polygons become point hinges at vertices of the inscribed polygons (Sections 2.4.1 and 3.4).

From a rigorous viewpoint a skeletal flexagon is the dual of an edge flexagon only in certain positions. In these positions only the degree of freedom corresponding to an edge hinge is operative in a point hinge. These positions can be identified by examining paper strips used as approximations to point hinges, provided that the axis of a paper strip bisects the vertex angles of the connected polygons. If the paper strips are simply folded about perpendiculars to their axes (Fig. 2.2) then the position is a dual, but if they are distorted (Fig. 1.6) then the position is not a dual. This distinction does not always have to be taken into account.

Some of the properties of fundamental skeletal flexagons are the same as those of precursor fundamental edge flexagons. In particular, the torsion per sector

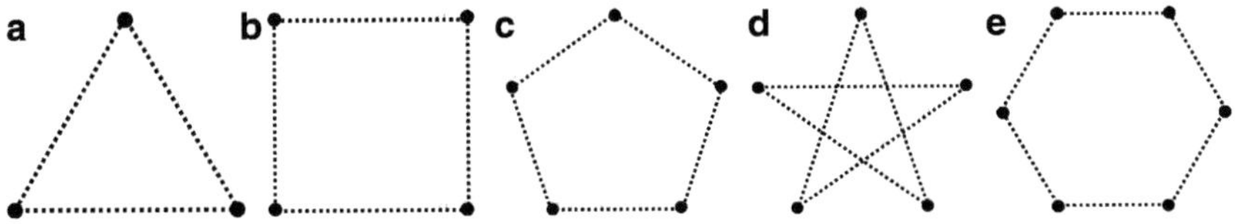

Fig. 5.1 Flexagon figures for first order fundamental even skeletal flexagons and fundamental point flexagons. (**a**) Triangle $S\langle 3, 3\rangle$. (**b**) Square $S\langle 4, 4\rangle$. (**c**) Pentagon $S\langle 5, 5\rangle$. (**d**) Pentagon $S\langle 5, 5/2\rangle$. (**e**) Hexagon $S\langle 6, 6\rangle$

is given by Equation (4.1) and fundamental skeletal flexagons are solitary flexagons. The flexagon symbol and sector symbol for a first order fundamental even skeletal flexagon are the same as those for the precursor first order fundamental even edge flexagon. In a flexagon figure, the inscribed polygon is inscribed at the vertices of the circumscribing polygon, rather than at the midpoints of its edges. Figure 5.1 shows flexagon figures for all the possible first order fundamental even skeletal flexagons made from polygons with up to six edges (cf. Figs. 4.1 to 4.4). To avoid confusion, circumscribing polygons are shown with dots representing their vertices.

The extra degree of freedom in a point hinge (Section 1.2) means that a main position of a first order fundamental even skeletal flexagon can always be laid flat, and also that it is not rigid when confined to a plane. The extra degree of freedom also means that the leaves in a pat are not constrained to overlap exactly, although in practice the use of short paper strips as approximations to point hinges does provide some stability. In the descriptions of first order fundamental even skeletal flexagons below, it is assumed that leaves are constrained to overlap exactly in main and intermediate positions. It is also assumed that main positions, intermediate positions, and flexes used are all constrained to have S-fold rotational symmetry, where S is the number of sectors.

5.2.2 *First Order Fundamental Triangle Even Skeletal Flexagons*

The first order fundamental triangle even skeletal flexagons described in this section are $2\langle 3, 3\rangle$ and $3\langle 3, 3\rangle$. Some of the properties of the two flexagons are given in Table 5.1 (cf. Table 4.3). The torsion per sector is 1. The flexagon figure is shown in Fig. 5.1a.

The net for the first order fundamental triangle even skeletal flexagon $2\langle 3, 3\rangle$ is shown in Fig. 5.2a. This was derived by using the net for the first order fundamental triangle even edge flexagon $2\langle 3, 3\rangle$ (Fig. 4.13) as a precursor, and constructing its dual. As assembled, the flexagon is in principal main position 1(2) which is, in appearance, a flat regular vertex ring of four equilateral triangles (Fig. 1.6). In this position the flexagon is not a dual of the first order fundamental triangle even edge

Table 5.1 Properties of first order fundamental triangle even skeletal flexagons. Principal cycles are in bold

Flexagon symbol	Typical main position	Cycle type	Number of cycles	Sector symbol
2⟨3, 3⟩	**1(2)**	**3-cycle**	1	**⟨3, 3, 2, 1⟩**
3⟨3, 3⟩	**1(2)**	**3-cycle**	1	**⟨3, 3, 2, 1⟩**

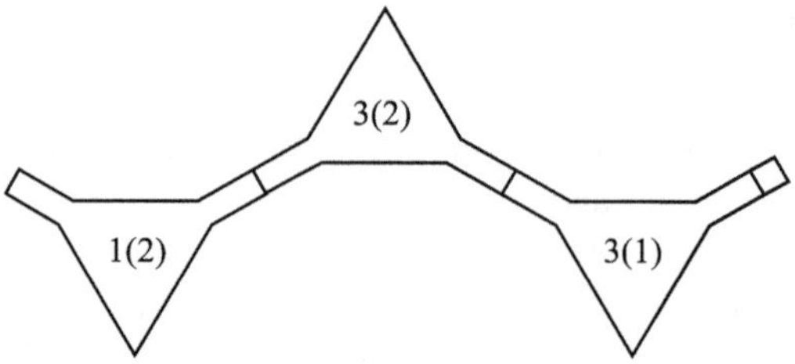

Fig. 5.2 Net for the following triangle flexagons. (**a**) First order fundamental even skeletal flexagon 2⟨3, 3⟩. Two copies needed. (**b**) First order fundamental even skeletal flexagon 3⟨3, 3⟩. Three copies needed. (**c**) Fundamental point flexagon 1⟨3, 3⟩. One copy needed. Fold until leaves numbered 1 are visible. (**d**) Degenerate triangle point flexagon type AD. One copy needed. Fold until leaves numbered 1 are visible

flexagon 2⟨3, 3⟩. The principal 3-cycle shown in the intermediate position map can be traversed by using the twofold pinch flex. The intermediate position map is the same as that for the first order fundamental triangle even edge flexagons *S*⟨3, 3⟩ (Fig. 4.7). Intermediate positions are equilateral triangle vertex pairs connected by single point hinges (Fig. 1.8b). Unlike the first order fundamental triangle even edge flexagon 2⟨3, 3⟩ (Section 4.2.3), the principal 3-cycle can be traversed without disconnecting a hinge, refolding the flexagon, and reconnecting the hinge. The flexagon can also be flexed as the augmented fundamental triangle point flexagon 1⟨3, 3, 6⟩ (Section 5.5.2). This is a transformation between flexagons.

The net for the first order fundamental triangle even skeletal flexagon 3⟨3, 3⟩ is shown in Fig. 5.2b. This was derived by using the net for the trihexaflexagon (Fig. 4.14) as a precursor and constructing its dual. As assembled, the flexagon is in principal main position 1(2) which is a flat regular vertex ring of six equilateral triangles, with the triangles pointing outwards (Fig. 5.3a). The principal 3-cycle shown in the intermediate position map can be traversed by using the threefold pinch flex. The intermediate position map is the same as that for the first order fundamental triangle even edge flexagons *S*⟨3, 3⟩ (Fig. 4.7). Only one degree of freedom of the point hinges is needed, and during flexing the flexagon is always a dual of the trihexaflexagon. In effect, the lines across the paper strips linking the leaves are used as short edge hinges. Paper clips are needed to keep the flexagon under control.. Intermediate positions are equilateral triangle vertex triples with a common vertex (Fig. 2.2). Intermediate positions can also be opened into flat regular vertex rings of six equilateral triangles, with the triangles pointing inwards (Fig. 5.3b). This requires both degrees of freedom of the point hinges, and in this

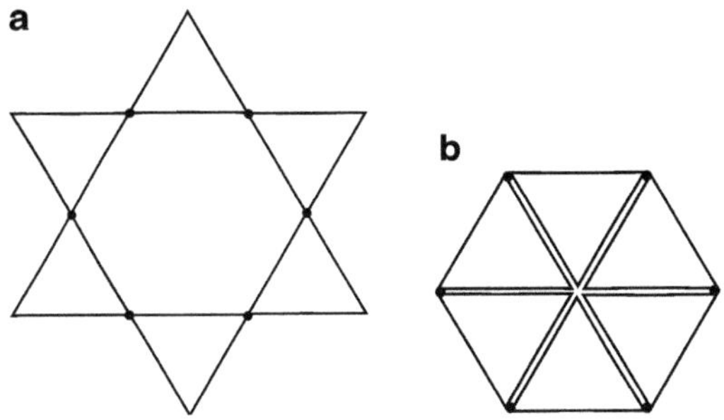

Fig. 5.3 Flat regular vertex rings of six equilateral triangles. (**a**) Pointing outwards. (**b**) Pointing inwards

position the flexagon is not a dual of the trihexaflexagon. In the figure triangles are slightly separated for clarity. Principal main positions can be rotated continuously between the two forms shown in Fig. 5.3 (cf. Section 2.4.2). Both degrees of freedom of the point hinges are used.

5.2.3 A First Order Fundamental Square Even Skeletal Flexagon

The first order fundamental square even skeletal flexagon described in this section is 2⟨4, 4⟩. Some of its properties are given in Table 5.2 (cf. Table 4.4). The torsion per sector is 2. The flexagon figure is shown in Fig. 5.1b.

The net for the first order fundamental square even skeletal flexagon 2⟨4, 4⟩ is shown in Fig. 5.4a. This was derived by using the first order fundamental square even edge flexagon 2⟨4, 4⟩ (Figs. 1.2 and 4.16a) as a precursor and constructing its dual. As assembled, the flexagon is in principal main position 1(2), which is, in appearance, a flat regular vertex ring of four squares with point hinges adjacent (Fig. 2.23c). The principal 4-cycle shown in the intermediate position map can be traversed by using the twofold pinch flex. The intermediate position map is the same as that for the first order fundamental square even edge flexagons *S*⟨4, 4⟩ (Fig. 4.8). Intermediate positions are square vertex pairs connected by single point hinges. Only one degree of freedom of the point hinges is needed, and during flexing the flexagon is always a dual of the first order fundamental square even edge flexagon 2⟨4, 4⟩. The extra degree of freedom in point hinges means that principal main positions can be manoeuvred, while constrained to a plane, into the less symmetrical position shown in Fig. 2.23d. In this position the flexagon is not a dual of the first order fundamental square even edge flexagon 2⟨4, 4⟩.

Intermediate positions can also be opened into subsidiary main positions, which are box vertex rings of four squares (Fig. 2.23a). These can be turned inside out by rotating continuously via a flat regular vertex ring of four squares with point hinges

Table 5.2 Properties of a first order fundamental square even skeletal flexagon. The principal cycle is in bold

Flexagon symbol	Typical main position	Cycle type	Number of cycles	Sector symbol
2⟨4, 4⟩	**1(2)**	**4-cycle**	**1**	**⟨4, 4, 3, 1⟩**
2⟨4, 4⟩	1(3)	None	–	⟨4, 4, 2, 2⟩

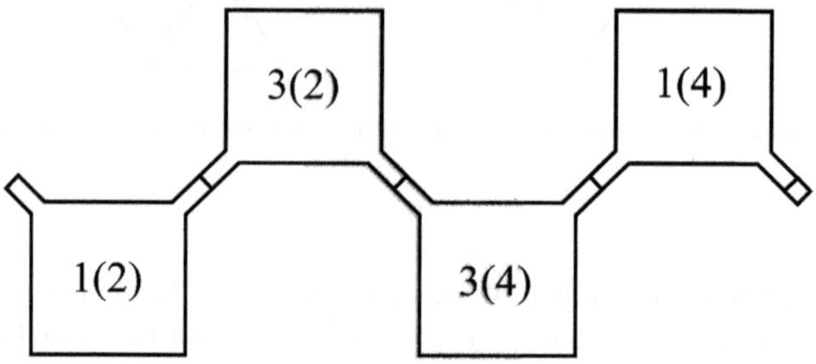

Fig. 5.4 Net for the following square flexagons. (**a**) First order fundamental even skeletal flexagon 2⟨4, 4⟩. Two copies needed. (**b**) Fundamental point flexagon 1⟨4, 4⟩. One copy needed. Fold until leaves numbered 1 are visible

diagonally opposite (Fig. 2.23b, cf. Section 2.4.2 and previous section). Both degrees of freedom of the point hinges are used. The flexagon can also be flexed as the augmented fundamental square point flexagon 1⟨4, 4, 8⟩ (Section 5.5.3). This is a transformation between flexagons.

5.3 Fundamental Point Flexagons

5.3.1 General Properties and Unagons

Fundamental point flexagons are made from fundamental vertex nets (Section 3.4), and a standard face numbering sequence (Section 4.1.1) is used. They are a special case of fundamental skeletal flexagons (Section 5.2.1), with the number of sectors reduced to one. Hence, they are solitary flexagons. All fundamental point flexagons can be flexed by using the simple flex and this is their characteristic flex. Main positions are, in appearance, polygon vertex pairs linked by pairs of point hinges (Fig. 1.8a), and intermediate positions are, in appearance, single polygons. The associated polygon for a fundamental point flexagon is the same as that for its net so its Schläfli symbol is $\{c\}$ (Section 3.1).

In intermediate positions, but not in any other position, the dual of a fundamental point flexagon is a fundamental unagon. A fundamental unagon is made from a first order fundamental edge net (Section 3.2), and it is a first order fundamental even edge flexagon (Section 4.2.1) with the number of sectors reduced to one. It can exist only in an intermediate position which is, in appearance, a single polygon. Fundamental unagons cannot be flexed, so they are not flexagons. The flexagon symbol for the precursor flexagon becomes the unagon symbol, with one outside the brackets to show that there is one sector. The associated polygon for a fundamental unagon is the same as that for the precursor flexagon. Gardner (1965, 2008) recognised the existence of unagons, and suggested possible names for a unagon made from equilateral triangles. This is the fundamental triangle unagon $1\langle 3,3\rangle$.

The net for the fundamental pentagon unagon $1\langle 5, 5\rangle$ is shown in Fig. 4.23c. This was derived by using the net for the for first order fundamental pentagon even edge flexagon $2\langle 5, 5\rangle$ Fig. 4.23a as a precursor, and reducing the number of sectors from two to one. As assembled, the unagon is in intermediate position 1. This is, in appearance, a single regular pentagon. Other face numbers can be seen only by bending the leaves.

Some of the properties of fundamental point flexagons are analogous to those of first order fundamental even edge flexagons. Flexagon symbols and sector symbols (Section 4.2.5.1) are appropriate. A flexagon symbol $1\langle s, c\rangle$ completely defines a fundamental point flexagon except for its size and whether the torsion is positive or negative. The number of sectors is 1, $\{s\}$ is the Schläfli symbol for the constituent polygons, and $\langle c\rangle$ is the fundamental vertex net symbol. The torsion per sector, which is also the total torsion, is given by Equation (4.1). In general, fundamental point flexagons are more stable than first order fundamental edge flexagons, and are much more stable than fundamental skeletal flexagons. Some of the properties of all the possible fundamental point flexagons made from polygons with up to 12 edges are given in Table 5.3 (cf. Table 4.1).

A fundamental point flexagon can have only one type of main position appearance and only one type of cycle. This is because of restricted flexing options; the simple flex is possible only where pairs of point hinges used in the flex are adjacent in an intermediate position. Thus, some cycles and corresponding main positions, present in first order fundamental even edge flexagons, are absent in corresponding fundamental point flexagons. When main positions are absent, lines representing them are absent from intermediate position maps. In other words, the intermediate position map for a fundamental point flexagon may be a subset of that for the corresponding first order fundamental even edge flexagon.

To design the net for any desired fundamental point flexagon, start by selecting the appropriate first order fundamental edge net in Table 5.3. Next, make the number of leaves in the net equal to the number of edges on the constituent polygons. Finally, apply the appropriate standard face numbering sequence (Section 4.1.1). Nets for some examples are given below.

5.3.2 The Fundamental Triangle Point Flexagon

The fundamental triangle point flexagon $1\langle 3, 3\rangle$ described in this section is the only fundamental triangle point flexagon (Table 5.3). The torsion is 1. The flexagon

Table 5.3 Properties of fundamental point flexagons. A standard face numbering sequence (Section 4.1.1) is used

Leaf type	Flexagon symbol	Fundamental vertex net	Cycle type	Sector symbol	Torsion
Triangle	1⟨3, 3⟩	⟨3⟩	3-cycle	⟨3, 3, 2, 1⟩	1
Square	1⟨4, 4⟩	⟨4⟩	4-cycle	⟨4, 4, 3, 1⟩	2
Pentagon	1⟨5, 5⟩	⟨5⟩	5-cycle	⟨5, 5, 4, 1⟩	3
Pentagon	1⟨5, 5/2⟩	⟨5/2⟩	5-cycle	⟨5, 5/2, 4, 1⟩	1
Hexagon	1⟨6, 6⟩	⟨6⟩	6-cycle	⟨6, 6, 5, 1⟩	4
Heptagon	1⟨7, 7⟩	⟨7⟩	7-cycle	⟨7, 7, 6, 1⟩	5
Heptagon	1⟨7, 7/2⟩	⟨7/2⟩	7-cycle	⟨7, 7/2, 6,1⟩	3
Heptagon	1⟨7, 7/3⟩	⟨7/3⟩	7-cycle	⟨7, 7/3, 6,1⟩	1
Octagon	1⟨8, 8⟩	⟨8⟩	8-cycle	⟨8, 8, 7, 1⟩	6
Octagon	1⟨8, 8/3⟩	⟨8/3⟩	8-cycle	⟨8, 8/3, 7,1⟩	2
Enneagon	1⟨9, 9⟩	⟨9⟩	9-cycle	⟨9, 9, 8, 1⟩	7
Enneagon	1⟨9, 9/2⟩	⟨9/2⟩	9-cycle	⟨9, 9/2, 8,1⟩	5
Enneagon	1⟨9, 9/4⟩	⟨9/4⟩	9-cycle	⟨9, 9/3, 8,1⟩	1
Decagon	1⟨10, 10⟩	⟨10⟩	10-cycle	⟨10, 10, 9, 1⟩	8
Decagon	1⟨10, 10/3⟩	⟨10/3⟩	10-cycle	⟨10, 10/3, 9, 1⟩	4
11-gon	1⟨11, 11⟩	⟨11⟩	11-cycle	⟨11, 11, 10, 1⟩	9
11-gon	1⟨11, 11/2⟩	⟨11/2⟩	11-cycle	⟨11, 11/2, 10, 1⟩	7
11-gon	1⟨11, 11/3⟩	⟨11/3⟩	11-cycle	⟨11, 11/3, 10, 1⟩	5
11-gon	1⟨11. 11/4⟩	⟨11/4⟩	11-cycle	⟨11, 11/4, 10, 1⟩	3
11-gon	1⟨11, 11/5⟩	⟨11/5⟩	11-cycle	⟨11, 11/5, 10, 1⟩	1
Dodecagon	1⟨12, 12⟩	⟨12⟩	12-cycle	⟨12, 12, 11, 1⟩	10
Dodecagon	1⟨12, 12/5⟩	⟨12/5⟩	12-cycle	⟨12, 12/5, 11, 1⟩	2

figure is shown in Fig. 5.1a. Its net is shown in Fig. 5.2c. As assembled, the flexagon is in intermediate position 1, which is, in appearance, a single equilateral triangle. To flex to principal main position 1(2), rotate the single leaf marked 1(2) through 180° about a pair of point hinges. Both degrees of freedom of the point hinges are needed. Principal main positions are equilateral triangle vertex pairs connected by pairs of point hinges (Fig. 1.8a). To reach intermediate position 2, fold together the pair of leaves numbered 1. By repeating this version of the twofold pinch flex, called the simple flex, the principal 3-cycle shown in the intermediate position map can be traversed. The intermediate position map is the same as that for the first order fundamental triangle even edge flexagons *S*⟨3, 3⟩ (Fig. 4.7).

5.3.3 The Fundamental Square Point Flexagon

The fundamental square point flexagon 1⟨4, 4⟩ described in this section is the only possible fundamental square point flexagon (Table 5.3). The torsion is 2. The flexagon figure is shown in Fig. 5.1b. Its net is shown in Fig. 5.4b. As assembled, the flexagon is in intermediate position 1, which is, in appearance, a single square. The principal 4-cycle shown in the intermediate position map (Fig. 5.5) can be traversed

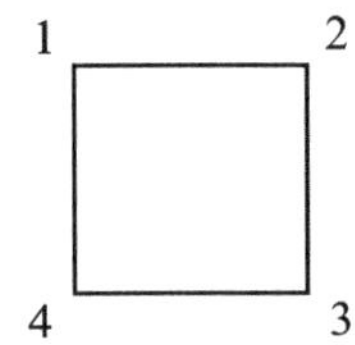

Fig. 5.5 Intermediate position map for the fundamental square point flexagon 1⟨4, 4⟩

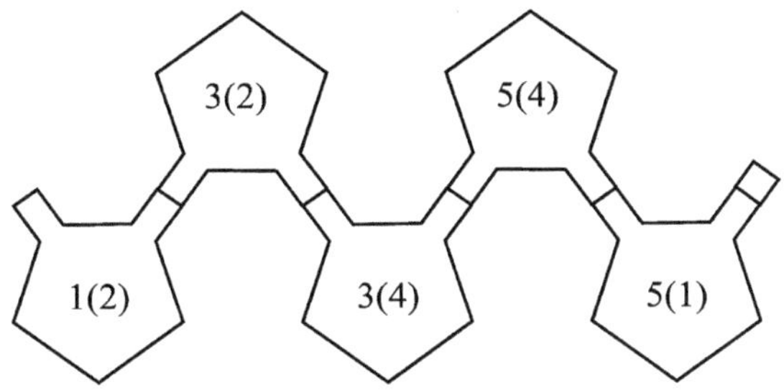

Fig. 5.6 Net for the fundamental pentagon point flexagon 1⟨5, 5⟩. One copy needed. Fold until leaves numbered 1 are visible

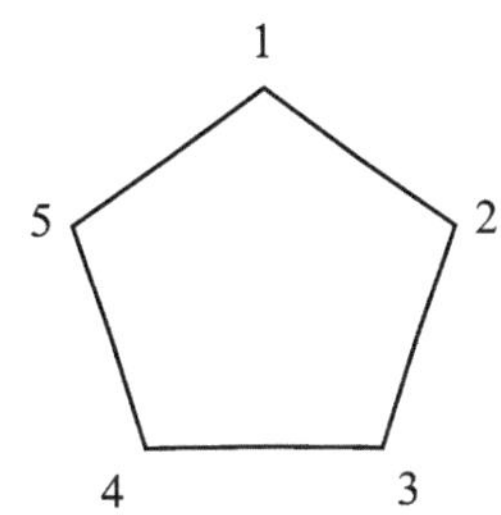

Fig. 5.7 Intermediate position map for the fundamental pentagon point flexagon 1⟨5, 5⟩

by using the simple flex, with one leaf turned over in each intermediate position. Principal main positions are square vertex pairs connected by pairs of point hinges (Fig. 3.23). Subsidiary main positions corresponding to box positions in the first order fundamental square even edge flexagon 2⟨4, 4⟩ (Section 4.2.4) are absent.

5.3.4 Fundamental Pentagon Point Flexagons

There are two fundamental pentagon point flexagons, 1⟨5, 5⟩ and 1⟨5, 5/2⟩ (Table 5.3), and these are described in this section. The torsion is 3 for 1⟨5, 5⟩ and one for 1⟨5, 5/2⟩. Their flexagon figures are shown in Fig. 5.1c and d. The net for the fundamental pentagon point flexagon 1⟨5, 5⟩ is shown in Fig. 5.6. As assembled, the flexagon is in intermediate position 1, which is, in appearance, a single regular pentagon. The principal 5-cycle shown in the intermediate position map (Fig. 5.7) can be traversed by using the simple flex, with one leaf turned over in each intermediate position. Principal main positions are pentagon vertex pairs connected by pairs of point hinges. The subsidiary 5-cycle, which is present in the first

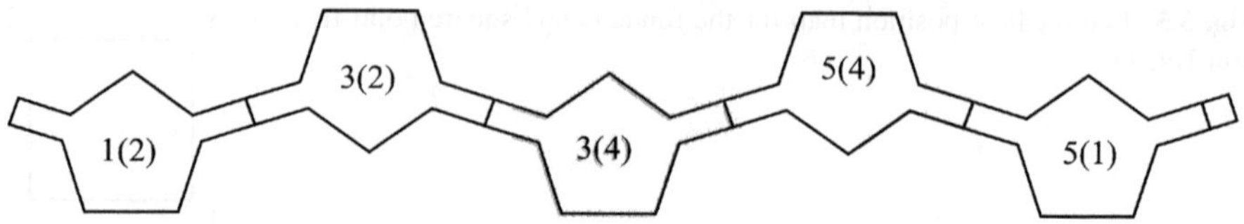

Fig. 5.8 Net for the fundamental pentagon point flexagon 1⟨5, 5/2⟩. One copy needed. Fold until leaves numbered 1 are visible

Fig. 5.9 Intermediate position map for the fundamental pentagon point flexagon 1⟨5, 5/2⟩

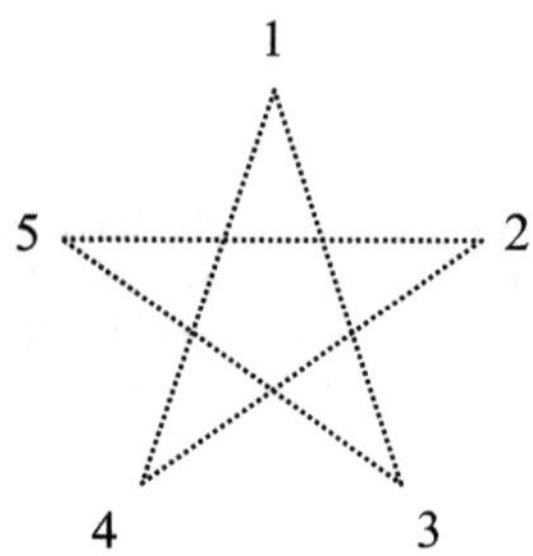

order fundamental pentagon even edge flexagon 2⟨5, 5⟩ (Section 4.2.6, Fig. 4.9), is absent together with corresponding main positions. The dual of the fundamental point flexagon, in intermediate positions, is the fundamental pentagon unagon 1⟨5, 5⟩ (Section 5.3.1).

The net for the fundamental pentagon point flexagon 1⟨5, 5/2⟩ is shown in Fig. 5.8. As assembled, the flexagon is in intermediate position 1, which is a single regular pentagon. The subsidiary 5-cycle shown in the intermediate position map (Fig. 5.9) can be traversed by using the simple flex, with two leaves turned over in each intermediate position. Subsidiary main positions are pentagon vertex pairs connected by pairs of point hinges. The principal 5-cycle, present in the first order fundamental pentagon even edge flexagon 2⟨5, 5/2⟩ (Section 4.2.6, Fig. 4.9), is absent together with corresponding main positions.

5.3.5 The Fundamental Hexagon Point Flexagon

The fundamental hexagon point flexagon 1⟨6, 6⟩ described in this section is the only possible fundamental hexagon point flexagon (Table 5.3). The torsion is 4. The flexagon figure is shown in Fig. 5.1e. Its net is shown in Fig. 5.10. As assembled, the flexagon is in intermediate position 1, which is, in appearance, a single regular hexagon. The principal 6-cycle shown in the intermediate position map (Fig. 5.11) can be traversed by using the simple flex, with one leaf turned over in each intermediate position. In principal main positions the flexagon is a hexagon vertex pair connected by pairs of point hinges. The two subsidiary 3-cycles, present in the first order fundamental hexagon even edge flexagon 2⟨6, 6⟩ (Section 4.2.7,

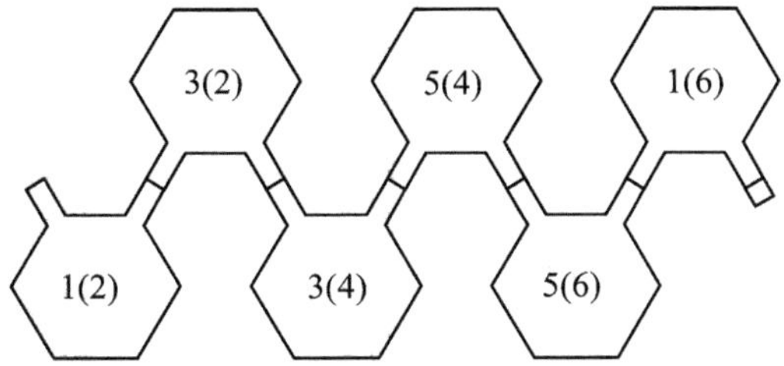

Fig. 5.10 Net for the fundamental hexagon point flexagon 1⟨6, 6⟩. One copy needed. Fold until leaves numbered 1 are visible

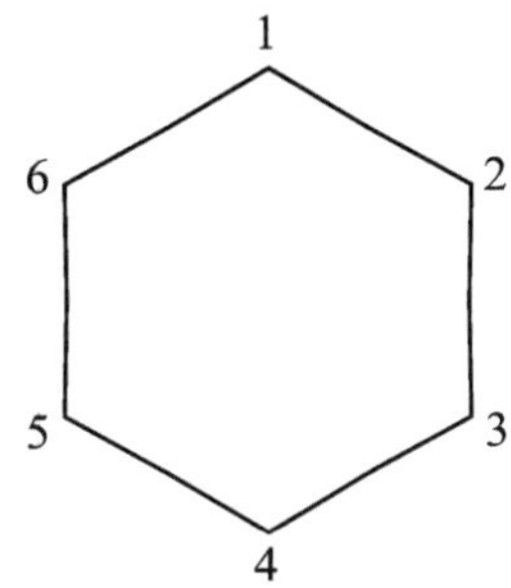

Fig. 5.11 Intermediate position map for the fundamental hexagon point flexagon 1⟨6, 6⟩

Fig. 4.10), are absent together with corresponding main positions. Subsidiary main positions corresponding to box positions in the first order fundamental hexagon even edge flexagon 2⟨6, 6⟩ are also absent.

5.4 Interleaved Fundamental Point Flexagons

5.4.1 General Properties

Flexagons are usually assembled by folding together pairs of leaves that have the same face number. In particular, fundamental point flexagons (Section 5.3.1) are assembled in this way. However, there are interleaved point flexagons in which leaves are interleaved during assembly (Pook 2007a; Sherman 2007b, c). These include interleaved fundamental point flexagons. Interleaved fundamental point flexagons are made from fundamental vertex nets (Section 3.4). The net symbol ⟨c⟩ must be a whole number. Thus, the fundamental vertex net ⟨5⟩ (Fig. 3.20a) can be used, but not ⟨5/2⟩ (Fig. 3.20b). All interleaved fundamental point flexagons can be flexed by using the simple flex and this is their characteristic flex. The nets are renumbered versions of nets for fundamental point flexagons (Section 5.3.1), so interleaved fundamental point flexagons are solitary flexagons. In intermediate positions, but not in any other position, the dual of an interleaved fundamental point flexagon is an interleaved fundamental unagon. These exist only in intermediate positions and cannot be flexed.

The associated polygon for an interleaved fundamentals point flexagon is a regular star polygon which has the same number of vertices as the constituent polygons. There is a distinct type of interleaved fundamental point flexagon for each type of regular star polygon. Hence, interleaved fundamental point flexagons can be enumerated by enumerating possible regular star polygons. Associated polygons are used to construct face numbering sequences for interleaved fundamental point flexagons. These sequences ensure that faces appear in cyclic order.

The flexagon symbol for a fundamental point flexagon, $1\langle s, c\rangle$, (Section 5.3.1) becomes $1\langle s, c, a\rangle$ for an interleaved fundamental point flexagon, where $\{a\}$ is the Schläfli symbol for the associated polygon. Flexagon figures are not appropriate for interleaved fundamental point flexagons. The torsion of an interleaved fundamental point flexagon is given by

$$T_S = s - 2A \tag{5.1}$$

where A is the denominator of a (cf. Equation (4.1)).

Some properties of all possible interleaved fundamental point flexagons made from polygons with up to 12 edges are given in Table 5.4 (cf. Table 5.3). Some fundamental point flexagons (Section 5.3.1) are included for comparison. These are not interleaved. The face numbering sequences are as applied to nets. The conven-

Table 5.4 Properties of interleaved fundamental point flexagons

Leaf type	Flexagon symbol	Fundamental vertex net	Face numbering sequence	Torsion
Pentagon	1⟨5, 5, 5/2⟩	⟨5⟩	1/2, 4/3, 5/1, 3/2, 4/5	1
Heptagon	1⟨7, 7, 7/2⟩	⟨7⟩	1/2, 4/3, 5/6, 1/7, 2/3, 5/4, 6/7	3
Heptagon	1⟨7, 7, 7/3⟩	⟨7⟩	1/2, 5/4, 7/1, 4/3, 6/7, 3/2, 5/6	1
Octagon	1⟨8, 8, 8/3⟩	⟨8⟩	1/2, 5/4, 7/8, 3/2, 5/6, 1/8, 3/4, 7/6	2
Enneagon	1⟨9, 9, 9/2⟩	⟨9⟩	1/2, 4/3, 5/6, 8/7, 9/1, 3/2, 4/5, 7/6, 8/9	5
Enneagon	1⟨9, 9, 9/4⟩	⟨9⟩	1/2, 6/5, 9/1, 5/4, 8/9, 4/3, 7/8, 3/2, 6/7	1
Decagon	1⟨10, 10, 10/3⟩	⟨10⟩	1/2, 5/4, 7/8, 1/10, 3/4, 7/6, 9/10, 3/2, 5/6, 9/8	4
11-gon	1⟨11, 11, 11/2⟩	⟨11⟩	1/2, 4/3, 5/6, 8/7, 9/10, 1/11, 2/3, 5/4, 6/7, 9/8, 10/11	7
11-gon	1⟨11, 11, 11/3⟩	⟨11⟩	1/2, 5/4, 7/8, 11/10, 2/3, 6/5, 8/9, 1/11, 3/4, 7/6, 9/10	5
11-gon	1⟨11, 11, 11/4⟩	⟨11⟩	1/2, 6/5, 9/10, 3/2, 6/7, 11/10, 3/4, 8/7, 11/1, 5/4, 8/9	3
11-gon	1⟨11, 11, 11/5⟩	⟨11⟩	1/2, 7/6, 11/1, 6/5, 10/11, 5/4, 9/10, 4/3, 8/9, 3/2, 7/8	1
Dodecagon	1⟨12, 12, 12/5⟩	⟨12⟩	1/2, 7/6, 11/12, 5/4, 9/1, 3/2, 7/8, 1/12, 5/6, 11/10, 3/4, 9/8	2

tion used (Section 4.1.1) is that numbers before slashes are for the upper faces, and those after slashes are for the lower faces. Interleaved fundamental point flexagons are usually stable.

Derivation of the face numbering sequences is straightforward. For example, for the interleaved fundamental pentagon point flexagon 1⟨5, 5, 5/2⟩, start by numbering the vertices of the constituent polygon, which is a regular convex pentagon, in cyclic order, to give 1/2, 2/3, 3/4, 4/5, 5/1. Next, inscribe a regular star pentagon in the regular convex pentagon, with the same vertices. Re-arrange the numbers in the order in which they appear at the vertices of the regular star pentagon to give 1/2, 3/4, 5/1, 2/3, 4/5. Finally, invert every other pair of numbers to give. 1/2, 4/3, 5/1, 3/2, 4/5. This is the desired face numbering sequence.

To design the net for any desired interleaved fundamental point flexagon, start by selecting the appropriate first order fundamental edge net in Table 5.4. Next, make the number of leaves in the net, equal to the number of edges on the constituent polygons. Finally, apply the face numbering sequence listed in the table. Nets for two examples are given below.

There is an interleaved fundamental point flexagon that corresponds to each fundamental point flexagon in the sense that both have the same associated polygon. Thus, the associated polygon for both the interleaved fundamental pentagon point flexagon 1⟨5, 5, 5/2⟩, and the fundamental pentagon point flexagon 1⟨5, 5/2⟩, is the regular star pentagon {5/2} (Fig. 3.1). The torsion and face numbering sequences for both flexagons are the same. Restricted flexing options (Section 5.3.1) mean that some cycles and corresponding main positions, present in first order fundamental even edge flexagons, are absent in corresponding interleaved fundamental point flexagons.

5.4.2 The Interleaved Fundamental Pentagon Point Flexagon

The interleaved fundamental pentagon point flexagons, 1⟨5, 5, 5/2⟩ described in this section is the only possible interleaved fundamental pentagon point flexagon (Table 5.4). The torsion is 1. Its net is shown in Fig. 5.12. It is assembled as follows. The same general method is used for other interleaved point flexagons. Position the

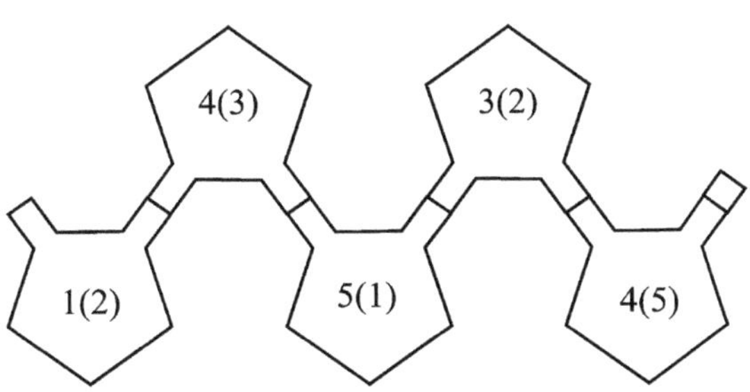

Fig. 5.12 Net for the interleaved fundamental pentagon point flexagon 1⟨5, 5, 5/2⟩. One copy needed. Interleave during assembly, with leaves numbered 1 left visible

leaves in turn, working along the net. Start at the leaf numbered 1/2, and fold the leaf numbered 4/3 under it so that leaf numbers 2 and 3 are adjacent. Then fold the leaf numbered 5/1 so that numbers 4 and 5 are adjacent. Next, interleave the leaf numbered 3/2 between the leaves numbered 1/2 and 4/3 so that the numbers 2, and also the numbers 3, are adjacent. Finally, interleave the leaf numbered 4/5 between the leaves numbered 4/3 and 5/1 so that the numbers 4, and also the numbers 5, are adjacent, and join the ends of the net.

As assembled, the flexagons are in intermediate position 1, which is, in appearance, a single regular pentagon. The principal 5-cycle shown in the intermediate position map can be traversed by using the simple flex, with one leaf turned over in each intermediate position. The intermediate position map is the same as that for the fundamental pentagon point flexagon $1\langle 5, 5\rangle$ (Fig. 5.7). The subsidiary 5-cycle, present in the first order fundamental pentagon even edge flexagon $2\langle 5, 5\rangle$ (Section 4.2.6, Fig. 4.9), is absent together with corresponding main positions. In main positions the flexagon is a pentagon vertex pair connected by a pair of point hinges.

5.4.3 An Interleaved Fundamental Enneagon Point Flexagon

There are two interleaved fundamental enneagon point flexagons, $1\langle 9, 9, 9/2\rangle$ and $1\langle 9, 9, 9/4\rangle$ (Table 5.4). $1\langle 9, 9, 9/2\rangle$ is described in this section. The torsion is 5. Its net is shown in Fig. 5.13. Despite the large number of edges on the leaves, the flexagon is stable and easy to handle. As assembled, the flexagon is in intermediate position 1, which is, in appearance, a single regular enneagon. The principal 9-cycle shown in the intermediate position map (Fig. 5.14) can be traversed by using the simple flex, with one leaf turned over in each intermediate position. Subsidiary cycles, and corresponding main positions, present in first order fundamental enneagon even edge flexagons, are absent. In main positions, the flexagon is an enneagon vertex pair connected by a pair of point hinges.

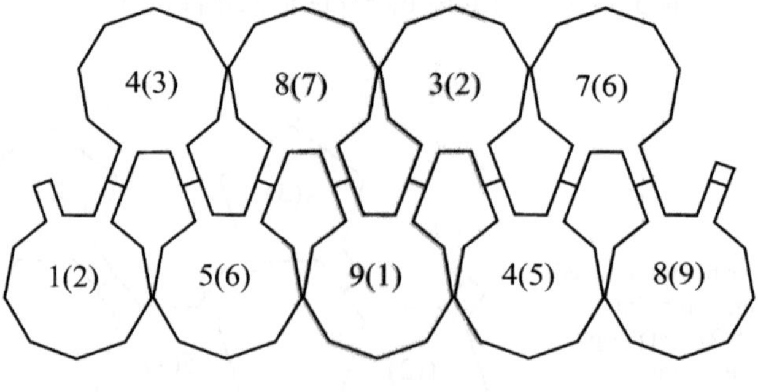

Fig. 5.13 Net for the interleaved fundamental enneagon point flexagon $1\langle 9, 9, 9/2\rangle$. One copy needed. Interleave during assembly, with leaves numbered 1 left visible

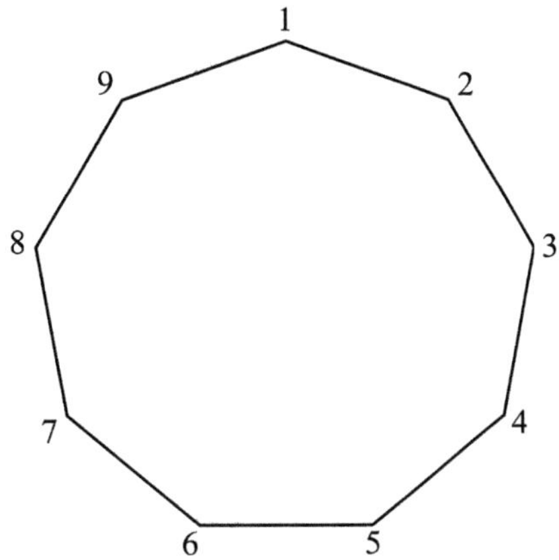

Fig. 5.14 Intermediate position map for the interleaved fundamental enneagon point flexagon 1⟨9, 9, 9/2⟩

5.5 Augmented Fundamental Point Flexagons

5.5.1 General Properties

An augmented point flexagon is a point flexagon in which the leaves have been replaced by leaves with a smaller number of edges. This number is a factor of the number of edges on the original leaves. Augmented point flexagons are made from fundamental vertex nets (Section 3.4). They can also be constructed by linking fundamental point flexagons using end links (Section 11.3.1). All augmented fundamental point flexagons can be flexed by using the simple flex and this is their characteristic flex. The notation used is mostly the same as that for interleaved fundamental point flexagons (Section 5.4.1). Sector symbols (Section 4.2.5) are used to describe the pat structure. Because of the relationship to fundamental point flexagons, they are solitary flexagons. Augmented fundamental point flexagons are usually easier to handle than their fundamental point flexagon equivalents. In intermediate positions, and a limited number of main positions, the dual of an augmented point flexagon is an augmented fundamental unagon. Unusually for unagons, a limited amount of flexing is possible, so they can be described as deficient flexagons.

In a fundamental point flexagon (Section 5.3.1), there is only one cycle that can be traversed, and the number of main positions in the cycle is equal to the number of edges on the constituent polygons. In an augmented fundamental point flexagon there is always more than one type of cycle that can be traversed. The number of principal main positions in a principal cycle is a multiple of the number of edges on the constituent polygons, and is the same as the number of edges on the associated polygon, which is a regular convex polygon. Some properties of all possible augmented fundamental point flexagons with associated polygons of up to 12 edges are given in Table 5.5. Nets for two examples are given below.

In an intermediate position map for a first order fundamental even edge flexagon lines of different lengths represent main positions with different appearances (Section 4.2.1). However, in intermediate position maps for augmented fundamental point flexagons lines of different lengths sometimes represent main positions with the same appearance. Intermediate positions of augmented fundamental point flexagons have more than one point hinge at each vertex. These do not overlap and there is only one possible method of assembly. Restricted flexing options (Section 5.3.1) mean that some cycles and corresponding main positions, present in first order fundamental even edge flexagons, are absent in corresponding augmented fundamental point flexagons.

5.5.2 An Augmented Fundamental Triangle Point Flexagon

The augmented fundamental triangle point flexagon 1⟨3, 3, 6⟩ is described in this section. It is the simplest augmented fundamental triangle point flexagon (Table 5.5), and also the simplest augmented fundamental point flexagon. Some of its properties are given in Table 5.6. There are some analogies with the properties of first order fundamental hexagon even edge flexagons (Table 4.6). The torsion is 2. Its dual, the fundamental augmented triangle unagon 1⟨3, 3, 6⟩, has the same pat structure as the first order fundamental triangle even edge flexagon 2⟨3, 3⟩ (Section 4.2.3). Hence, it is a deficient flexagon and a transformation between flexagons is possible.

Table 5.5 Properties of augmented fundamental point flexagons. Standard face numbering sequences (Section 4.1.1) are used

Leaf type	Associated polygon	Flexagon symbol	Fundamental vertex net	Number of cycle types	Torsion
Triangle	Hexagon	1⟨3, 3, 6⟩	⟨3⟩	2	2
Triangle	Enneagon	1⟨3, 3, 9⟩	⟨3⟩	4	3
Triangle	Dodecagon	1⟨3, 3, 12⟩	⟨3⟩	5	4
Square	Octagon	1⟨4, 4, 8⟩	⟨4⟩	2	4
Square	Dodecagon	1⟨4, 4, 12⟩	⟨4⟩	4	6
Pentagon	Decagon	1⟨5, 5, 10⟩	⟨5⟩	2	6
Pentagon	Decagon	1⟨5, 5/2, 10⟩	⟨5/2⟩	2	2
Hexagon	Dodecagon	1⟨6, 6, 12⟩	⟨6⟩	2	8

Table 5.6 Properties of the augmented fundamental triangle point flexagon 1⟨3, 3, 6⟩. The associated polygon is a hexagon. Fundamental vertex net ⟨3⟩ is used. The principal cycle is in bold

Flexagon symbol	Typical main position	Cycle type	Number of cycles	Sector symbol
1⟨3, 3, 6⟩	**1(2)**	**6-cycle**	**1**	**⟨3, 3, 5, 1⟩**
1⟨3, 3, 6⟩	1(3)	3-cycle	2	⟨3, 3, 4, 2⟩
1⟨3, 3, 6⟩	1(4)	None	–	⟨3, 3, 3, 3⟩

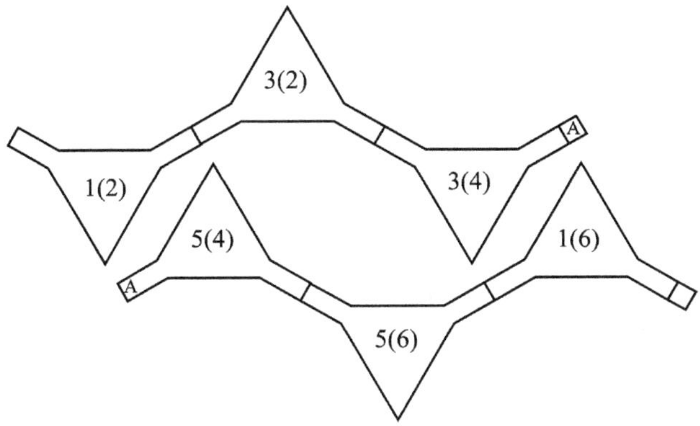

Fig. 5.15 Net for the augmented fundamental triangle point flexagon 1⟨3, 3, 6⟩. One copy needed. Join the two parts of the net at A-A. Fold until leaves numbered 1 are visible

The net for the augmented fundamental triangle point flexagon 1⟨3, 3, 6⟩ is shown in Fig. 5.15. This was be derived by using the fundamental hexagon point flexagon 1⟨6, 6⟩ (Fig. 5.10) as a precursor and replacing the regular hexagons by equilateral triangles. As assembled, the flexagon is in intermediate position 1, which is, in appearance, a single equilateral triangle. The principal 6-cycle shown in the intermediate position map can be traversed by using the simple flex. The intermediate position map is the same as that for the first order fundamental hexagon even edge flexagons *S*⟨6, 6⟩ (Fig. 4.10). The two subsidiary 3-cycles shown can also be traversed by using the simple flex. In these three cycles, main positions are equilateral triangle vertex pairs linked by pairs of point hinges (Fig. 1.8a). Subsidiary main positions 1(4), 2(5) and 3(6) are not in cycles and are triangle vertex pairs linked by single point hinges (Fig. 1.8b). By turning over an appropriate number of leaves in a simple flex at an intermediate position, any of the other five intermediate positions can be reached directly.

The pat structure of subsidiary main positions 1(4), 2(5) and 3(6) is identical to that of intermediate positions of the first order fundamental triangle even skeletal flexagon 2⟨3, 3⟩ (Section 5.2.2). The dual marked net, shown in Fig. 5.16, has numbers for flexing as the skeletal fundamental triangle flexagon and letters for flexing as the augmented fundamental triangle point flexagon 1⟨3, 3, 6⟩. This is a transformation between flexagons. Some other skeletal triangle flexagons can also be flexed as triangle point flexagons.

5.5.3 *An Augmented Fundamental Square Point Flexagon*

The augmented fundamental square point flexagon 1⟨4, 4, 8⟩ is described in this section. It is the simplest augmented fundamental square point flexagon. Some of its properties are given in Table 5.7 Also see Table 5.5. There are some analogies

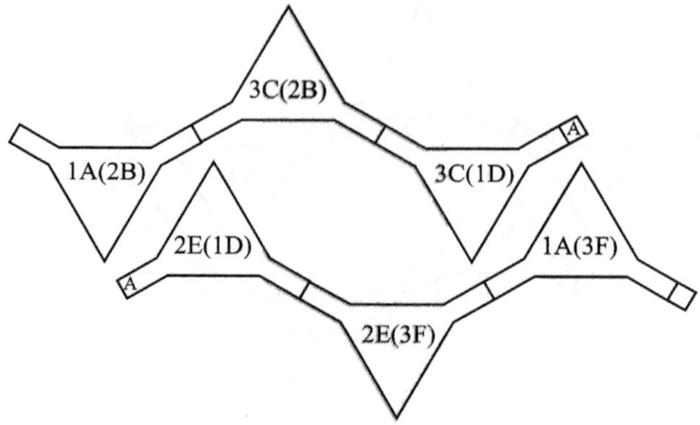

Fig. 5.16 Dual marked net for the augmented fundamental triangle point flexagon 1⟨3, 3, 6⟩ and the first order fundamental triangle even skeletal flexagon 2⟨3, 3⟩. One copy needed. Join the two parts of the net at A-A. Fold until leaves numbered 1 are visible

Table 5.7 Properties of the augmented fundamental square point flexagon 1⟨4, 4, 8⟩. The associated polygon is an octagon. Fundamental vertex net ⟨4⟩ is used. The principal cycle is in bold

Flexagon symbol	Typical main position	Cycle type	Number of cycles	Sector symbol
1⟨4, 4, 8⟩	**1(2)**	**8-cycle**	**1**	**⟨4, 4, 7, 1⟩**
1⟨4, 4, 8⟩	1(4)	8-cycle	1	⟨4, 4, 5, 3⟩
1⟨4, 4, 8⟩	1(5)	None	–	⟨4, 4, 4, 4⟩

with the properties of first order fundamental octagon even edge flexagons (Table 4.7). The torsion is 4. Its dual, the fundamental augmented square unagon 1⟨4, 4, 8⟩ has the same pat structure as the first order fundamental even edge flexagon 2⟨4, 4⟩ (Section 4.2.4). Hence, it is a flexagon and transformation between flexagons is possible.

The net for the augmented fundamental square point flexagon 1⟨4, 4, 8⟩ is shown in Fig. 5.17. This was derived by using the fundamental octagon point flexagon1⟨8, 8⟩ (Table 5.3) as a precursor, and replacing the regular octagons by squares. As assembled, the flexagon is in intermediate position 1, which is, in appearance, a single square. The principal 8-cycle shown in the intermediate position map (Fig. 5.18, solid lines) can be traversed by using the simple flex. The subsidiary 8-cycle (dotted lines) can be traversed also by using the simple flex. In these cycles, main positions are square vertex pairs linked by pairs of point hinges (Fig. 3.23). Subsidiary main positions 1(5), 2(6), 3(7) and 4(8) are not in cycles, and are square vertex pairs linked by one point hinge. The two subsidiary 4-cycles, and corresponding main positions, present in first order fundamental octagon even edge flexagons (Fig. 4.12, Table 4.7), are absent. The flexagon can also be flexed as the first order fundamental square even skeletal flexagon 2⟨4, 4⟩ (Section 5.2.3), and transformation between flexagons is possible.

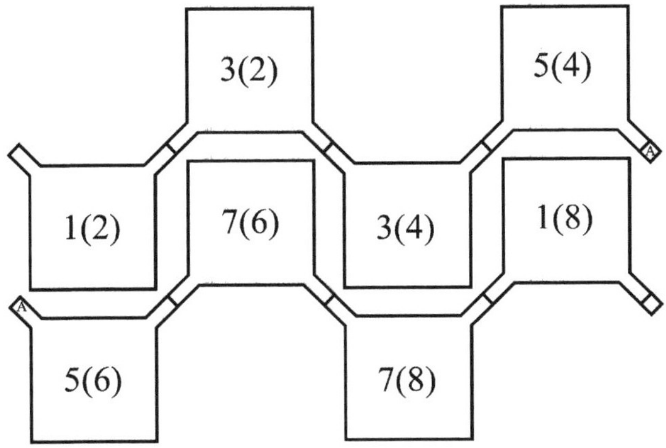

Fig. 5.17 Net for the augmented fundamental square point flexagon 1⟨4, 4, 8⟩. One copy needed. Join the two parts of the net at A-A. Fold until leaves numbered 1 are visible

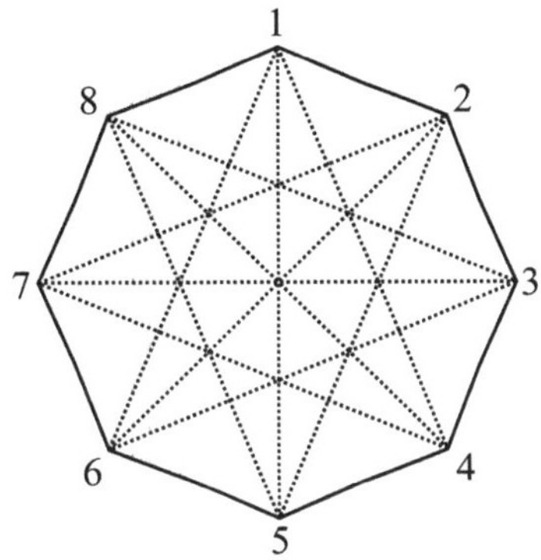

Fig. 5.18 Intermediate position map for the augmented fundamental square point flexagon 1⟨4, 4, 8⟩

5.6 Augmented Interleaved Fundamental Point Flexagons

5.6.1 General Properties

Augmented interleaved fundamental point flexagons have the same relationship to interleaved fundamental point flexagons (Section 5.4.1) as augmented fundamental point flexagons (Section 5.5.1) have to fundamental point flexagons (Section 5.3.1). They are solitary flexagons, and differ from augmented fundamental point flexagons in that their associated polygons are regular star polygons rather than regular convex polygons. All augmented interleaved fundamental point flexagons can be flexed by using the simple flex and this is their characteristic flex. Some properties of all the possible augmented interleaved fundamental point flexagons, with associated polygons of up to 12 edges. are given in Table 5.8. Face numbering sequences were derived using the method described in Section 5.4.1.

Table 5.8 Properties of augmented interleaved fundamental point flexagons

Leaf type	Flexagon symbol	Fundamental vertex net	Associated polygon	Face numbering sequence	Torsion
Triangle	1⟨3, 3, 9/2⟩	⟨3⟩	Enneagon (9/2)	1/2, 4/3, 5/6, 8/7, 9/1, 3/2, 4/5, 7/6, 8/9	3
Triangle	1⟨3, 3, 9/4⟩	⟨3⟩	Enneagon (9/4)	1/2, 6/5, 9/1, 5/4, 8/9, 4/3, 7/8, 3/2, 6/7	1
Triangle	1⟨3, 3, 12/5⟩	⟨3⟩	Dodecagon (12/5)	1/2, 7/6, 11/12, 5/4, 9/1, 3/2, 7/8, 1/12, 5/6, 11/10, 3/4, 9/8	4
Square	1⟨4, 4, 8/3⟩	⟨4⟩	Octagon (8/3)	1/2, 7/6, 3/4, 1/8, 5/6, 3/2, 7/8, 5/4	1
Square	1⟨4, 4, 12/5⟩	⟨4⟩	Dodecagon (12/5)	1/2, 7/6, 11/12, 5/4, 9/1, 3/2, 7/8, 1/12, 5/6, 11/10, 3/4, 9/8	2
Pentagon	1⟨5, 5, 10/3⟩	⟨5⟩	Decagon (10/3)	1/2, 5/4, 7/8, 1/10, 3/4, 7/6, 9/10, 3/2, 5/6, 9/8	6
Hexagon	1⟨6, 6, 12/5⟩	⟨6⟩	Dodecagon (12/5)	1/2, 7/6, 11/12, 5/4, 9/1, 3/2, 7/8, 1/12, 5/6, 11/10, 3/4, 9/8	6

To design the net for any desired augmented interleaved fundamental point flexagon, start by selecting the appropriate first order fundamental edge net in Table 5.8. Next, make the number of leaves in the net equal to the number of edges on the constituent polygons. Finally, apply the face numbering sequence listed in the table. Nets for some examples are given below.

In an intermediate position map for a first order fundamental even edge flexagon lines of different lengths represent main positions with different appearances (Section 4.2.1). However, in intermediate position maps for augmented interleaved fundamental point flexagons lines of different lengths sometimes represent main positions with the same appearance.

Intermediate positions of augmented interleaved fundamental point flexagons have more than one point hinge at each vertex. These overlap at a vertex, and there is more than one possible method of arranging the hinges during assembly. As a result, an augmented interleaved fundamental point flexagon can be either a knotted band or an unknotted band. The complexities of knot theory (Adams 1994) make enumeration of possible forms difficult. However, the intermediate position map is the same for all the forms of an augmented interleaved fundamental point flexagon. In paper models, with paper strips used as approximations to point hinges, augmented interleaved fundamental point flexagons have a neater appearance, and are easier to handle, if hinges are nested during assembly, but this is not always possible. When nesting is possible, there is only one method of assembly in which all the hinges are nested, and this is regarded as the correct method. When nesting is not possible there is no one correct method of assembly.

In intermediate positions, but not in any other position, the dual of an augmented interleaved fundamental point flexagon is an augmented interleaved fundamental unagon. These are theoretically possible, but as far as is known paper models of these unagons are impossible to assemble. Theoretically, they can exist only in intermediate positions, and cannot be flexed. Duals of augmented interleaved fundamental point flexagons are not possible if there are un-nested hinges so there are no corresponding unagons. Restricted flexing options (Section 5.3.1) mean that some cycles and corresponding main positions, present in first order fundamental even edge flexagons, are absent in corresponding augmented fundamental point flexagons.

5.6.2 Augmented Interleaved Fundamental Triangle Point Flexagons

The augmented interleaved fundamental triangle point flexagons 1⟨3, 3, 9/2⟩ and 1⟨3, 3, 9/4⟩ are described in this section. They are the simplest augmented interleaved fundamental triangle point flexagons. Some of their properties are given in Table 5.9. Also see Table 5.8. There are some analogies with the properties of first order fundamental enneagon even edge flexagons (Table 4.1). The torsion is 3 for 1⟨3, 3, 9/2⟩ and one for 1⟨3, 3, 9/4⟩.

The net for the augmented interleaved fundamental triangle point flexagon1⟨3, 3, 9/2⟩ is shown in Fig. 5.19. This was derived by using the interleaved fundamental enneagon point flexagons 1⟨9, 9, 9/2⟩ (Fig. 5.13) as a precursor, and replacing the regular enneagons by equilateral triangles. If assembled with nested hinges, then the flexagon is a knotted band. This is a trefoil knot (Fig. 5.20), which is the simplest knot (Adams 1994). The flexagon can be assembled also as an unknotted band, but there are then some un-nested hinges. The correct method of assembly is with nested hinges, and is as follows. Start at leaf number 1/2 and fold leaf 4/3 under it so that numbers 2 and 3 are adjacent. Next, fold leaf 5/6 under leaf 4/3 so that numbers 4 and 5 are adjacent. Follow with leaf 8/7 so that numbers 6 and 7 are adjacent, and leaf 9/1 so that numbers 8 and 9 are adjacent. Then interleave leaf 3/2 between leaves 1/2 and 4/3 so that the number 2s, and the number 3s, are adjacent. Ensure that the two adjacent hinges are nested. Next, interleave leaf 4/5 between leaves 4/3 and 5/6 so that the 4s, and the 5s, are adjacent. Then interleave leaf 7/6 between leaves 5/6 and 8/7 so that the 6s, and the 7s, are adjacent. Finally, interleave leaf 8/9 between leaves 8/7 and 9/1 so that the 8s and the 9s are adjacent. Ensure that the two adjacent hinges are nested, and join the ends. The same general method of assembly is used for other point flexagons with nested hinges.

Table 5.9 Properties of augmented interleaved fundamental triangle point flexagons. Fundamental vertex net ⟨3⟩ is used. Principal cycles are in bold

Flexagon symbol	Associated polygon	Typical main position	Cycle type	Number of cycles	Sector symbols
1⟨3, 3, 9/2⟩	**Enneagon ⟨9/2⟩**	**1(2)**	**9-cycle**	**1**	**⟨3, 3, 8, 1⟩**
1⟨3, 3, 9/2⟩	Enneagon ⟨9/2⟩	1(4)	3-cycle	3	⟨3, 3, 6, 3⟩
1⟨3,3, 9/4⟩	Enneagon ⟨9/4⟩	1(2)	9-cycle	1	⟨3, 3, 8, 1⟩

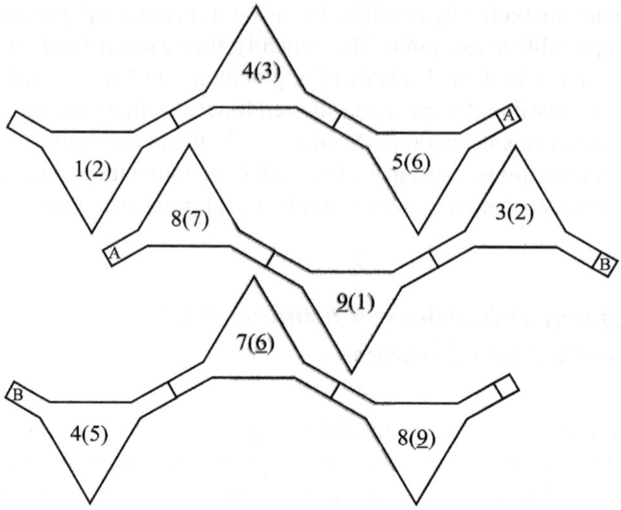

Fig. 5.19 Net for the augmented interleaved fundamental triangle point flexagon 1⟨3, 3, 9/2⟩. One copy needed. Join the three parts of the net at A-A and B-B. Interleave during assembly, nesting hinges, and with leaves numbered 1 left visible. See text

Fig. 5.20 A trefoil knot

As assembled, the flexagon is in intermediate position 1, which is, in appearance, a single equilateral triangle. The principal 9-cycle shown in the intermediate position map (Fig. 5.21) can be traversed by using the simple flex. Three subsidiary 3-cycles can be traversed by using the simple flex also. Main positions are, in appearance, triangle vertex pairs linked by pairs of point hinges (Fig. 1.8a). The two subsidiary 9-cycles, and corresponding main positions, present in first order fundamental enneagon even edge flexagons, are absent.

The net for the augmented interleaved fundamental triangle point flexagon1⟨3, 3, 9/4⟩ is shown in Fig. 5.22. Assembly and flexing are difficult. Hinges cannot be nested so there is no one correct method of assembly (previous section). The flexagon can be assembled as either a knotted band or an un-knotted band. As assembled the flexagon is in intermediate position 1, which is approximately a single equilateral triangle, with three point hinges at each vertex (Fig. 5.23a). In the photograph the hinges have been arranged as neatly as possible. The principal 9-cycle shown in the intermediate position map can be traversed by using the simple flex. The intermediate position map is the same as that for the interleaved fundamental enneagon point flexagon 1⟨9, 9, 9/2⟩ (Fig. 5.14). Principal main positions are approximately equi-

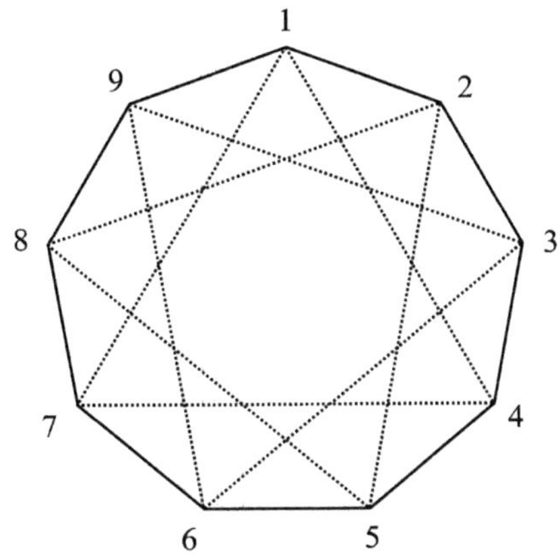

Fig. 5.21 Intermediate position map for the augmented interleaved fundamental triangle point flexagon 1⟨3, 3, 9/2⟩

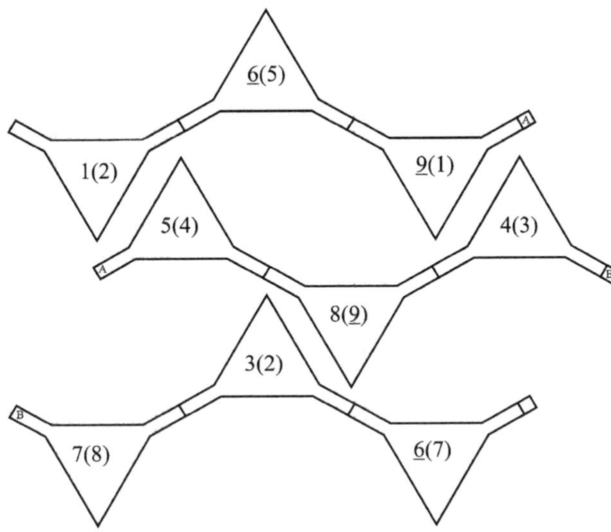

Fig. 5.22 Net for the augmented interleaved fundamental triangle point flexagon 1⟨3, 3, 9/4⟩. One copy needed. Join the three parts of the net at A-A and B-B. Interleave during assembly, with hinges adjacent, and with leaves numbered 1 left visible

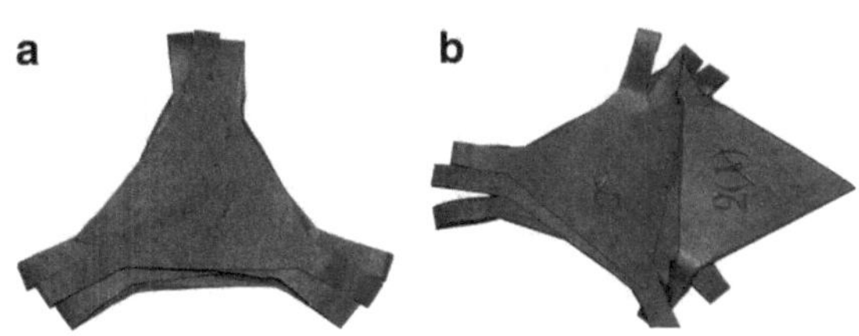

Fig. 5.23 A flexagon as approximately the following. Point hinges approximated by paper strips and cannot be nested. (**a**) An equilateral triangle. (**b**) An equilateral triangle vertex pair connected by pairs of point hinges

lateral triangle vertex pairs linked by pairs of point hinges (Fig. 5.23b). The five subsidiary cycles, and corresponding main positions, present in first order fundamental enneagon even edge flexagons, are absent.

5.6.3 *An Augmented Interleaved Fundamental Square Point Flexagon*

The augmented interleaved fundamental square point flexagon 1⟨4, 4, 8/3⟩, described in this section, is the simplest augmented interleaved fundamental square point flexagon. It is also the simplest augmented interleaved fundamental point flexagon Some of its properties are given in Table 5.10. Also see Table 5.8. There are some analogies with properties of first order fundamental octagon even edge flexagons (Table 4.7). The torsion is 1.

The net for the augmented interleaved fundamental square point flexagon 1⟨4, 4, 8/3⟩ is shown in Fig. 5.24. This was derived by using the interleaved fundamental octagon point flexagon 1⟨8, 8, 8/3⟩ (Table 5.4) as a precursor, and replacing the regular octagons by squares. As assembled, the flexagon is in intermediate position 1, which is, in appearance, a single square. Some hinges are nested and the flexagon is a knotted band. This is a trefoil knot, which is the simplest knot (Adams 1994). The flexagon can be assembled as an unknotted band, but there are then some unnested hinges. The principal 8-cycle shown in the intermediate position map (Fig. 5.25) can be traversed by using the simple flex. Principal main positions are square vertex pairs linked by pairs of point hinges (Fig. 3.23). The three subsidiary

Table 5.10 Properties of the augmented interleaved fundamental square point flexagon 1⟨4, 4, 8/3⟩. Fundamental vertex net ⟨4⟩ is used. The principal cycle is in bold

Flexagon symbol	Associated polygon	Typical main position	Cycle type	Number of cycles	Sector symbol
1⟨4, 4, 8/3⟩	**Octagon (8/3)**	**1(2)**	**8-cycle**	**1**	**⟨4, 4, 7, 1⟩**

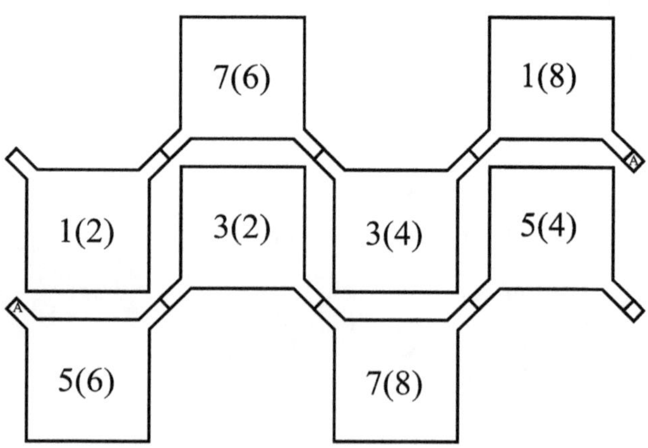

Fig. 5.24 Net for the augmented interleaved fundamental square point flexagon 1⟨4, 4, 8/3⟩. One copy needed. Join the two parts of the net at A-A. Interleave during assembly, nesting hinges, and with leaves numbered 1 left visible

Fig. 5.25 Intermediate position map for the augmented interleaved fundamental square point flexagon 1⟨4, 4, 8/3⟩

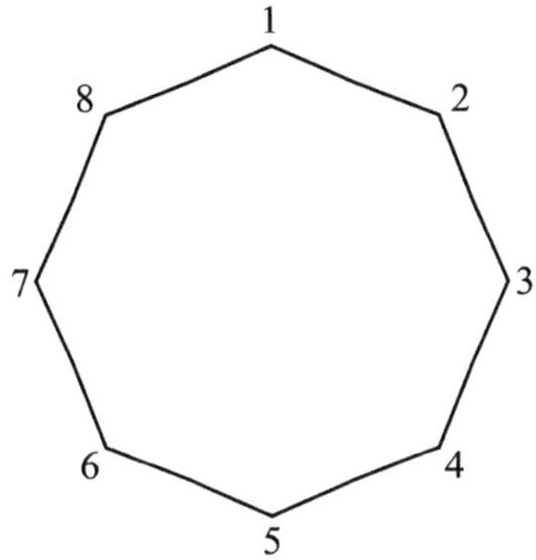

cycles, and corresponding main positions, present in first order fundamental octagon even edge flexagons, are absent.

References

Adams CC (1994) The Knot Book; an Elementary Introduction to the Mathematical Theory of Knots. W.H. Freeman, New York

Gardner M (1965) Mathematical Puzzles and Diversions. Penguin, Harmondsworth

Gardner M (2008) Hexaflexagons, Probability Paradoxes, and the Tower of Hanoi. Cambridge University Press, Cambridge

Pook LP (2007a) Interleaved point flexagons. Flexagon Lovers Group posting 317. http://tech.groups.yahoo.com/group/Flexagon_Lovers/. Accessed 8 May 2007

Pook LP (2007b) Point flexagons. Flexagon Lovers Group posting 297. http://tech.groups.yahoo.com/group/Flexagon_Lovers/. Accessed 25 April 2007

Sherman S (2007a) Point flexagons. http://loki3.com/flex/point-flexagon.html. Accessed 17 April 2007

Sherman S (2007b) Point flexagons. Flexagon Lovers Group posting 309. http://tech.groups.yahoo.com/group/Flexagon_Lovers/. Accessed 1 May 2007

Sherman S (2007c) Point flexagons. Flexagon Lovers Group posting 314. http://tech.groups.yahoo.com/group/Flexagon_Lovers/. Accessed 6 May 2007

Chapter 6
Fundamental Compound Edge Flexagons

6.1 Introduction

Compound edge flexagons are even edge flexagons with main positions that are, in appearance, compound edge rings of regular convex polygons (Section 2.2.4). In a compound edge ring, alternate hinge angles (Fig. 1.3) are the same and, in general, alternate polygons are the same distance from the centre of the ring. Compound edge flexagons with at least some flat main positions were briefly described by Conrad and Hartline (1962) and in more detail by Pook (2003). A compound edge ring of $2n$ regular convex polygons, and corresponding compound edge flexagons, can be divided into n identical sectors each containing two polygons. All the possible flat non overlapping, compound edge rings, containing up to ten polygons, with up to 12 edges on the constituent polygons are listed in Table 2.5. Possible non flat compound edge rings of four regular polygons, with up to eight edges on the constituent polygons, are listed in Table 2.6.

Fundamental compound edge flexagons are made from first order fundamental edge nets (Section 3.2), and a standard face numbering sequence is used (Section 4.1.1). They are regular cycle flexagons in which all the main positions of a cycle have the same appearance and the same pat structure. They have some analogies with first order fundamental even edge flexagons (Section 4.2.1). In particular, they are solitary flexagons and the topological invariants are the same. Leaves usually have to be bent during flexing, although there are exceptions. Most fundamental compound edge flexagons are too difficult to flex to be aesthetically satisfying, and are therefore of only theoretical interest. Various flexes are used but there is no characteristic flex that can be used for all fundamental compound edge flexagons.

There are fundamental compound vertex flexagons with main positions that are, in appearance, compound vertex rings of regular convex polygons (Section 2.4.1) but they are very unstable and are not discussed. Nets included in this chapter are restricted to fundamental compound edge flexagons that are reasonably easy to handle.

L. Pook, *Serious Fun with Flexagons,* Solid Mechanics and Its Applications 164, DOI 10.1007/978-90-481-2503-6_6, © Springer Science+Business Media B.V. 2009

6.2 General Properties

Possible fundamental compound edge flexagons can be enumerated by fan folding an appropriate first order fundamental edge net (Section 3.2) onto a sector of a compound ring. For a particular sector and a particular first order fundamental edge net there are always, neglecting enantiomorphs, two distinct solutions. These two solutions are shown as sector diagrams in Fig. 6.1 for the first order fundamental square edge net ⟨4⟩ (Fig. 3.3) fan folded onto a sector of the flat compound ring of eight squares (Fig. 1.11).

The flexagon symbol for a first order fundamental even edge flexagon is $S\langle s, c\rangle$ where S is the number of sectors, $\{s\}$ is the Schläfli symbol for the constituent polygons, and ⟨c⟩ is the first order fundamental edge net symbol (Section 4.2.1). The flexagon symbol for a fundamental compound edge flexagon is S⟨s, c, a⟩ where S is the number of sectors, {s} is the Schläfli symbol for the constituent polygons, ⟨c⟩. is the first order fundamental edge net symbol, {a} is the Schläfli symbol for the associated polygon. The associated polygon is a regular convex polygon with the same number of edges as there are leaves per sector. The two distinct sector diagrams mean that fundamental compound edge flexagons always occur in pairs, and a geometric consequence is that, in a pair, the average value of a is always equal to s. Flexagon figures are not appropriate for fundamental compound edge flexagons.

The compound edge ring of eight squares (Fig. 1.11) has four sectors There are three leaves per sector in Fig. 6.1a and five in Fig. 6.1b. so the flexagon symbols for the corresponding fundamental square compound edge flexagons are 4⟨4, 4, 3⟩ and 4⟨4, 4, 5⟩. The associated polygon for 4⟨4, 4, 3⟩ is an equilateral triangle so it is a triangular compound edge flexagon. In full, it is a triangular fundamental square compound edge flexagon. The noun 'square' refers to the constituent polygons and the adjective 'triangular' to the associated polygon. Similarly, 4⟨4, 4, 5⟩ is a pentagonal compound edge flexagon, and in full a pentagonal fundamental square compound edge flexagon. To avoid confusion 'square-like' is used later as the adjectival form of 'square'.

Table 6.1 shows, for each of the flat compound edge rings listed in Table 2.5, all the possible fundamental compound edge flexagons with at least some main positions that are flat compound rings. Most are non overlapping compound rings but one overlapping compound ring is included.

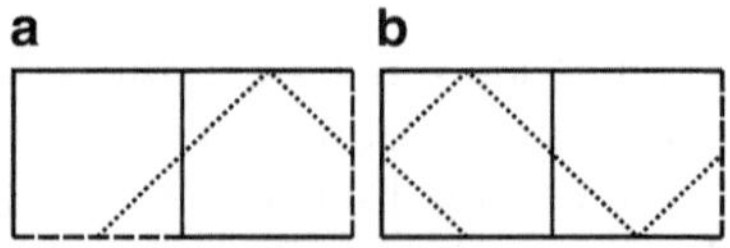

Fig. 6.1 Sector diagrams for fundamental square compound edge flexagons. (**a**) 4⟨4, 4, 3⟩. (**b**) 4⟨4, 4, 5⟩ (Les Pook, Flexagons inside out, 2003, © Cambridge University Press 2003, reprinted with permission)

Table 6.1 Flat compound edge rings and corresponding compound edge flexagons

Polygon type	Number in ring	Ring symbol	Flexagon symbols for corresponding flexagons
Square	8	4(0°, 90°)	4⟨4, 4, 3⟩, 4⟨4, 4, 5⟩
Pentagon	10	5(−36°, 108°)	5⟨5, 5, 3⟩, 5⟨5, 5/2, 4⟩, 5⟨5, 5/2, 6⟩, 5⟨5, 5, 7⟩
Hexagon	4	2(60°, 120°)	2⟨6, 6, 5⟩, 2⟨6, 6, 7⟩
Hexagon	6	3(0°, 120°)	3⟨6, 6, 4⟩, 3⟨6, 6, 8⟩
Octagon	4	2(45°, 135°)[a]	2⟨8, 8, 6⟩, 2⟨8, 8/3, 6⟩, 2⟨8, 8, 10⟩, 2⟨8. 8/3, 10⟩
Octagon	8	4(0°, 90°)	4⟨8, 8, 6⟩, 4⟨8, 8/3, 6⟩, 4⟨8, 8, 10⟩, 4⟨8, 8/3, 10⟩
Enneagon	6	3(20°, 100°)	3⟨9, 9/4, 5⟩, 3⟨9, 9, 7⟩, 3⟨9, 9/2, 8⟩, 3⟨9, 9/2, 10⟩, 3⟨9, 9, 11⟩, 3⟨9, 9/4, 13⟩
Decagon	4	2(36°, 144°)[a]	2⟨10, 10, 7⟩, 2⟨10, 10/3, 9⟩, 2⟨10, 10/3, 11⟩, 2⟨10, 10, 13⟩
Decagon	4	2(72°, 108°)	2⟨10, 10/3, 7⟩, 2⟨10, 10, 9⟩, 2⟨10, 10, 11⟩, 2⟨10, 10/3, 13⟩
Decagon	10	5(−36°, 108°)	5⟨10, 10, 6⟩, 5⟨10, 10/3, 8⟩, 5⟨10, 10/3, 12⟩, 5⟨10, 10, 14⟩
Decagon	10	5(0°, 72°)	5⟨10, 10/3, 6⟩, 5⟨10, 10, 8⟩, 5⟨10, 10, 12⟩, 5⟨10, 10/3, 14⟩
Dodecagon	4	2(30°, 150°)[a]	2⟨12, 12, 8⟩, 2⟨12, 12/5, 8⟩, 2⟨12, 12, 16⟩, 2⟨12, 12/5, 16⟩
Dodecagon	4	2(60°, 120°)	2⟨12, 12, 10⟩, 2⟨12, 12/5, 10⟩, 2⟨12, 12, 14⟩, 2⟨12, 12/5, 14⟩
Dodecagon	6	3(0°, 120°)	3⟨12, 12, 8⟩, 3⟨12, 12/5, 8⟩, 3⟨12, 12, 16⟩, 3⟨12, 12/5, 16⟩
Dodecagon	8	4(−30°, 120°)	4⟨12, 12, 7⟩, 4⟨12, 12/5, 11⟩, 4⟨12, 12/5, 13⟩, 4⟨12, 12, 17⟩,
Dodecagon	8	4(0°, 90°)	4⟨12, 12, 9⟩, 4⟨12, 12/5, 9⟩, 4⟨12, 12, 15⟩, 4⟨12, 12/5, 15⟩
Dodecagon	8	4(30°, 60°)	4⟨12, 12/5, 5⟩, 4⟨12, 12, 11⟩, 4⟨12, 12, 13⟩, 4⟨12, 12/5, 19⟩

[a]Overlapping ring.

A noticeable feature of the dynamic behaviour of most compound edge flexagons is that intermediate positions are not clearly defined. Nevertheless, intermediate position maps can be used to characterise their dynamic properties. A fundamental compound edge flexagon with the same associated polygon as a first order fundamental even edge flexagon, either has the same intermediate position map as the first order fundamental even edge flexagon, or an intermediate position map that is a subset. The latter arises because types of cycles, and corresponding main positions, that are present in first order fundamental even edge flexagons, may be absent from corresponding fundamental compound edge flexagons. This can happen for two reasons. First, main positions may be absent because of the limitation on the maximum permissible number of leaves in a pat (Section 4.2.5.1). Secondly, main positions may be absent because corresponding polygon rings are impossible (Section 2.2.4). If main positions are absent, then lines representing them are absent from intermediate position maps. In other words, the intermediate position map for a fundamental compound edge flexagon becomes a subset of that for the corresponding first order fundamental even edge flexagon.

A type of main position that is present in a fundamental compound edge flexagon has the same pat structure, and the same sector symbol, as the corresponding type of main positions in a first order fundamental even edge flexagon. Some properties of all the possible fundamental compound edge flexagons with at least some main positions that are flat non overlapping rings, and with up to eight leaves per sector, are given in Table 6.2. Two flexagons with main positions that are flat overlapping rings are included. There is no simple expression for the torsion per sector of a fundamental compound edge flexagon.

Table 6.2 Properties of fundamental compound edge flexagons with flat main positions

Associated polygon	Leaf type	Flexagon symbol	Fundamental edge net	Torsion per sector
Triangle	Square	4⟨4, 4, 3⟩	⟨4⟩	1
Triangle	Pentagon	5⟨5, 5, 3⟩	⟨5⟩	1
Square	Pentagon	5⟨5, 5/2, 4⟩	⟨5/2⟩	0
Square	Hexagon	3⟨6, 6, 4⟩	⟨6⟩	2
Pentagon	Square	4⟨4, 4, 5⟩	⟨4⟩	3
Pentagon	Hexagon	2⟨6, 6, 5⟩	⟨6⟩	3
Pentagon	Enneagon	3⟨9, 9/4, 5⟩	⟨9/4⟩	1
Pentagon	Dodecagon	4⟨12, 12/5, 5⟩	⟨12/5⟩	1
Hexagon	Pentagon	5⟨5, 5/2, 6⟩	⟨5/2⟩	2
Hexagon	Octagon	2⟨8, 8, 6⟩[a]	⟨8⟩	4
Hexagon	Octagon	4⟨8, 8, 6⟩	⟨8⟩	4
Hexagon	Octagon	2⟨8, 8/3, 6⟩	⟨8/3⟩	2
Hexagon	Octagon	4⟨8, 8/3, 6⟩	⟨8/3⟩	2
Hexagon	Decagon	5⟨10, 10, 6⟩	⟨10⟩	4
Hexagon	Decagon	5⟨10, 10/3, 6⟩	⟨10/3⟩	3
Heptagon	Pentagon	5⟨5, 5, 7⟩	⟨5⟩	3
Heptagon	Hexagon	2⟨6, 6, 7⟩	⟨6⟩	3
Heptagon	Enneagon	3⟨9, 9, 7⟩	⟨9⟩	5
Heptagon	Decagon	2⟨10, 10, 7⟩	⟨10⟩	5
Heptagon	Decagon	2⟨10, 10/3, 7⟩	⟨10/3⟩	3
Heptagon	Dodecagon	4⟨12, 12, 7⟩	⟨12⟩	5
Octagon	Hexagon	3⟨6, 6, 8⟩	⟨6⟩	4
Octagon	Enneagon	3⟨9, 9/2, 8⟩	⟨9/2⟩	4
Octagon	Decagon	5⟨10, 10, 8⟩	⟨10⟩	6
Octagon	Decagon	5⟨10, 10/3, 8⟩	⟨10/3⟩	4
Octagon	Dodecagon	2⟨12, 12, 8⟩[a]	⟨12⟩	6
Octagon	Dodecagon	3⟨12, 12, 8⟩	⟨12⟩	6
Octagon	Dodecagon	2⟨12, 12/5, 8⟩[a]	⟨12/5⟩	2
Octagon	Dodecagon	3⟨12, 12/5, 8⟩	⟨12/5⟩	2

[a] Overlapping ring.

6.3 Triangular Fundamental Compound Edge Flexagons

6.3.1 Some Properties

There are two triangular compound edge flexagons with flat main positions listed in Table 6.2. These are the triangular fundamental square compound edge flexagon 4⟨4, 4, 3⟩ and the triangular fundamental pentagon compound edge flexagon 5⟨5, 5, 3⟩. Some of their properties are given in Table 6.3. There are some analogies with the properties of first order fundamental triangle even edge flexagons (Table 4.3). The torsion per sector is 1. The sector diagrams are shown in Figs. 6.1a and 6.2.

6.3.2 A Fundamental Square Compound Edge Flexagon

The net for the triangular fundamental square compound edge flexagon 4⟨4, 4, 3⟩ is shown in Fig. 6.3. The flexagon is unusual for a fundamental compound edge flexagon in that it can be flexed without bending the leaves and, in the sequence used, there are clearly defined intermediate positions.

As assembled, the flexagon is in principal main position 2(1) which is, in appearance, a flat compound edge ring of eight squares (Fig. 1.11). The principal 3-cycle shown in the intermediate position map can be traversed by using a version of the twofold pinch flex. The intermediate position map is the same as that for the first order fundamental triangle even edge flexagons *S*⟨3, 3⟩ (Fig. 4.7). Starting from principal main position 2(1), and using parallel hinges, pinch together two diagonally opposite pairs of pats so that leaves numbered 1 are folded together. The new pats have leaves numbered 2 visible on both faces. The in between position (Fig. 6.4a) is with the pats partially pinched together. Then fold the new pats outwards to reach intermediate position 2. This is an irregular overlapping even edge ring of eight squares with mixed up face numbers (Fig. 6.4b). In the photograph, the eighth square is covered by the central square. The new pats, numbered 2 on both faces, are at the corners of the intermediate position and are the key pats which identify the intermediate position. To complete the flex and reach principal main position 3(2), reverse the first part of the flex, using hinges that are at right angles to those used in the first part, and so on round the principal 3-cycle.

Table 6.3 Properties of triangular fundamental compound edge flexagons. Principal cycles are in bold

Leaf type	Flexagon symbol	Typical main position	Cycle type	Number of cycles	Main position type	Ring symbol	Sector symbol	Curvature
Square	**4⟨4, 4, 3⟩**	**1(2)**	**3-cycle**	**1**	**Flat**	**4(0°, 90°)**	**⟨4, 4, 2, 1⟩**	**0°**
Pentagon	**5⟨5, 5, 3⟩**	**1(2)**	**3-cycle**	**1**	**Flat**	**5(−36°, 108°)**	**⟨5, 5, 2, 1⟩**	**0°**

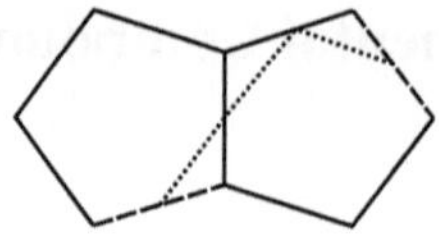

Fig. 6.2 Sector diagram for the fundamental pentagon compound edge flexagon 5⟨5, 5, 3⟩ (Les Pook, Flexagons inside out, 2003, © Cambridge University Press 2003, reprinted with permission)

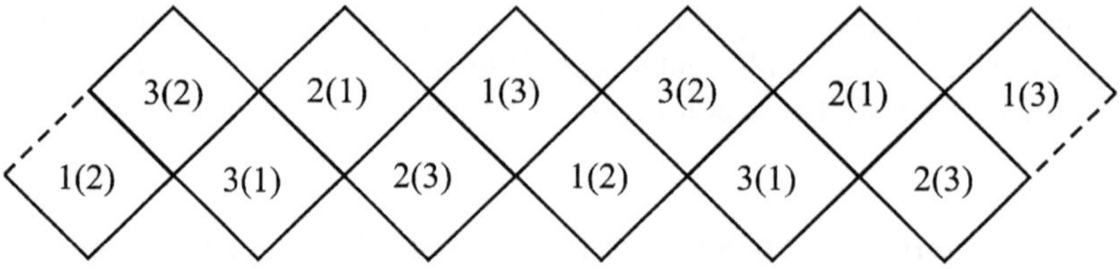

Fig. 6.3 Net for the fundamental square compound edge flexagon 4⟨4, 4, 3⟩. One copy needed

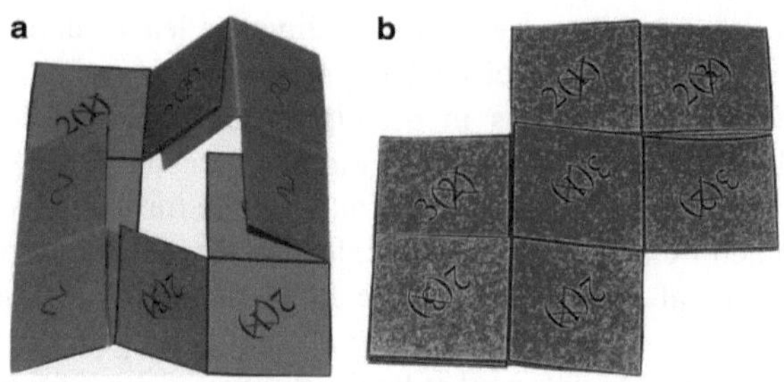

Fig. 6.4 The fundamental square compound edge flexagon 4⟨4, 4, 3⟩. (a) In between position. (b) Intermediate position, a flat overlapping irregular even edge ring of eight squares

The principal 3-cycle cycle can be traversed by using a twist flex also. Starting from main position 2(1), to flex to main position 3(2), hold a diametrically opposite pair of leaves numbered 3(2), turn them over and pull them gently apart so that they are at diagonally opposite corners of the flexagon. The intermediate position is not clearly defined.

6.3.3 A Fundamental Pentagon Compound Edge Flexagon

The net for the triangular fundamental pentagon compound edge flexagon 5⟨5, 5, 3⟩. is shown in Fig. 6.5. As assembled, the flexagon is in principal main position 2(1), which is, in appearance, a flat compound edge ring of ten regular pentagons (Fig. 6.6). A twist flex is used to traverse the principal 3-cycle shown in the intermediate position map. The intermediate position map is the same as that for the first order

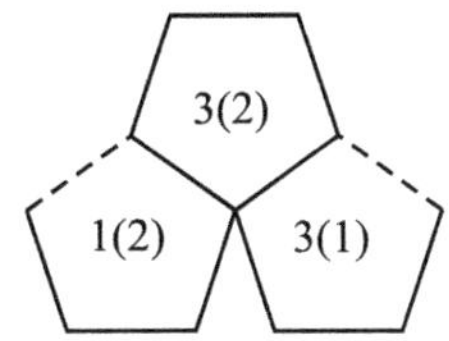

Fig. 6.5 Net for the fundamental pentagon compound edge flexagon 5⟨5, 5, 3⟩. Five copies needed

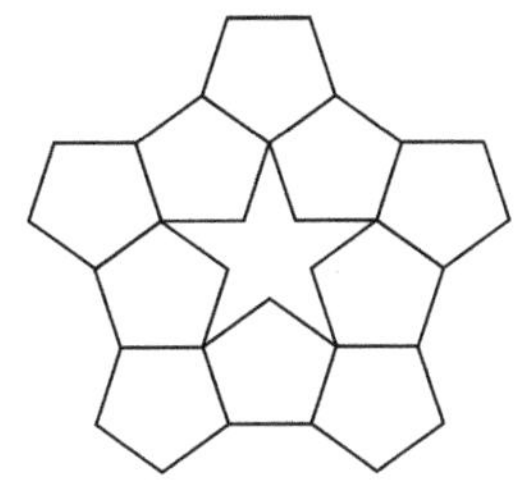

Fig. 6.6 Flat compound edge ring of ten regular pentagons (Les Pook, Flexagons inside out, 2003, © Cambridge University Press 2003, reprinted with permission)

fundamental triangle even edge flexagons *S*⟨3, 3⟩ (Fig. 4.7). Starting from principal main position 2(1), twist the five single leaves at the corners of principal main position 2(1), so that pairs of leaves numbered 1 are folded together, to reach principal main position 3(2) and so round the principal 3-cycle. Provided that fivefold rotational symmetry is approximately maintained during flexing, it is not necessary to hold all the single leaves, and only slight bending of the leaves is needed. The intermediate position is not clearly defined.

6.4 A Square-Like Fundamental Compound Edge Flexagon

6.4.1 Some Properties

There are two square-like compound edge flexagons with flat main positions listed in Table 6.2. One of these is the square-like fundamental hexagon compound edge flexagon 3⟨6, 6, 4⟩. Some of its properties are given in Table 6.4. There are some analogies with the properties of a first order square even edge flexagon (Table 4.4). The torsion per sector is 2. The sector diagram is shown in Fig. 6.7a.

6.4.2 A Fundamental Hexagon Compound Edge Flexagon

The net for the square-like fundamental hexagon compound edge flexagon 3⟨6, 6, 4⟩ is shown in Fig. 6.8. As assembled, the flexagon is in principal main position 2(1) which is, in appearance, a flat compound edge ring of six regular hexagons (Fig. 6.9a). To flex into principal main position 2(3), hold leaves marked 2(3) and

Table 6.4 Properties of the square-like fundamental compound edge flexagon. The principal cycle is in bold

Leaf type	Flexagon symbol	Typical main position	Cycle type	Number of cycles	Main position type	Ring symbol	Sector symbol	Curvature
Hexagon	**3⟨6, 6, 4⟩**	**1(2)**	**4-cycle**	**1**	**Flat**	**3(0°, 120°)**	**⟨6, 6, 3, 1⟩**	**0°**
Hexagon	3⟨6, 6, 4⟩	1(3)	None	–	Antibox	3(−60°, 60°)	⟨6, 6, 2, 2⟩	360°

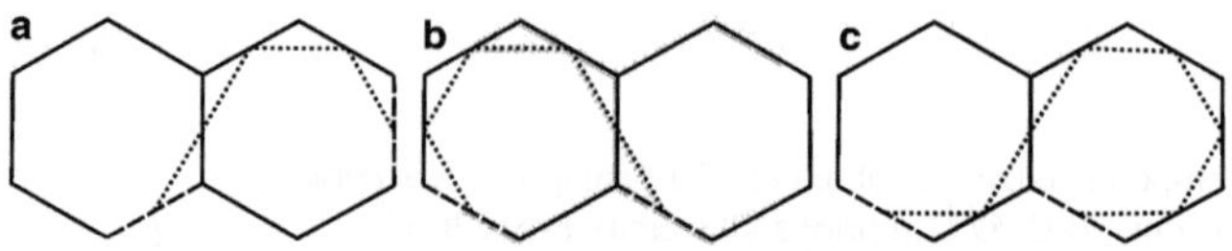

Fig. 6.7 Sector diagrams for fundamental hexagon compound edge flexagons. (**a**) 3⟨6, 6, 4⟩ (**b**) 2⟨6, 6, 5⟩. (**c**) 2⟨6, 6, 7⟩. Figures 6.7b and c: (Les Pook, Flexagons inside out, 2003, © Cambridge University Press 2003, reprinted with permission)

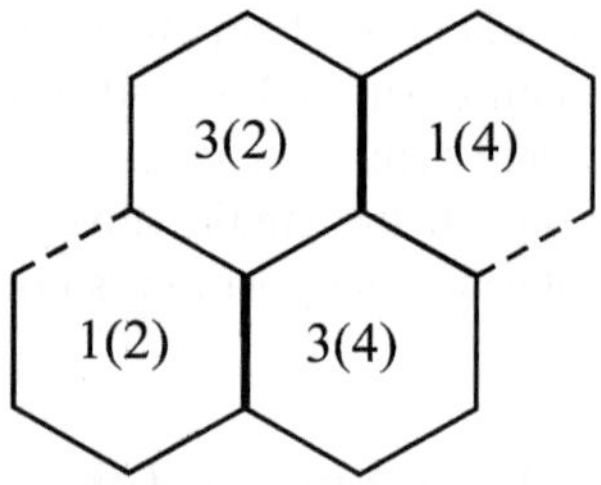

Fig. 6.8 Net for the fundamental hexagon compound edge flexagon 3⟨6, 6, 4⟩. Three copies needed

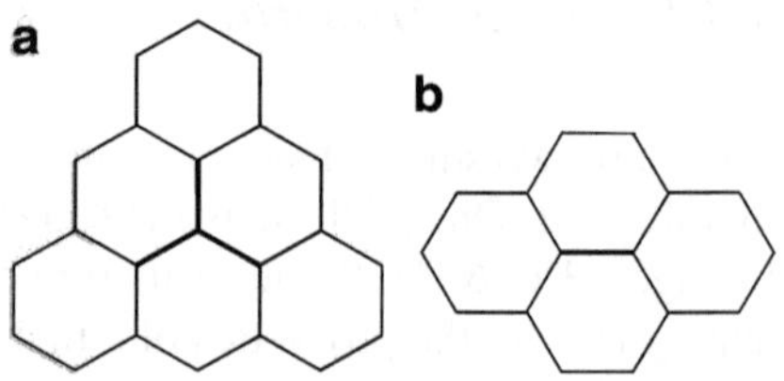

Fig. 6.9 Flat compound edge rings of regular hexagons. (**a**) Six hexagons. (**b**) Four hexagons (Les Pook, Flexagons inside out, 2003, © Cambridge University Press 2003, reprinted with permission)

pull them gently outwards in a twist flex, and so on round the principal 4-cycle shown in the intermediate position map. The intermediate position map is the same as that for the first order fundamental square even edge flexagons $S\langle 4, 4\rangle$ (Fig. 4.8). Intermediate positions are not clearly defined.

It is also possible, by using twist flexes, to traverse the principal 4-cycle by flexing via subsidiary main positions. These are antibox edge rings of six regular hexagons (Fig. 2.13). Starting from principal main position 1(2), work round the flexagon in a twist flex by folding leaves numbered 1 together in pairs, ensuring that each pat consists of a folded pile of two leaves, to reach subsidiary main position 2(4). Then work round the flexagon in another twist flex by folding pairs of leaves numbered 4 together to reach principal main position 2(3). Each part of these twist flexes is a local flex because some pats are left unchanged. If these are not carried out correctly then the flexagon can become muddled, that is, in an unwanted position.

6.5 Pentagonal Fundamental Compound Edge Flexagons

6.5.1 *Some Properties*

There are four pentagonal compound edge flexagons with flat main positions listed in Table 6.2. Two of these are the pentagonal fundamental square compound edge flexagon 4⟨4, 4, 5⟩ and the pentagonal fundamental hexagon compound edge flexagon 2⟨6, 6, 5⟩. Some of their properties are given in Table 6.5. There are some analogies with the properties of first order fundamental pentagon even edge flexagons (Table 4.5). The torsion per sector is 3. The sector diagrams are shown in Figs. 6.1b and 6.7b.

6.5.2 *A Fundamental Square Compound Edge Flexagon*

The net for the pentagonal fundamental square compound edge flexagon 4⟨4, 4, 5⟩ is shown in Fig. 6.10. As assembled, the flexagon is in subsidiary main position 1(3) which is, in appearance, a flat compound edge ring of eight squares (Fig. 1.11). The subsidiary 5-cycle shown in the intermediate position map can be traversed by using a twist flex. The intermediate position map is the same as that for the fundamental pentagon point flexagon 1⟨5, 5/2⟩ (Fig. 5.9). Starting at subsidiary main position 1(3), open the pats at diagonally opposite corners at leaves numbered 5, and fold the pairs of leaves numbered 1 together to reach intermediate position 3. This is clearly defined and is an irregular box edge ring of eight squares (Fig. 6.11), with leaves numbered 3 on the outside of the ring. In the photograph the newly folded pats are held together with paper clips. To complete the twist flex to subsidiary main position 3(5), open the pats that were at the other pair of corners at leaves numbered 5, and fold the pairs of leaves numbered 1 together, and so on round the subsidiary 5-cycle. The second part of the twist flex is difficult. The flexagon is easily muddled, that is put into an unwanted position, and it can be difficult to see how to return to a wanted position.

The principal 5-cycle, and corresponding main positions, present in first order fundamental pentagon even edge flexagons (Section 4.2.6), are absent because the permissible number of leaves in a pat is exceeded (Sections 4.2.5.1 and 6.2).

Table 6.5 Properties of pentagonal fundamental compound edge flexagons. The principal cycles is in bold

Leaf type	Flexagon symbol	Typical main position	Cycle type	Number of cycles	Main position type	Ring symbol	Sector symbols	Curvature
Square	4⟨4, 4, 5⟩	1(2)	5-cycle	1	Flat	4(0°, 90°)	⟨4, 4, 3, 2⟩	0°
Hexagon	**2⟨6, 6, 5⟩**	**1(2)**	**5-cycle**	**1**	**Flat**	**2(60°, 120°)**	**⟨6, 6, 4, 1⟩**	**0°**
Hexagon	2⟨6, 6, 5⟩	1(3)	5-cycle	1	Slant	2(0°, 60°)	⟨6, 6, 3, 2⟩	240°

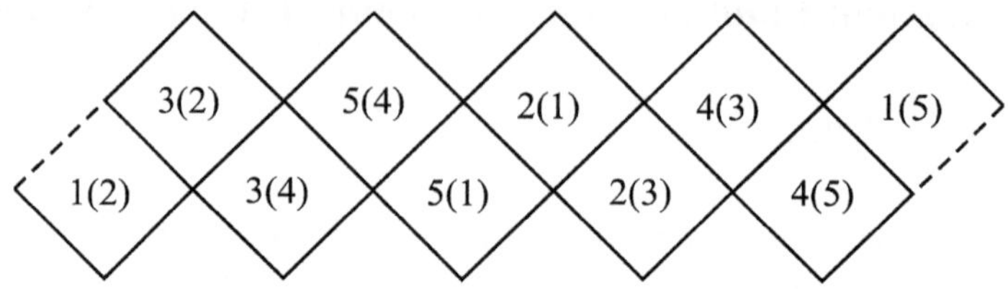

Fig. 6.10 Net for the fundamental square compound edge flexagon 4⟨4, 4, 5⟩. Two copies needed. Fold until leaves numbered 1 and 3 are visible

6.5.3 A Fundamental Hexagon Compound Edge Flexagon

The net for the pentagonal fundamental hexagon compound edge flexagon 2⟨6, 6, 5⟩ is shown in Fig. 6.12. As assembled, it is in principal main position 1(2) which is, in appearance, a flat compound edge ring of four regular hexagons (Fig. 6.9b). The principal 5-cycle shown in the intermediate position map can be traversed by using a twist flex. The intermediate position map is the same as that for the first order fundamental pentagon even edge flexagons *S*⟨5, 5⟩ and *S*⟨5, 5/2⟩ (Fig. 4.9). Starting from principal main position 1(2), hold the groups of three leaves with numbers 1 and 3 on the top and bottom. Then slide the groups of leaves past each other, ensuring that pairs of leaves numbered 1 fold together, to reach principal main position 2(3) and so on round the principal 5-cycle. The intermediate position is not clearly defined. The principal 5-cycle can also be traversed by using a two-fold pinch flex. Intermediate positions are clearly defined, but hinges are not in line so leaves have to be bent during flexing. They also have to be bent to avoid intersection.

Another type of twist flex can be used to flex to subsidiary main positions. These are slant compound edge rings of four regular hexagons (Fig. 2.14). Starting from principal main position 1(2) fold together pairs of leaves numbered 2 to reach subsidiary main position 1(3), or pairs of leaves numbered 1 to reach subsidiary main position 2(5). The intermediate position is not clearly defined. The high curvature (240°) means that subsidiary main positions cannot be turned inside out, so the subsidiary 5-cycle shown in the intermediate position map cannot be traversed directly, but it can be traversed via principal main positions.

Fig. 6.11 Intermediate position of the fundamental square compound edge flexagon 4⟨4, 4, 5⟩

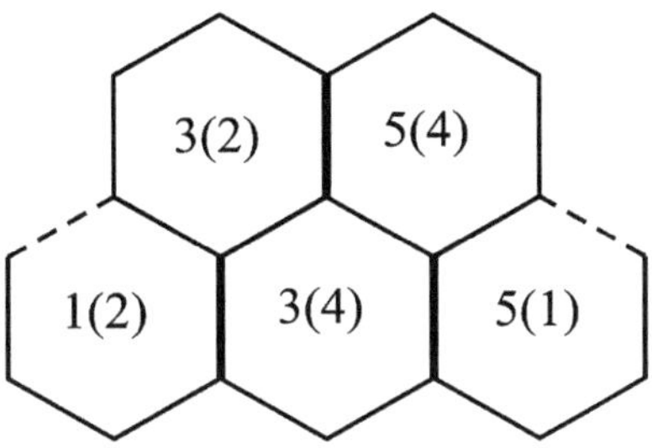

Fig. 6.12 Net for the fundamental hexagon compound edge flexagon 2⟨6, 6, 5⟩. Two copies needed

6.6 A Hexagonal Fundamental Compound Edge Flexagon

6.6.1 Some Properties

There are seven hexagonal compound edge flexagons with flat main positions listed in Table 6.2. One of these is the hexagonal fundamental octagon compound edge flexagon 2⟨8, 8, 6⟩. Some of its properties are given in Table 6.6. There are some analogies with first order fundamental hexagon even edge flexagons (Table 4.6). The torsion per sector is 4. The sector diagram is shown in Figs. 6.13.

6.6.2 A Fundamental Octagon Compound Edge Flexagon

The net for the hexagonal fundamental octagon compound edge flexagon 2⟨8, 8, 6⟩ is shown in Fig. 6.14. The leaves have been truncated to irregular hexagons to simplify assembly, but the appearance of the flexagon is satisfactory. The resulting truncated flexagon is a partial overlap flexagon and could also be called an irregular hexagon even edge flexagon.

As assembled, the flexagon is in principal main position 2(1) which is, in appearance, nominally a flat overlapping compound edge ring of four regular octagons (Fig. 6.15a). The pats at the ends of the flexagon are single leaves. The principal 6-cycle shown in the intermediate position map can be traversed by using a twist

Table 6.6 Properties of a hexagonal fundamental octagon compound edge flexagon. The principal cycle is in bold

Leaf type	Flexagon symbol	Typical main position	Cycle type	Number of cycles	Main position type	Ring symbol	Sector symbol	Curvature
Octagon	**2⟨8, 8, 6⟩**	**1(2)**	**6-cycle**	**1**	**Flat**[a]	**2(45°, 135°)**	**⟨8, 8, 5, 1⟩**	**0°**
Octagon	2⟨8, 8, 6⟩	1(3)	3-cycle	2	Slant	2(0°, 90°)	⟨8, 8, 4, 2⟩	180°
Octagon	2⟨8, 8, 6⟩	1(4)	None	–	Antibox	2(–45°, 45°)	⟨8, 8, 3, 3⟩	360°

[a]Overlapping ring

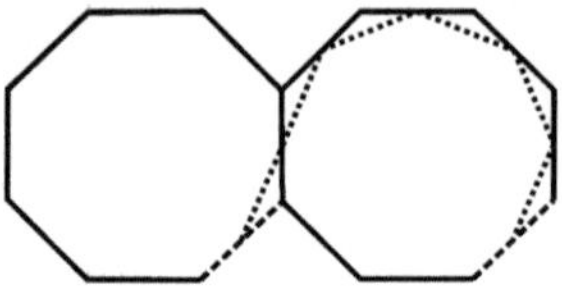

Fig. 6.13 Sector diagram for the fundamental octagon compound edge flexagon 2⟨8, 8, 6⟩

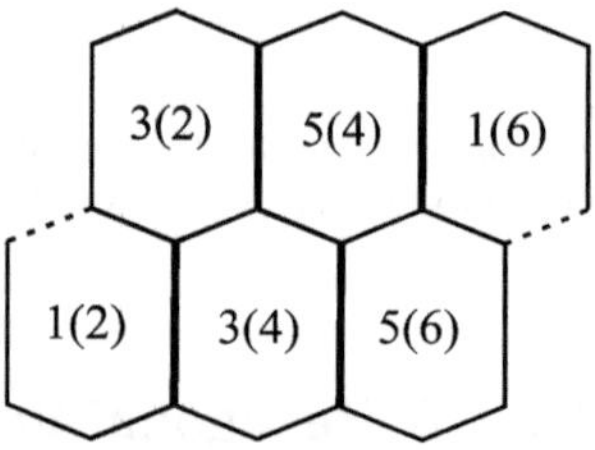

Fig. 6.14 Net for the fundamental octagon (irregular hexagon) compound edge flexagon 2⟨8, 8, 6⟩. Two copies needed

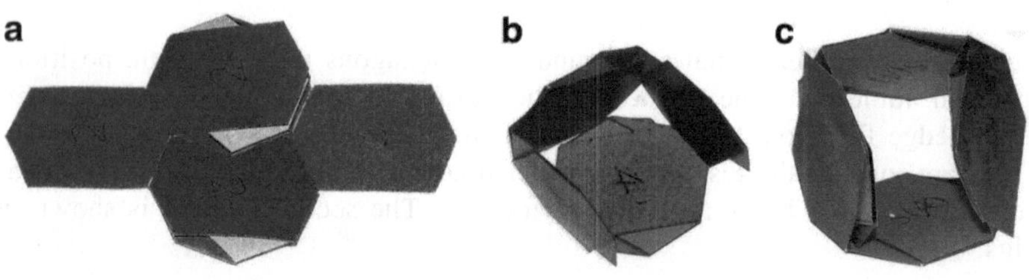

Fig. 6.15 A flexagon as compound edge rings of four regular octagons (irregular hexagons). (**a**) Flat overlapping. (**b**) Slant. (**c**) Antibox

flex. The intermediate position map is the same as that for the first order fundamental hexagon even edge flexagons *S*⟨6, 6⟩ (Fig. 4.10). Starting from principal main position 2(1), unfold a single leaf numbered 2 from the top of each of the two overlapping pats and pull gently, easing the overlapping pats into principal main position 2(3), and so on round the principal 6-cycle.

Subsidiary main positions can be reached by using different twist flexes. To reach subsidiary main position 2(4) from principal main position 2(1), unfold a pair of leaves with visible numbers 2 and 4 from each of the two overlapping pats, and ease the flexagon into subsidiary main position 2(4). This is, nominally, a slant compound edge ring of four regular octagons (Fig. 6.15b). The high curvature

(180°) means that it cannot be turned inside out, so the two subsidiary 3-cycles shown in the intermediate position map cannot be traversed directly, but they can be traversed via principal main positions. Similarly, to reach subsidiary main position 2(5) which is, nominally an antibox compound edge ring of four regular octagons (Fig. 6.15c), unfold groups of three leaves.

6.7 Heptagonal Fundamental Compound Edge Flexagons

6.7.1 Some Properties

There are six heptagonal compound edge flexagons with flat main positions listed in Table 6.2 which have seven leaves per sector. Two of these are the heptagonal fundamental hexagon compound edge flexagon 2⟨6, 6, 7⟩ and the heptagonal fundamental decagon compound edge flexagon 2⟨10, 10/3, 7⟩. Some of their properties are given in Table 6.7. There are some analogies with first order fundamental heptagon even edge flexagons. (Table 4.1). The torsion per sector is 3. The sector diagrams are shown in Figs. 6.7c and 6.16.

6.7.2 A Fundamental Hexagon Compound Edge Flexagon

The net for the heptagonal fundamental hexagon compound edge flexagon 2⟨6, 6, 7⟩ is shown in Fig. 6.17. As assembled, it is in first subsidiary main position 3(1) which is, in appearance, a flat compound edge ring of four regular hexagons (Fig. 6.9b). The first subsidiary 7-cycle shown by dotted lines in the intermediate position map (Fig. 6.18) can be traversed by using a twist flex. Starting from first subsidiary main

Table 6.7 Properties of heptagonal fundamental compound edge flexagons. The principal cycle is in bold

Leaf type	Flexagon symbol	Typical main position	Cycle type	Number of cycles	Main position type	Ring symbol	Sector symbols	Curvature
Hexagon	2⟨6, 6, 7⟩.	1(3)	7-cycle[a]	1	Flat	2(60°, 120°)	⟨6, 6, 5, 2⟩	0°
Hexagon	2⟨6, 6, 7⟩.	1(4)	7-cycle[b]	1	Slant	2(0°, 60°).	⟨6, 6, 4, 3⟩	240°
Decagon	**2⟨10, 10/3, 7⟩**	**1(2)**	**7-cycle**	**1**	**Flat**	**2(72°, 108°)**	**2⟨10, 10/3, 6, 1⟩**	**0°**
Decagon	2⟨10, 10/3, 7⟩	1(3)	7-cycle[a]	1	Slant	2(8°, 18°)	2⟨10, 10/3, 5, 2⟩	308°

[a]First subsidiary cycle.
[b]Second subsidiary cycle.

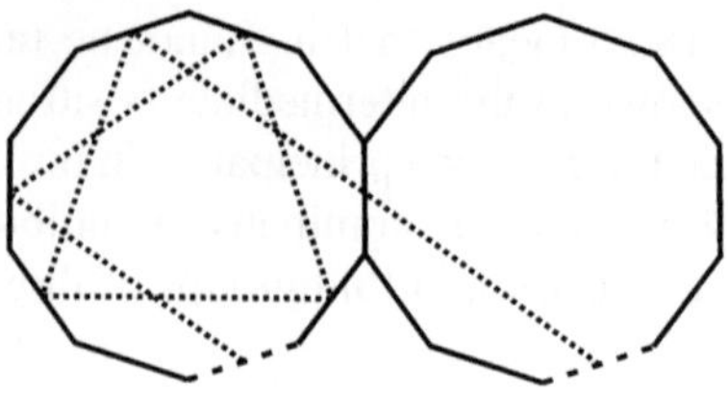

Fig. 6.16 Sector diagram for the fundamental compound decagon edge flexagon 2⟨10, 10/3, 7⟩

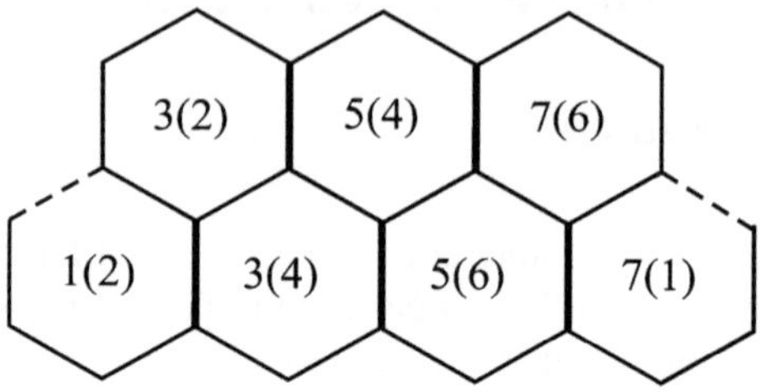

Fig. 6.17 Net for the fundamental hexagon compound edge flexagon 2⟨6, 6, 7⟩. Two copies needed. Fold until leaves numbered 1 and 3 are visible

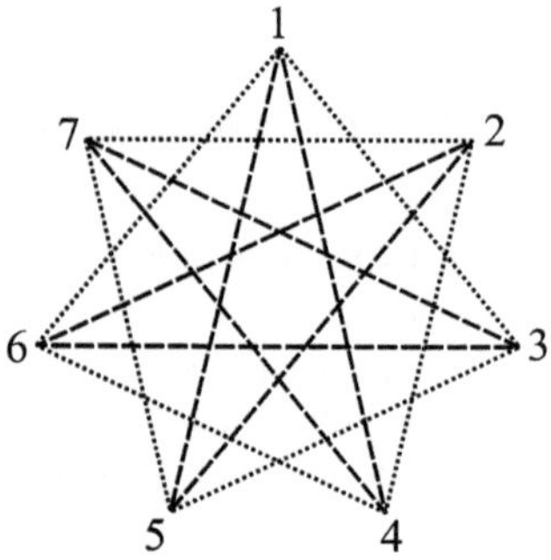

Fig. 6.18 Intermediate position map for the fundamental hexagon compound edge flexagon 2⟨6, 6, 7⟩. Dotted lines are the first subsidiary cycle and dashed lines the second subsidiary cycle

position 3(1), separate the two end pats at the leaves numbered 5. Hold the folded pairs of leaves with leaves numbered 3 and 5 visible and twist gently to reach first subsidiary main position 5(3), with the two pairs of leaves becoming pats at the sides of the flexagon, and so round the first subsidiary 7-cycle.

Second subsidiary main positions can be reached by using a different twist flex. To reach second subsidiary main position 1(4) from first subsidiary main position 3(2), lift the two leaves numbered 3(4) and fold them so that pairs of leaves numbered 3 are folded together. The second subsidiary main position is a slant compound edge ring of four regular hexagons (Fig. 2.14). The high curvature (240°) means that it cannot be turned inside out so the second subsidiary 7-cycle shown by dashed lines in the intermediate position map cannot be traversed directly, but it can be traversed via first subsidiary main positions.

The principal 7-cycle, present in first order fundamental heptagon even edge flexagons is absent, together with corresponding main positions, because the permissible number of leaves in a pat is exceeded (Sections 4.2.5.1 and 6.2).

6.7.3 A Fundamental Decagon Compound Edge Flexagon

The net for the heptagonal fundamental decagon compound edge flexagon 2⟨10, 10/3, 7⟩ is shown in Fig. 6.19. The flexagon is unstable but fairly easy to handle. As assembled, it is in principal main position 1(2) which is, in appearance, a flat compound edge ring of four regular decagons (Fig. 6.20). The principal 7-cycle shown by solid lines in the intermediate position map (Fig. 6.21) can be tra-

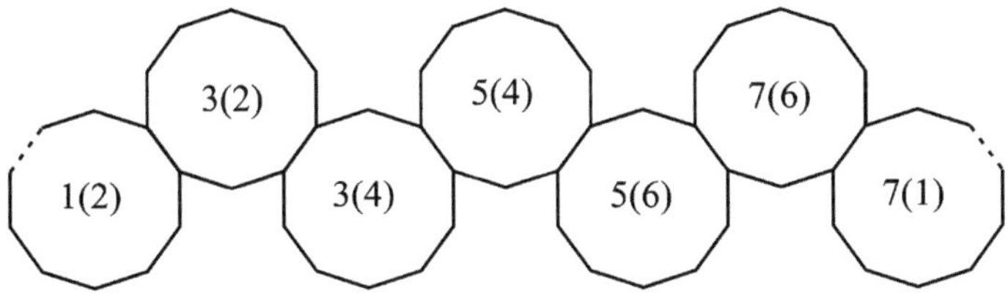

Fig. 6.19 Net for the fundamental decagon compound edge flexagon 2⟨10, 10/3, 7⟩. Two copies needed

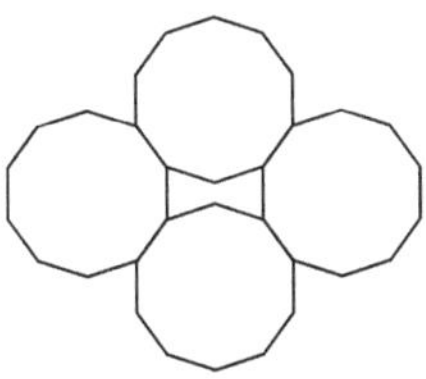

Fig. 6.20 Flat compound edge ring of four regular decagons

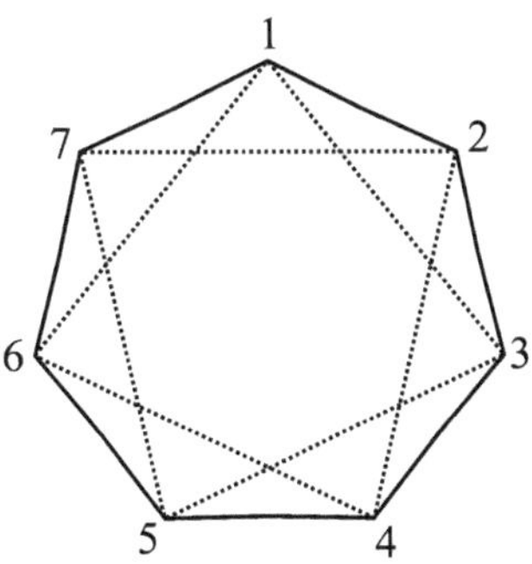

Fig. 6.21 Intermediate position map for the fundamental decagon compound edge flexagon 2⟨10, 10/3, 7⟩

Fig. 6.22 A flexagon as a slant compound edge ring of four regular decagons

versed by using a twofold pinch flex. The hinges are not in line, so leaves have to be bent during flexing. Intermediate positions are clearly defined, but leaves do not overlap exactly.

First subsidiary main positions can be reached by using a twist flex. To reach first subsidiary main position 1(3) from principal main position 1(2), lift the two leaves numbered 2(3) and fold them so that pairs of leaves numbered 2 are folded together. The first subsidiary main position is a slant compound ring of four regular decagons (Fig. 6.22). The high curvature (308°) means that this cannot be turned inside out, so the first subsidiary 7-cycle, shown by dotted lines in the intermediate position map, cannot be traversed directly, but it can be traversed via principal main positions. Similarly, by flexing via first subsidiary main position, the principal 7-cycle can be traversed without bending the leaves.

The second subsidiary 7-cycle, present in first order fundamental heptagon even edge flexagons, and corresponding main positions, are absent because the corresponding polygon ring is impossible (Sections 2.2.4 and 6.2).

References

Conrad AS, Hartline DK (1962) *Flexagons.* RIAS Technical Report 62-11. Baltimore, MD: RIAS

Pook LP (2003) Flexagons Inside Out. Cambridge University Press, Cambridge

Chapter 7
Irregular Cycle Flexagons

7.1 Introduction

In a regular cycle flexagon made from regular convex polygons all the main positions of a cycle have the same appearance and the same pat structure. Among others, fundamental even edge flexagons (Section 4.2.1) and fundamental point flexagons (Section 5.3.1) are made from regular convex polygons and are regular cycle flexagons. They are solitary flexagons. In an irregular cycle flexagon made from regular convex flexagons the pat structure, but not the appearance, of main positions changes as a cycle is traversed. The irregular cycle flexagons described in this chapter can be regarded as variants of regular cycle flexagons, and are all solitary flexagons. The characteristic feature of an irregular cycle flexagon is that its associated polygon, and hence the inscribed portion of its flexagon figure, is an irregular polygon with the same vertices as a regular polygon.

Three families of irregular cycle flexagons are described. The first is irregular cycle even edge flexagons. Every irregular cycle even edge flexagons has a corresponding first order fundamental even edge flexagon with the same intermediate position map. The topological invariants (Section 4.2.1) are the same. In general, nets for irregular cycle even edge flexagons are irregular, that is they are not first order fundamental edge nets (Section 3.2). Standard face numbering sequences (Section 4.1.1) are not appropriate for irregular cycle even edge flexagons. Possible irregular cycle even edge flexagons can be enumerated by counting possible irregular polygons that have the same vertices as regular polygons (Conrad and Hartline 1962). These irregular polygons are possible associated polygons for irregular cycle even edge flexagons. The one possible irregular quadrilateral, and possible irregular pentagons and irregular hexagons are shown in Figs. 7.1–7.3. All have intersecting edges. Rotations and reflections are not regarded as distinct. The arbitrary type letters shown in Figs. 7.2 and 7.3 are used in descriptions of corresponding flexagons. The numbers of possible irregular polygons, with the same vertices as regular polygons, and hence numbers of irregular cycle even edge flexagons, increase rapidly with the number of edges on polygons. There are no irregular triangles, one irregular quadrilateral, two irregular pentagons and 11 irregular hexagons.

L. Pook, *Serious Fun with Flexagons*, Solid Mechanics and Its Applications 164,
DOI 10.1007/978-90-481-2503-6_7, © Springer Science+Business Media B.V. 2009

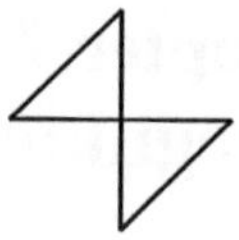

Fig. 7.1 The irregular quadrilateral with the same vertices as a square

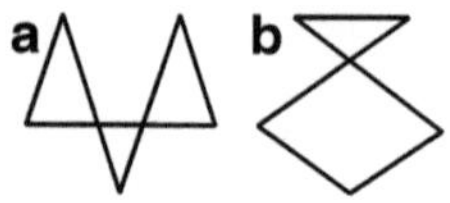

Fig. 7.2 Irregular pentagons with the same vertices as a regular pentagon. (a) Type A. (b) Type B

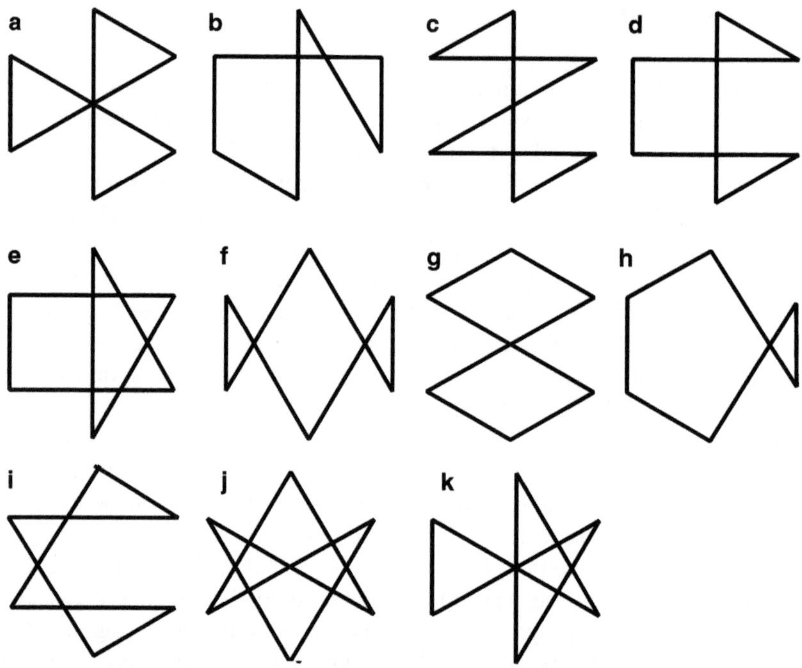

Fig. 7.3 Irregular hexagons with the same vertices as a regular hexagon. (a) Type A. (b) Type B. (c) Type C. (d) Type D. (e) Type E. (f) Type F. (g) Type G. (h) Type H. (i) Type I. (j) Type J. (k) Type K

The second family is irregular cycle interleaved point flexagons. The topological invariants for irregular cycle interleaved point flexagons are the same as those for fundamental point flexagons (Section 5.1). They are related to irregular cycle even edge flexagons because the possible irregular polygons shown in Figs. 7.1–7.3 are also possible associated polygons for irregular cycle interleaved point flexagons. They are made from fundamental vertex nets (Section 3.4), but standard face numbering sequences (Section 4.1.1) are not appropriate. Various transformations between flexagons are possible.

The third family is irregular cycle non interleaved point flexagons. The topological invariants for irregular cycle non interleaved point flexagons are the same as those for fundamental point flexagons (Section 5.1). Irregular cycle non interleaved point flexagons are duals of irregular cycle even edge *unagons* (cf. Section 5.3.1). Hence, irregular cycle non interleaved point flexagons correspond to irregular cycle even edge flexagons, and the possible irregular polygons shown in Figs. 7.1–7.3 are possible associated polygons for irregular cycle non interleaved point flexagons. There are apparently no irregular cycle compound edge flexagons, but this has not been proved.

7.2 Irregular Cycle Even Edge Flexagons

7.2.1 General Properties

The topological invariants for irregular cycle even edge flexagons are the same as those for first order fundamental even edge flexagons (Section 4.2.1). All irregular cycle even edge flexagon can be flexed by using a pinch flex. This is their characteristic flex. Every irregular cycle even edge flexagons has a corresponding first order fundamental even edge flexagon with the same main and intermediate position appearances, and the same intermediate position map. In general, nets for irregular cycle even edge flexagons are irregular, that is they are not first order fundamental edge nets (Section 3.2). Standard face numbering sequences (Section 4.1.1) are not appropriate for irregular cycle even edge flexagons. For the nets given below leaves are numbered using specially derived face numbering sequences which ensure that pairs of faces and intermediate positions appear in cyclic order. An irregular cycle even edge unagon is an irregular cycle even edge flexagon with the number of sectors reduced to one (cf. Section 5.3.1).

All flexagons exist as enantiomorphic pairs (Pook 2009). All first order fundamental even edge flexagons and some irregular cycle even edge flexagons are twisted bands. A flexagon that is a twisted band must have an enantiomorph because the torsion can be either positive or negative (Section 4.2.1). The two members of an enantiomorphic pair are not usually considered to be distinct types of flexagon (Section 1.2) but, if they are regarded as distinct, then the enantiomorph of a flexagon with numbered faces can be derived by interchanging the numbers on the faces of each leaf of its net (Section 1.4.1). From a topological viewpoint it is not possible to transform between the two enantiomorphs because this is a nonperformable operation. Some irregular cycle even edge flexagons are untwisted bands, that is their torsion is zero. From a topological viewpoint this means that there is no enantiomorph (Pook 2003). However, interchanging the numbers on each leaf of a net results in a flexagon that is an enantiomorph. If leaf bending is allowed, transforming between the forms with these enantiomorphic face numberings is a performable operation that, in practice, can sometimes be achieved by using a band flex.

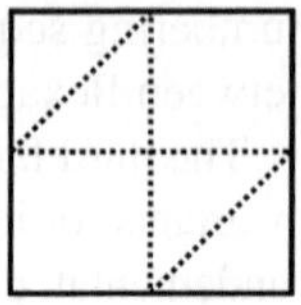

Fig. 7.4 Flexagon figure for irregular cycle square even edge flexagons (Les Pook, Flexagons inside out, 2003, © Cambridge University Press 2003, reprinted with permission)

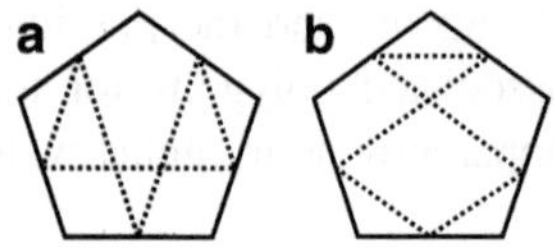

Fig. 7.5 Flexagon figures for irregular cycle pentagon even edge flexagons. (a) Type A. (b) Type B (Les Pook, Flexagons inside out, 2003, © Cambridge University Press 2003, reprinted with permission)

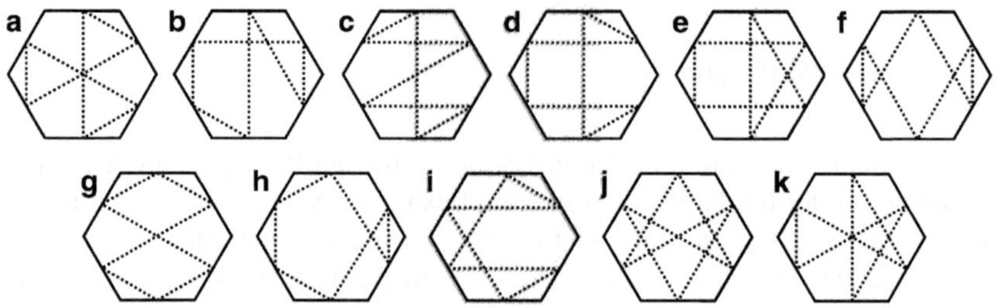

Fig. 7.6 Flexagon figures for irregular cycle hexagon even edge flexagons. (a) Type A. (b) Type B. (c) Type C. (d) Type D. (e) Type E. (f) Type F. (g) Type G. (h) Type H. (i) Type I. (j) Type J. (k) Type K (Les Pook, Flexagons inside out, 2003, © Cambridge University Press 2003, reprinted with permission)

There are no irregular cycle triangle even edge flexagons. Flexagon figures for irregular cycle square even edge flexagons, irregular cycle pentagon even edge flexagons, and irregular cycle hexagon even edge flexagons are shown in Figs. 7.4–7.6. These were derived by using the irregular polygons shown in Figs. 7.1–7.3 as the inscribed polygons of flexagon figures, and the same type letters are used. Figures 7.1–7.3 are the associated polygons of corresponding irregular cycle even edge flexagons. Irregular cycle even edge flexagons are identified by descriptions, including the number of sectors and type letters where appropriate.

7.2.2 Derivation of Nets

The net for an even edge flexagon, which is defined in terms of its flexagon figure, may be derived by using the following practical procedure. In effect, this reverses the method used to construct a flexagon figure (Section 4.2.5.3). The first order fundamental square even edge flexagon $2\langle 4, 4\rangle$ (Section 4.2.4) is used as an example.

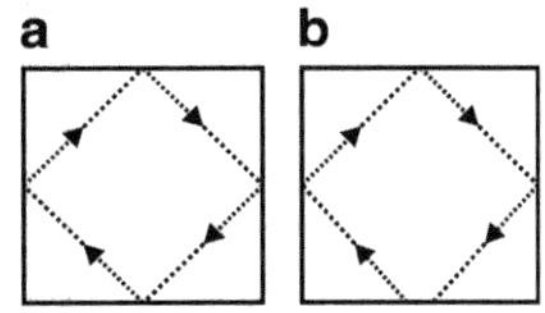

Fig. 7.7 Flexagon figure for the first order fundamental square even edge flexagon 2⟨4, 4⟩. (**a**) Arbitrary direction selected. (**b**) Disconnected at an arbitrary vertex (Les Pook, Flexagons inside out, 2003, © Cambridge University Press 2003, reprinted with permission)

This procedure, with due attention to detail, was used to derive the nets for irregular cycle even edge flexagons given below. Face numbering sequences were derived by traversing a cycle of the completed flexagon, and numbering faces in sequence as they appeared. Nets including face numbering sequences can, of course, be derived theoretically. but in practice the practical procedure is easier.

Start by choosing an arbitrary direction around the inscribed polygon, a square, of the flexagon figure (Fig. 4.2), as shown in Fig. 7.7a. Next, disconnect the square at an arbitrary vertex, as shown at the at the bottom of Fig. 7.7b. The circumscribing polygon of the flexagon figure is a square, so the leaves of the flexagon are squares. The inscribed polygon in the flexagon figure has four edges, so for each sector of an intermediate position a pat consisting of a folded pile of four leaves is required. Prepare a pile of four separate paper leaves. Figure 7.7b indicates how the four individual paper leaves are to be hinged together. Starting at the bottom of the pile, mark the lower edge of the first leaf, then join the first and second leaves with transparent adhesive tape at the left hand edges, the second and third leaves at the upper edges, the third and fourth leaves at the right hand edges, and mark the lower edge of the fourth leaf. Unfold the fan folded pile to get the net for one sector. With practice, pairs of pre-hinged leaves can be used. To construct a flexagon with two sectors, make a second identical pat, turn it over and join the top leaves, then the bottom leaves. The resulting net is shown in Figs. 1.2 and 4.16. Reversing the arbitrary direction around the inscribed polygon results in the enantiomorph. Thus, there are always two solutions, which are an enantiomorphic pair, but the two enantiomorphs are not usually regarded as distinct flexagons. (Section 1.2).

7.2.3 The Irregular Cycle Square Even Edge Flexagon

There is one two sector irregular cycle square even edge flexagon. Some of its properties are given in Table 7.1 (cf. Table 4.4). The torsion is zero. It is the simplest flexagon that is an untwisted band. The flexagon figure is shown in Fig. 7.4.

The net for the two sector irregular cycle square even edge flexagon is shown in Fig. 7.8. Face numbers and hinge letters have been chosen to simplify comparison with the first order fundamental square even edge flexagon 2⟨4, 4⟩ (Section 4.2.4). As assembled, the flexagon is in principal main position 2(1). This is, in appearance, a flat regular even edge ring of four squares (Fig. 1.1b). The principle 4-cycle

Table 7.1 Properties of the two sector irregular cycle square even edge flexagon. The principal cycle is in bold

Typical main position	Cycle type	Number of cycles	Main position type	Ring symbol	Curvature
1(2)	**4-cycle**	**1**	**Flat**	**4(90°)**	**0°**
1(3)	None	–	Box	4(0°)	360°

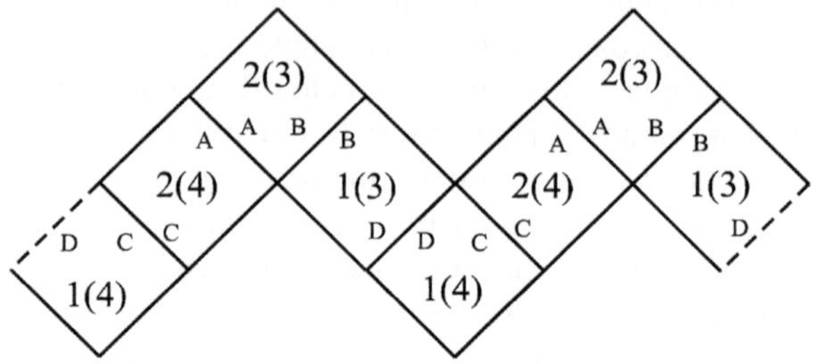

Fig. 7.8 Net for the two sector irregular cycle square even edge flexagon. One copy needed

shown in the intermediate position map, and the Tuckerman diagram, can be traversed by using the twofold pinch flex. The intermediate position map and Tuckerman diagram are the same as those for the first order fundamental square even edge flexagons $S\langle 4, 4\rangle$ (Figs. 4.8 and 4.17). Intermediate positions are square edge pairs (Fig. 1.15) They can be opened into box positions, which are box edge rings of four squares (Fig. 2.5).

The full map (Fig. 7.9) shows in more detail than the Tuckerman diagram what happens when the flexagon is flexed around the principal 4-cycle, in the direction of the arrows, by using the twofold pinch flex (cf. Section 4.2.5.3). As assembled, the flexagon is in principal main position 2(1). To flex to principal main position 3(2) fold the flexagon in two, along hinge line D-D, to reach intermediate position 2, with leaves numbered 1 concealed. Keep the hinge line uppermost, and unfold about hinge line B-B to reach principal main position 3(2). To traverse the complete principal 4-cycle in the direction of the arrows, continue flexing, changing hinges as appropriate at each position. If the flexagon is turned over, the cycle may be traversed in the reverse direction. This sequence is identical to that for the first order fundamental square even edge flexagon $2\langle 4, 4\rangle$.

In the principal 4-cycle, alternate principal main positions have the same pat structure, provided that enantiomorphs are not regarded as distinct. In principal main positions 3(2) and 1(4), alternate pats are a single leaf and a folded pile of three leaves, and these two positions appear identical on the full map, but they are actually an enantiomorphic pair. In principal main positions 2(1) and 4(3), each pat is a folded pile of two leaves and the two positions are completely identical in that they are not an enantiomorphic pair.

A dual marked net is shown in Fig. 7.10. Figures are used for faces of the enantiomorph shown in Fig. 7.8 and letters for the other enantiomorph. As assembled, the flexagon is in principal main position 2(1), and the intermediate position map shown in Fig. 4.8 can be traversed. Starting from principal main position 2(1), the flexagon can be transformed into principal main position B(A) by using a band flex in which the

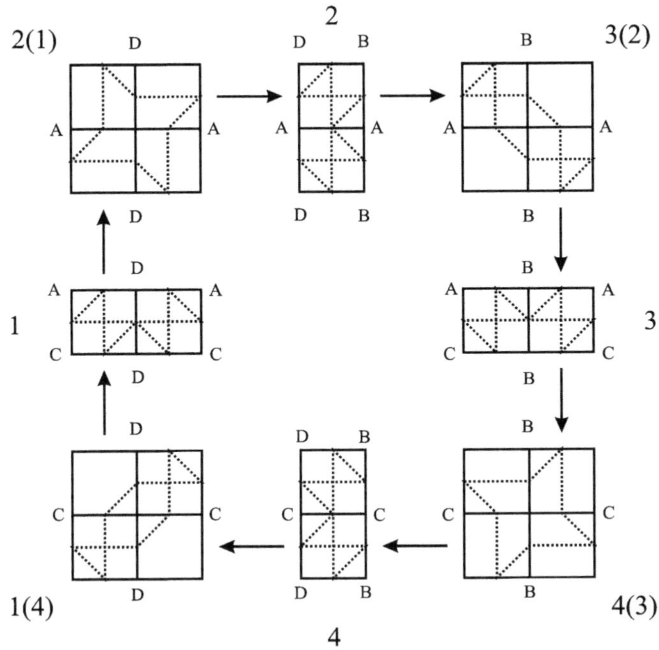

Fig. 7.9 Full map for the principal cycle of the two sector irregular cycle square even edge flexagon

Fig. 7.10 Dual marked net for the two sector irregular cycle square even edge flexagon.

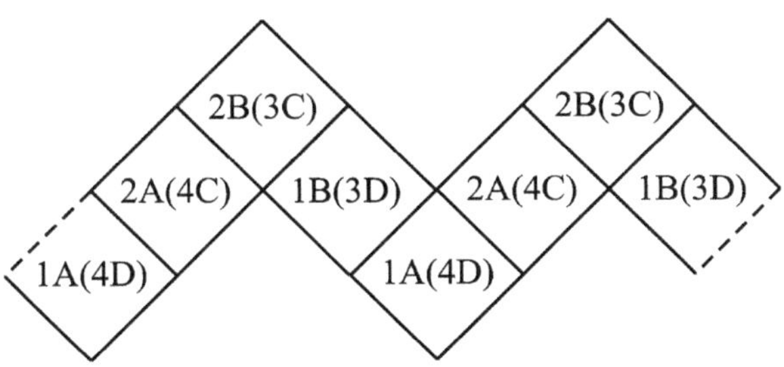

Fig. 7.11 Intermediate position map for the lettered enantiomorph of the two sector irregular cycle square even edge flexagon

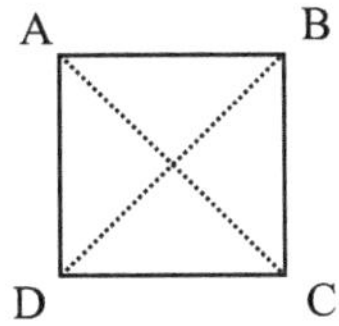

flexagon is collapsed into an open band, and then closed into principal main position B(A) by folding together pairs of leaves marked C and pairs of leaves marked D. This is a transformation between flexagons. If the folding is done simultaneously this particular band flex becomes a twist flex. After the transformation the intermediate position map for the lettered enantiomorph (Fig. 7.11) can be traversed.

7.2.4 An Irregular Cycle Pentagon Even Edge Flexagon

There are two sector irregular cycle pentagon even edge flexagons, types A and B. One of these, type B, is described in this section. Some of its properties are given in Table 7.2 (cf. Table 4.5). The torsion per sector is 1. The flexagon figure is shown in Fig. 7.5b.

The net for the two sector irregular cycle pentagon even edge flexagon type B is shown in Fig. 7.12. As assembled, the flexagon is in intermediate position 1, which is, in appearance, a pentagon edge pair. Apart from differences in pat structure, the dynamic behaviour of the flexagon is the same as that of the first order fundamental pentagon even edge flexagon 2⟨5, 5⟩ (Section 4.2.6). The principal 5-cycle shown on the intermediate position map can be traversed by using the twofold pinch flex. The intermediate position map is the same as that for the first order fundamental pentagon even edge flexagon 2⟨5, 5⟩ (Fig. 4.9). Principal main positions are skew regular even edge rings of four regular pentagons (Fig. 2.7). Intermediate positions can also be opened into subsidiary main position, which are slant regular even edge rings of four regular pentagons (Fig. 4.24). The high curvature (216°) means that these cannot be turned inside out so the subsidiary 5-cycle shown in the intermediate position map cannot be traversed directly. All the subsidiary main positions can be visited by flexing via principal main positions.

The flexagon diagrams for principal main positions, projected onto a plane (Fig. 7.13), show that three different types of pat structure appear as the flexagon is flexed around its principal 5-cycle. Subsidiary main positions also have three different types of pat structure.

Table 7.2 Properties of the two sector irregular cycle fundamental pentagon even edge flexagon type B. The principal cycle is in bold

Typical main position	Cycle type	Number of cycles	Main position type	Ring symbol	Curvature
1(2)	**5-cycle**	**1**	**Skew**	**4(108°)**	**–72°**
1(3)	5-cycle	1	Slant	4(36°)	216°

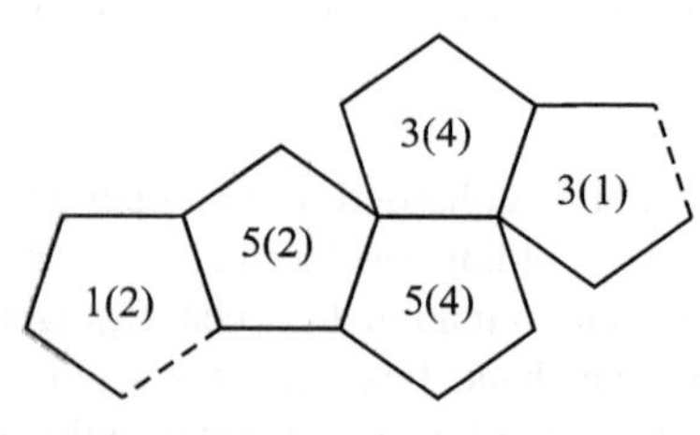

Fig. 7.12 Net for the two sector irregular pentagon even edge flexagon type B. Two copies needed. Fold until leaves numbered 1 are visible

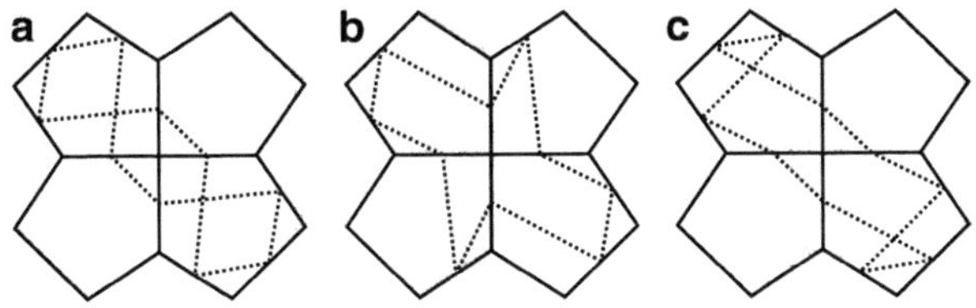

Fig. 7.13 Flexagon diagrams for the principal main positions of the two sector irregular cycle pentagon even edge flexagon type B. (**a**) 2(1). (**b**) 3(2) and 1(5). (**c**) 4(3) and 5(4)

7.2.5 Irregular Cycle Hexagon Even Edge Flexagons

There are 11 two sector irregular cycle hexagon even edge flexagons, types A-K. of these, types A and C are described in this section. Some of their properties are given in Table 7.3 (cf. Table 4.6). The torsion per sector is 2 for type A and zero for type C. Flexagon figures are shown in Fig. 7.6a and c.

Nets for the two sector irregular cycle hexagon even edge flexagons, type A and C are shown in Figs. 7.14 and 7.15. As assembled, the flexagons are in intermediate position 1, which are, in appearance, hexagon edge pairs. Apart from differences in pat structure, the dynamic behaviour of the flexagons is the same as that of the first order fundamental hexagon even edge flexagon 2⟨6, 6⟩ (Section 4.2.7). The principal6-cycle shown on the intermediate position map can be traversed by using the twofold pinch flex. The two sector irregular cycle hexagon even edge flexagon type C is typical of irregular cycle even edge flexagons in that the irregularity in pat structure is clearly noticeable while traversing the principal 6-cycle. The intermediate position map is the same as that for the first order fundamental hexagon even edge flexagon 2⟨6, 6⟩ (Fig. 4.10). Principal main positions are skew regular even edge rings of four regular hexagons (Fig. 1.5). Intermediate positions can also be opened into subsidiary main position which are slant regular even edge rings of four regular hexagons. The high curvature (120°) means that these cannot be turned inside out, so the two subsidiary 3-cycles shown in the intermediate position map cannot be traversed directly. All the subsidiary main positions can also be visited by flexing via principal main positions. Intermediate positions can be opened into box positions, which are box edge rings of four regular hexagons (Fig. 4.31a).

The flexagon diagrams for the principal main positions, projected onto a plane, of the two sector irregular cycle hexagon even edge flexagon type A (Fig. 7.16) show that there are two different types of pat structure. These appear alternately as the principal 6-cycle is traversed. The pat structure is the same in all the subsidiary main positions of a subsidiary 3-cycle, but the two subsidiary 3-cycles have different types of pat structure. The pat structure of all the box positions is the same.

Principal main positions of the two sector irregular cycle hexagon even edge flexagon type C have four different types of pat structure, two of which appear twice. Each of the subsidiary main positions in a subsidiary 3-cycle has a different pat structure. In the subsidiary 3-cycles there are, in total, four different types of pat structures, two of which appear in both subsidiary 3-cycles. Each of the three box position has a different type of pat structure.

Table 7.3 Properties of the two sector irregular cycle hexagon even edge flexagons types A and C. The principal cycle is in bold

Typical main position	Cycle type	Number of cycles	Main position type	Ring symbol	Curvature
1(2)	**6-cycle**	**1**	**Skew**	**4(120°)**	**–120°**
1(3)	3-cycle	2	Slant	4(60°)	120°
1(4)	None	–	Box	4(0°)	360°

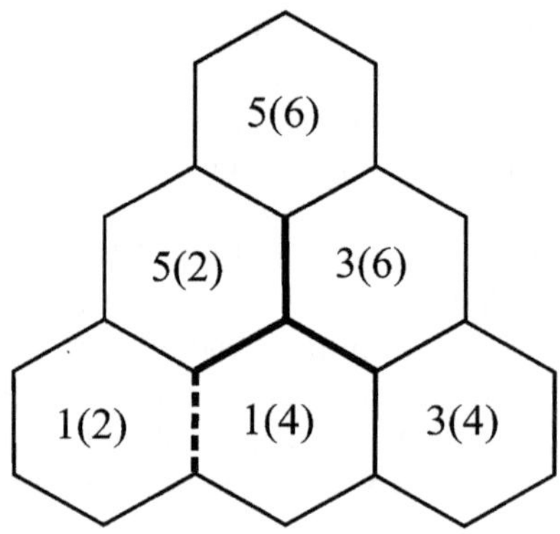

Fig. 7.14 Net for the two sector irregular cycle hexagon even edge flexagon type A. Two copies needed. Fold until leaves numbered 1 are visible

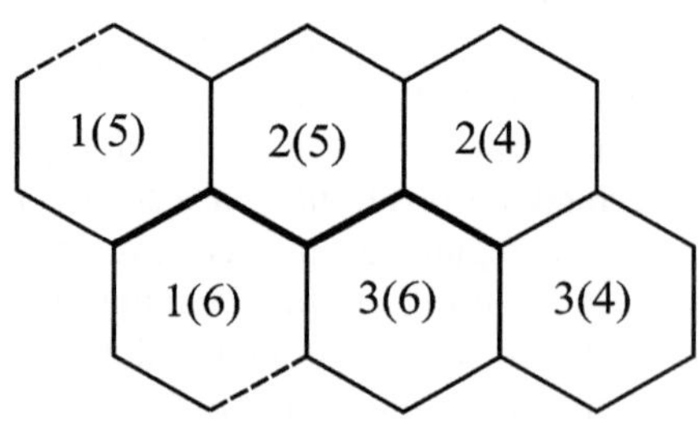

Fig. 7.15 Net for the two sector irregular cycle hexagon even edge flexagon type C. Two copies needed. Fold until leaves numbered 1 are visible

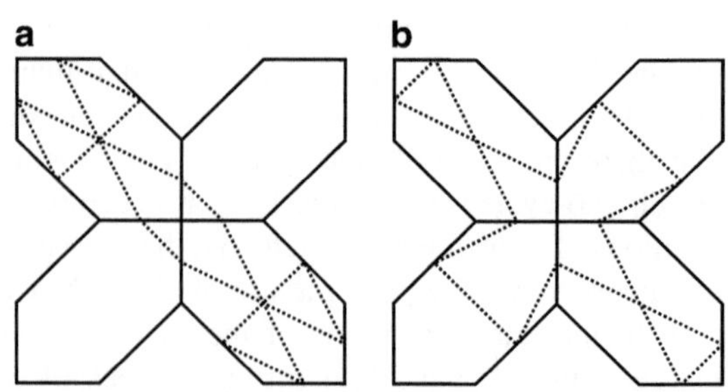

Fig. 7.16 Flexagon diagrams for the principal main positions of the two sector irregular cycle hexagon even edge flexagon type A. (**a**) 2(1), 4(3) and 6(5). (**b**) 3(2) 5(4) and 1(6)

7.3 Irregular Cycle Interleaved Point Flexagons

7.3.1 General Properties

The topological invariants for irregular cycle interleaved point flexagons are the same as those for fundamental point flexagons (Section 5.1). All irregular cycle interleaved point flexagons can be flexed by using the simple flex and this is their characteristic flex. Irregular cycle interleaved point flexagons are related to irregular cycle even edge flexagons (Section 7.2.1). This is because the irregular polygons shown in Figs. 7.1–7.3 are possible associated polygons for irregular cycle interleaved point flexagons. The pat structures of main positions and intermediate positions vary as a cycle is traversed. Irregular cycle interleaved point flexagons are made from fundamental vertex nets (Section 3.4), but standard face numbering sequences (Section 4.1.1) are not appropriate. Face numbering sequences which ensure that pairs of faces and intermediate positions appear in cyclic order can be derived from associated polygons using the method for interleaved fundamental point flexagons described in Section 5.4.1. Flexagon figures are not appropriate for interleaved fundamental point flexagons.

A feature of irregular cycle interleaved point flexagons is that a particular type has more than one possible distinct face numbering sequence. Each distinct face numbering sequence corresponds to a distinct form of an irregular cycle interleaved point flexagon. Enumeration of distinct face numbering sequences for a particular irregular cycle point flexagon requires consideration of symmetries of its associated polygon (Pook 2007). For example, a square has fourfold rotational symmetry and the irregular quadrilateral shown in Fig. 7.1 has both reflectional symmetry and twofold rotational symmetry so there are $4 \div 2 = 2$ distinct ways in which the quadrilateral can have the same vertices as a square. Hence, there are two distinct face numbering sequences, for the corresponding irregular cycle interleaved square point flexagon. Similarly, for each of the irregular pentagons shown in Fig. 7.2 there are five distinct face numbering sequences for corresponding irregular cycle interleaved pentagon point flexagons. For irregular cycle interleaved hexagon point flexagons, corresponding to the irregular hexagons shown in Fig. 7.3, there are two distinct face numbering sequences for type A, three for types F, G and J, six for types C, D, E, H, I and K, and 12 for type B. The number of distinct face numbering sequence doubles for types B and C because the associated hexagons do not have reflectional symmetry (Fig. 7.3b and c).

Various transformations between flexagons are possible. It is always possible to transform between the distinct forms of a particular type of irregular cycle interleaved point flexagon. It is sometimes possible to transform between different types of irregular cycle interleaved point flexagon, and also to transform between an irregular cycle interleaved point flexagon and an interleaved fundamental point flexagon (Section 5.4.1).

One distinct face numbering sequence is given in Tables 7.4–7.6 for each of the irregular cycle interleaved point flexagons described below. The convention given in Section 4.1.1 is used. Face numbering sequences were derived from associated polygons as described in Section 5.4.1. The dynamic properties of a particular type of irregular cycle interleaved point flexagon are the same for all the distinct face numbering sequences. Hence, when considering transformations between flexagons, it is sufficient to consider only one face numbering sequence.

Some irregular cycle interleaved point flexagons can be constructed without interleaving during assembly, but some have to be interleaved during assembly, as described in Sections 5.4.2 and 5.6.2. When interleaving is required this is noted in descriptions of individual flexagons.

Table 7.4 Properties of the irregular cycle interleaved square point flexagon and the fundamental square point flexagon 1⟨4, 4⟩. Fundamental vertex net ⟨4⟩ is used

Type	Rotational symmetry of associated quadrilateral	Distinct face numbering sequences	Typical face numbering sequence	Torsion
Irregular	Twofold	2	1/2, 3/2, 4/1, 4/3	0
1⟨4, 4⟩	Fourfold	1	1/2, 3/2, 3/4, 4/1	2

Table 7.5 Properties of irregular cycle pentagon interleaved point flexagons types A and B, the interleaved fundamental pentagon point flexagon 1⟨5, 5, 5/2⟩, and the fundamental pentagon point flexagon 1⟨5, 5⟩. Fundamental vertex net ⟨5⟩ is used

Type	Rotational symmetry of associated pentagon	Distinct face numbering sequences	Typical face numbering sequence	Torsion
A	Onefold	5	1/2, 5/4, 2/3, 4/3, 5/1	1
B	Onefold	5	1/2, 4/3, 4/5, 1/5, 2/3	1
1⟨5, 5, 5/2⟩	Fivefold	1	1/2, 4/3, 5/1, 3/2, 4/5	1
1⟨5, 5⟩.	Fivefold	1	1/2, 3/2, 3/4, 5/4, 5/1	3

Table 7.6 Properties of the irregular cycle interleaved hexagon point flexagon types A-K, and the fundamental hexagon point flexagon 1⟨6, 6⟩. Fundamental vertex net ⟨6⟩ is used

Type	Rotational symmetry of associated hexagon	Distinct face numbering sequences	Typical face numbering sequence	Torsion
A	Threefold	2	1/2, 3/2, 5/6, 1/6, 3/4, 5/4	2
B	Onefold	12	1/2, 4/3, 2/3, 1/6, 5/6, 5/4	2
C	Twofold	6	1/2, 5/4, 3/4, 6/5, 2/3, 1/6	0
D	Onefold	6	1/2, 3/2, 6/1, 6/5, 3/4, 5/4	0
E	Onefold	6	1/2, 4/3, 5/6, 1/6, 2/3, 5/4	2
F	Twofold	3	1/2, 4/3, 2/3, 5/4, 6/1, 6/5	0
G	Twofold	3	1/2, 3/2, 5/6, 5/4, 3/4, 1/6	0
H	Onefold	6	1/2, 4/3, 2/3, 5/4, 5/6, 1/6	2
I	Onefold	6	1/2. 3/2, 6/1, 5/4, 3/4, 6/5	0
J	Twofold	3	1/2, 4/3, 6/1, 5/4, 2/3, 6/5	0
K	Onefold	6	1/2, 4/3, 6/1, 6/5, 2/3, 5/4	0
1⟨6, 6⟩	Sixfold	1	1/2, 3/2. 3/4, 5/4, 5/6, 1/6	4

7.3.2 Interleaf Flexes

In general, point flexagons can be flexed around cycles by using the simple flex. Some point flexagons can also be flexed by using interleaf flexes. One necessary condition for an interleaf flex to be possible is that the point flexagon must be made from a net in which point hinges are at adjacent vertices on the polygons. For example, interleaf flexes are sometimes possible for pentagon point flexagons made using the fundamental pentagon vertex net ⟨5⟩ (Fig. 3.20a), but are never possible for those made from the fundamental pentagon vertex net ⟨5/2⟩ (Fig. 3.20b).

Intermediate positions of point flexagons are folded piles of polygons which are, in appearance, single polygons (Section 5.1). In an interleaf flex, two or more polygons (leaves) are unfolded, from one end an intermediate position, by rotating them through 180° and then folding them back in a different order. In the single interleaf flex the two leaves at one end of an intermediate position are unfolded, and then interchanged by folding back in the reverse order. In the double interleaf flex, two leaves at each end of the intermediate position are interchanged. Double interleaf flexes are easier to specify than single interleaf flexes. More complicated interleaf flexes can always be decomposed into a combination of single interleaf flexes and simple flexes.

Two leaves connected by a point hinge cannot be twisted relative to each other (Section 1.2). Thus, for a single interleaf flex to be possible, the two leaves being interchanged must not be connected by a point hinge. Hence, a second necessary condition for an interleaf flex to be possible is that the point flexagon must be interleaved.

7.3.3 The Irregular Cycle Interleaved Square Point Flexagon

There is one irregular cycle interleaved square point flexagon. It is the simplest possible interleaved point flexagon. Some of its properties are given in Table 7.4. Properties of the fundamental square point flexagon 1⟨4, 4⟩ (Section 5.3.3) are included for comparison. It is not possible to transform between the irregular cycle interleaved square point flexagon and the fundamental square point flexagon 1⟨4, 4⟩ because the latter is not interleaved and it has a different torsion.

The net for the irregular cycle interleaved square point flexagon, using one of the two distinct face numbering sequences, is shown in Fig. 7.17. As assembled, the flexagon is in intermediate position 1, which is, in appearance, a single square. Apart from differences in pat structure, its dynamic behaviour is the same as that of the fundamental square point flexagon 1⟨4, 4⟩. The principal 4-cycle shown in the intermediate position map can be traversed by using the simple flex. The intermediate position map is the same as that for the fundamental square point flexagon 1⟨4, 4⟩ (Fig. 5.5). Principal main positions are square vertex pairs linked by pairs of point hinges (Fig. 3.23).

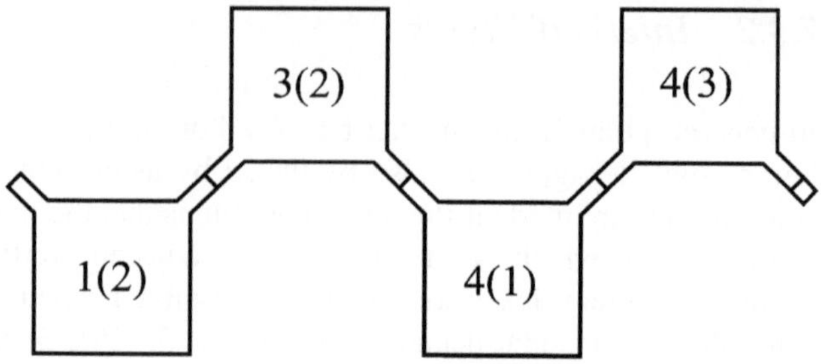

Fig. 7.17 Net for the irregular cycle interleaved square point flexagon. One copy needed. Interleave during assembly, with leaves numbered 1 left visible

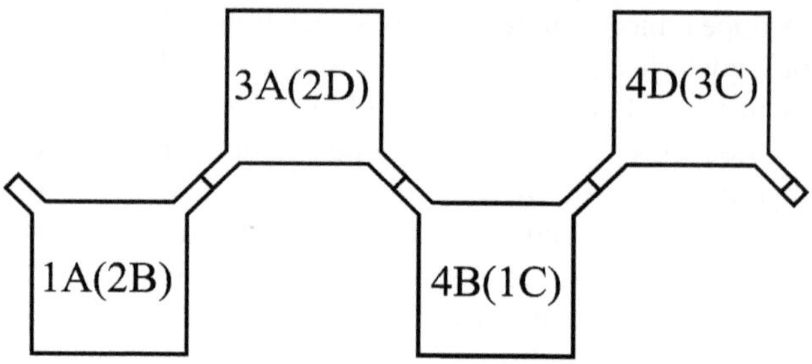

Fig. 7.18 Dual marked net for the irregular cycle interleaved square point flexagon. One copy needed. Interleave during assembly, with leaves numbered 1 left visible

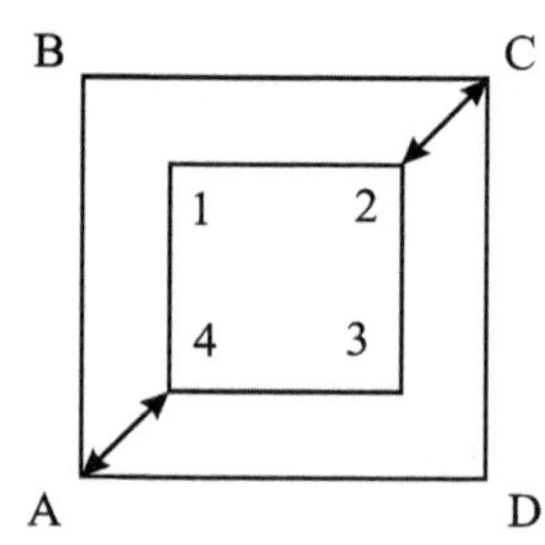

Fig. 7.19 Intermediate position map for the dual marked irregular cycle interleaved square point flexagon

A dual marked net is shown in Fig. 7.18. Numbers are used for one of the face numbering sequences and letters for the other. The lettered sequence was derived by carrying out a double interleaf flex and marking the faces as the principal 4-cycle was traversed. The markings are equivalent to shifting the face numbers two leaves along the net, so it is not a distinct sequence. Using a single interleaf flex and renumbering does lead to the other distinct face numbering sequence.

As assembled, the flexagon is in intermediate position 1, and the two visible leaves are lettered A and C. The numbered principal 4-cycle shown in the intermediate position map (Fig. 7.19) can be traversed by using the simple flex. Face letters are mixed up. The lettered principal 4-cycle shown in the intermediate position map can also be traversed by using the simple flex. Face numbers are mixed up.

To transform from the numbered principal 4-cycle to the lettered principal 4-cycle, use a double interleaf flex at either intermediate position 2 or intermediate position 4 to reach either intermediate position C or intermediate position A. The lines with pairs of arrowheads on the intermediate position map show where a double interleaf flex can be used to transform between the two principal 4-cycles. Transformations between flexagons are possible at two intermediate positions because the associated polygon has twofold rotational symmetry (Table 7.4). In other words, the transformation behaviour also has twofold rotational symmetry.

7.3.4 Irregular Cycle Interleaved Pentagon Point Flexagons

There are two irregular cycle interleaved pentagon point flexagons, types A and B. Some of their properties are given in Table 7.5. Properties of the interleaved fundamental pentagon point flexagon 1⟨5, 5, 5/2⟩ (Section 5.4.2) and the fundamental pentagon point flexagon 1⟨5, 5⟩ (Sections 5.3.4) are included for comparison. For convenience, an associated polygon that does not have rotational symmetry is described as having onefold rotational symmetry. Transformations between flexagons are possible between the irregular cycle interleaved pentagon point flexagons, types A and B, and also between them and the interleaved fundamental pentagon point flexagon 1⟨5, 5, 5/2⟩. However, the fundamental pentagon point flexagon 1⟨5, 5⟩ is not interleaved and has a different torsion, so transformations are not possible.

Nets for the irregular cycle interleaved pentagon point flexagons, types A and B, in each case using one of the five distinct face numbering sequences, are shown in Figs. 7.20 and 7.21. As assembled, the flexagons are in intermediate position 1, which are, in appearance, single regular pentagons. Apart from differences in pat

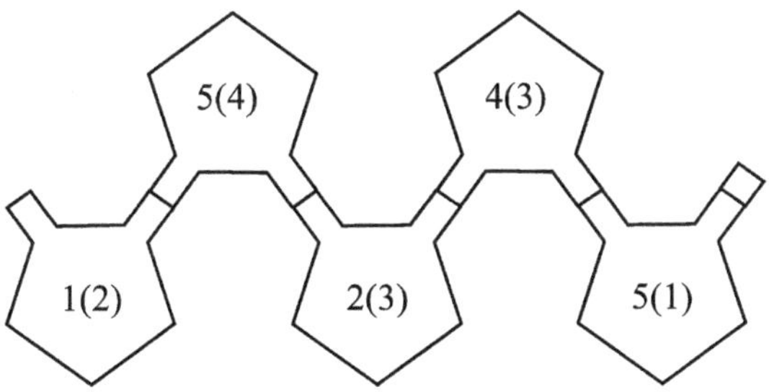

Fig. 7.20 Net for the irregular cycle interleaved pentagon point flexagon type A. One copy needed. Interleave during assembly, with leaves numbered 1 left visible

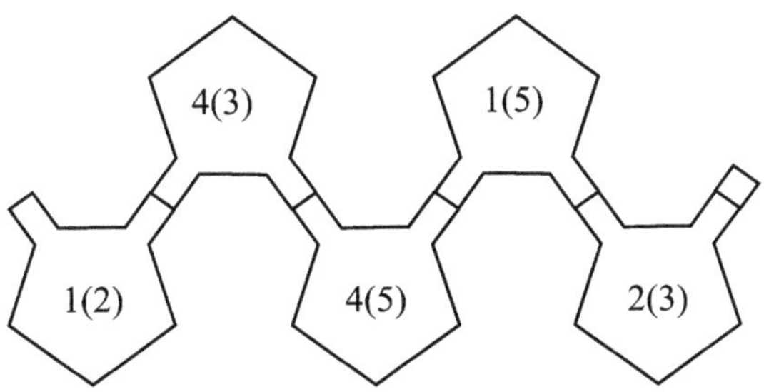

Fig. 7.21 Net for the irregular cycle interleaved pentagon point flexagon type B. One copy needed. Interleave during assembly, with leaves numbered 1 left visible

structure, their dynamic properties are the same as those of the interleaved fundamental pentagon point flexagon 1⟨5, 5, 5/2⟩ and the fundamental pentagon point flexagon 1⟨5, 5⟩. The principal 5-cycle shown in the intermediate position map can be traversed by using the simple flex. The intermediate position map is the same as that for the fundamental pentagon point flexagon 1⟨5, 5⟩ (Fig. 5.7). Principal main positions are pentagon vertex pairs linked by pairs of point hinges.

A single interleaf flex transforms the interleaved fundamental pentagon point flexagon 1⟨5, 5, 5/2⟩ into the irregular cycle interleaved pentagon flexagon type A. The associated polygon for 1⟨5, 5, 5/2⟩ has fivefold rotational symmetry (Table 7.5) so there are five distinct ways of carrying out the single interleaf flex, leading to the five distinct face numberings of type A. Similarly, the double interleaf flex leads to the five distinct face numberings of the irregular cycle interleaved pentagon point flexagon type B. In other words, the transformation behaviour also has fivefold rotational symmetry. These transformations are illustrated by the triple marked net for the irregular cycle interleaved pentagon point flexagons types A and B, and the interleaved fundamental pentagon point flexagon 1⟨5, 5, 5/2⟩ shown in Fig. 7.22. The Arabic numerals are the face numbering sequence for the interleaved fundamental pentagon point flexagon 1⟨5, 5, 5/2⟩, the Roman numerals are for one of the five distinct face numbering sequences for the irregular cycle interleaved pentagon point flexagon type A, and the letters correspond to one of the five distinct face numbering sequences for the irregular cycle interleaved pentagon point flexagon type B.

As assembled, the flexagon is in intermediate position 1 of the interleaved fundamental pentagon point flexagon 1⟨5, 5, 5/2⟩. The principal 5-cycle shown by Arabic numerals in the intermediate position map (Fig. 7.23) can be traversed by using the simple flex.

To transform to intermediate position v of the irregular cycle interleaved pentagon point flexagons type A, use a single interleaf flex at intermediate position 1, and interchange the two leaves marked with asterisks. This transformation is indicated by the line marked with an asterisk on the intermediate position map. The principal 5-cycle shown by Roman numerals can be traversed by using the simple flex. The transformation can be carried out in five distinct ways, leading to the five distinct face numbering sequences of the irregular cycle interleaved pentagon point flexagons type A. For clarity, face markings are shown only for one of these.

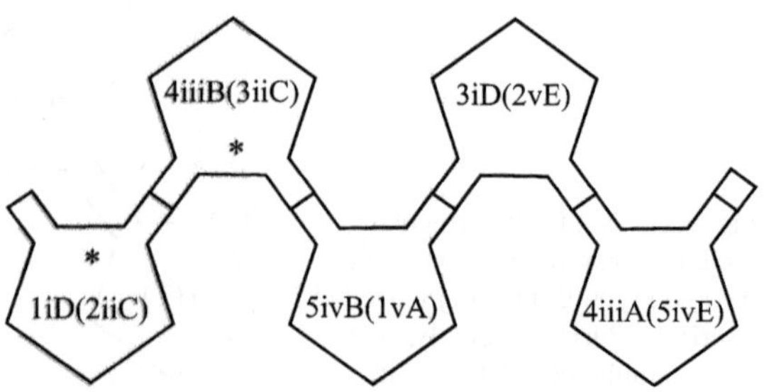

Fig. 7.22 Triple marked net for the interleaved fundamental pentagon point flexagon 1⟨5, 5, 5/2⟩ and the irregular cycle interleaved pentagon point flexagons, types A and B. One copy needed. Interleave during assembly, with leaves numbered 1 left visible

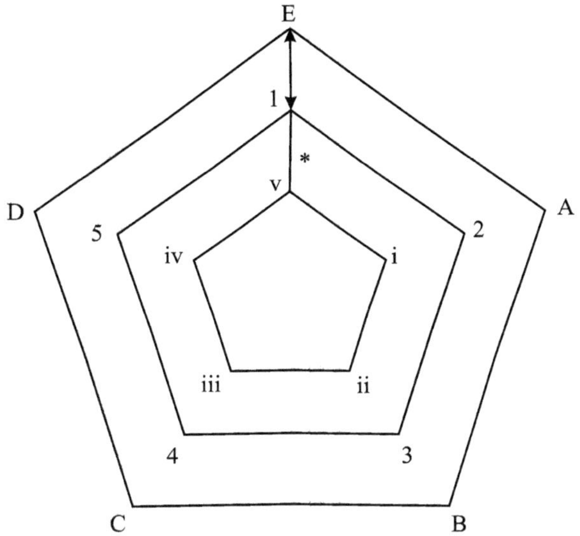

Fig. 7.23 Intermediate position map for the interleaved fundamental pentagon point flexagon 1⟨5, 5, 5/2⟩ (Arabic numerals), and the irregular cycle interleaved pentagon point flexagons, types A (Roman numerals) and B (letters)

To transform to intermediate position E of irregular cycle interleaved pentagon point flexagons type B, use a double interleaf flex at intermediate position 1. This transformation is indicated by the line with double arrowheads on the intermediate position map. The principal 5-cycle shown by letters can be traversed by using the simple flex. The transformation can be carried out in five distinct ways, leading to the five distinct face numbering sequences of the irregular cycle interleaved pentagon point flexagons type B. Face markings are shown only for one of these.

7.3.5 *Irregular Cycle Interleaved Hexagon Point Flexagons*

There are 11 irregular cycle interleaved hexagon point flexagons, types A-K. Some of their properties are given in Table 7.6. Properties of the fundamental hexagon point flexagon 1⟨6, 6⟩ (Section 5.3.5) are included for comparison. For convenience associated polygons that do not have rotational symmetry are described as having onefold rotational symmetry. Transformation between flexagons are possible within two groups of irregular cycle interleaved hexagon point flexagons. These groups are types A, B, E and H, which have torsion 2, and types C, D, F, G, I, J and K, which have zero torsion. Transformations with the fundamental hexagon point flexagon 1⟨6, 6⟩ are not possible because this is not interleaved and it has a different torsion. Interleaving during assembly is needed for types E, J and K.

Nets for the irregular cycle interleaved hexagon point flexagons, type A, C and J are shown in Figs. 7.24–7.26. As assembled, the flexagons are in intermediate position 1, which are, in appearance, single regular hexagons. Apart from differences in pat structure, their dynamic properties are the same as those of the fundamental

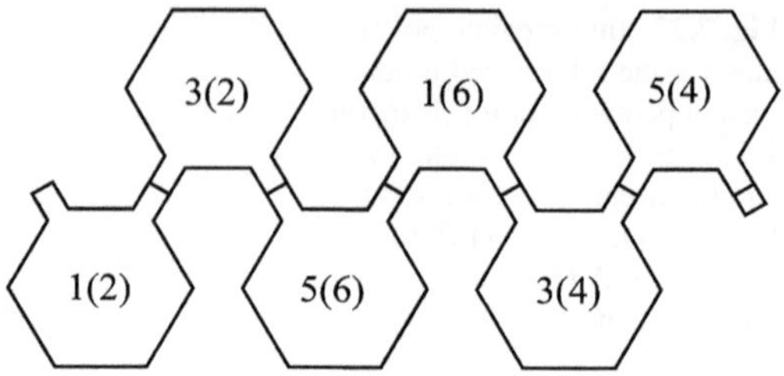

Fig. 7.24 Net for the irregular cycle interleaved hexagon point flexagon, type A. One copy needed. Interleave during assembly, with leaves numbered 1 left visible

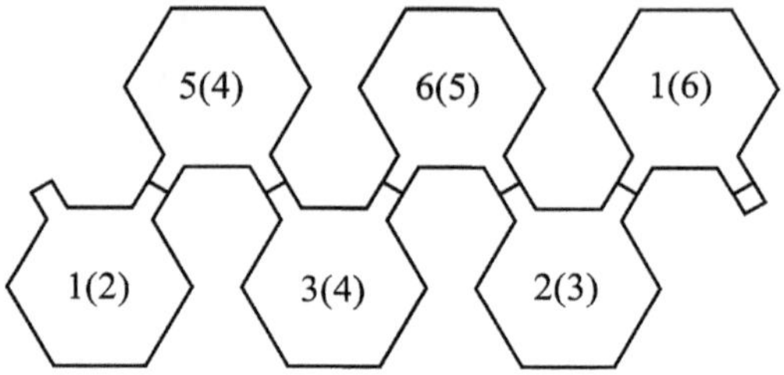

Fig. 7.25 Net for the irregular cycle interleaved hexagon point flexagon, type C. One copy needed. Interleave during assembly, with leaves numbered 1 left visible

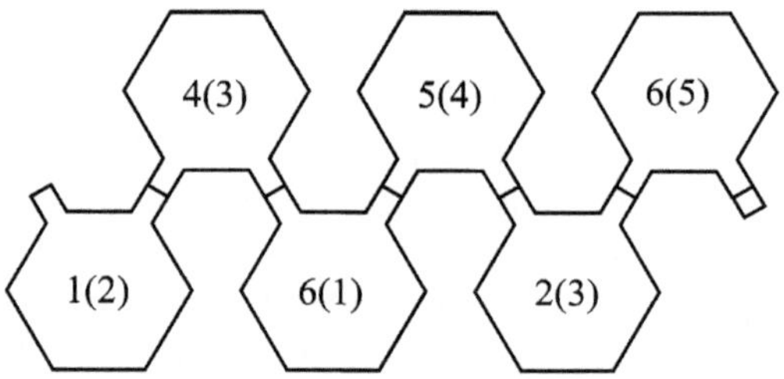

Fig. 7.26 Net for the irregular cycle interleaved hexagon point flexagon, type J. One copy needed. Interleave during assembly, with leaves numbered 1 left visible

hexagon point flexagon 1⟨6, 6⟩. The principal 6-cycle shown in the intermediate position map can be traversed by using the simple flex. The intermediate position map is the same as that for the fundamental hexagon point flexagon 1⟨6, 6⟩ (Fig. 5.11). Principal main positions are hexagon vertex pairs linked by pairs of point hinges.

The associated hexagon for the irregular cycle interleaved hexagon point flexagon type A has threefold rotational symmetry (Fig. 7.3a), so the transformation behaviour of the flexagon also has threefold rotational symmetry. In particular, the flexagon can be transformed into the irregular cycle interleaved hexagon point flexagon type E by making either a single interleaf flex or a double interleaf flex at intermediate position 2 or 4 or 8.

The associated hexagon for the irregular cycle interleaved hexagon point flexagon type C has twofold rotational symmetry (Fig. 7.3c), so transformation behaviour of the flexagon also has twofold rotational symmetry. In particular, it can be transformed into the irregular cycle interleaved hexagon point flexagon type F by making a single interleaf flex either at leaves numbered 1(2) and 2(3) or at leaves 4(5) and 5(6). It can be transformed also into the irregular cycle interleaved hexagon point flexagon type K by making a single interleaf flex either at leaves 2(3) and 3(4) or at leaves 5(6) and 6(1).

The associated hexagon for the irregular cycle interleaved hexagon point flexagon type J has twofold rotational symmetry (Fig. 7.3j), so transformation behaviour of the flexagon also has twofold rotational symmetry. The single interleaf flex is possible at all six pairs of adjacent leaves. This is the only irregular cycle interleaved hexagon point flexagon for which this is possible. In particular, it can be transformed into the irregular cycle interleaved hexagon point flexagon type I by making a single interleaf flex either at leaves numbered 2(3) and 3(4), or at leaves 5(6) and 6(1). Further, the flexagon can be transformed into the irregular cycle interleaved hexagon point flexagon type K by making a single interleaf flex at any of the other four pairs of adjacent leaves.

7.3.6 Augmented Irregular Cycle Interleaved Triangle Point Flexagons

An augmented point flexagon is a point flexagon in which the leaves have been replaced by leaves with a smaller number of edges, where this number of edges is a factor of the number of edges on the original leaves Section 5.5.1. Thus, an augmented irregular cycle interleaved triangle point flexagon is an irregular cycle interleaved hexagon point flexagon (previous section) in which the hexagons have been replaced by equilateral triangles. All augmented irregular cycle interleaved triangle point flexagons can be flexed by using the simple flex and this is their characteristic flex.

There are 11 irregular cycle interleaved hexagon point flexagons, types A-K. Therefore, there are 11 augmented irregular cycle interleaved triangle point flexagons, types A-K. Some of their properties are given in Table 7.7 (cf. Table 7.6). Properties of the augmented interleaved fundamental triangle point flexagon 1⟨3, 3, 6⟩ (Section 5.5.2) are included for comparison. No transformations between flexagons are possible. Interleaving during assembly is needed for types E, J and K. Types B, C, G. H, J and K have hinges that cannot be nested so there is no one correct method of assembly (Section 5.6.1).

The augmented irregular cycle interleaved triangle point flexagons, type A-K are all flexed using the simple flex. Intermediate positions of types A, D, E, F and I are, in appearance, single equilateral triangles. Most main positions are equilateral triangle vertex pairs linked by pairs of point hinges (Fig. 1.8a), but some are triangle vertex pairs linked by single point hinges (Fig. 1.8b). Intermediate positions of types B, C, G. H, J and K are approximately single triangles (Fig. 5.23a). Most main positions are approximately equilateral triangle vertex pairs linked by pairs of point hinges (Fig. 5.23b), but some are approximately triangle vertex pairs linked by single point hinges.

The dynamic properties are simplifications of those of the augmented interleaved fundamental triangle point flexagon 1⟨3, 3, 6⟩. These are shown in its intermediate position map, which is the same as that for the first order fundamental hexagon even edge flexagons *S*⟨6, 6⟩ (Fig. 4.10). The principal 6-cycle shown in this intermediate position map is present in all the augmented irregular cycle

Table 7.7 Properties of the augmented irregular cycle interleaved triangle point flexagons type A-K, and the augmented fundamental triangle point flexagon 1⟨3, 3, 6⟩. Fundamental vertex net ⟨3⟩ is used

Type	Rotational symmetry of associated hexagon	Distinct face numbering sequences	Typical face numbering sequence	Absent subsidiary main positions	Torsion
A	Threefold	2	1/2, 3/2, 5/6, 1/6, 3/4, 5/4	1(4,), 2(5), 3(6)	4
B	Onefold	12	1/2, 4/3, 2/3, 1/6, 5/6, 5/4	1(3), 3(5), 2(5), 3(6)	0
C	Twofold	6	1/2, 5/4, 3/4, 6/5, 2/3, 1/6	1(3), 2(4), 4(6), 5(1), 1(4,), 2(5)	2
D	Onefold	6	1/2, 3/2, 6/1, 6/5, 3/4, 5/4	4(6), 6(2), 1(4), 2(5)	2
E	Onefold	6	1/2, 4/3, 5/6, 1/6, 2/3, 5/4	1(3), 3(5), 1(4,), 2(5), 3(6)	0
F	Twofold	3	1/2, 4/3, 2/3, 5/4, 6/1, 6/5	1(3), 3(5), 4(6), 6(2), 3(5)	2
G	Twofold	3	1/2, 3/2, 5/6, 5/4, 3/4, 1/6	1(4), 2(5)	2
H	Onefold	6	1/2, 4/3, 2/3, 5/4, 5/6, 1/6	1(3), 3(5), 3(6)	0
I	Onefold	6	1/2. 3/2, 6/1, 5/4, 3/4, 6/5	4(6), 6(2), 1(4), 2(5)	2
J	Twofold	3	1/2, 4/3, 6/1, 5/4, 2/3, 6/5	1(3), 3(5), 4(6), 6(2), 1(4,), 2(5), 3(6)	0
K	Onefold	6	1/2, 4/3, 6/1, 6/5, 2/3, 5/4	1(3), 3(5), 4(6), 6(2), 1(4,), 2(5), 3(6)	2
1⟨3, 3, 6⟩	Threefold	1	1/2, 3/2. 3/4, 5/4, 5/6, 1/6	None	2

interleaved triangle point flexagons types A-K, but some of the subsidiary main positions are absent, as listed in Table 7.7. Hence, some subsidiary cycles are *incomplete cycles* that cannot be completely traversed.

For most of the flexagons listed in Table 7.7 an intermediate position map has the same rotational symmetry as the corresponding associated hexagon. One exception is the augmented irregular cycle interleaved triangle point flexagon type A where the intermediate position map has sixfold rotational symmetry (Fig. 7.27), and the associated hexagon has threefold rotational symmetry (Fig. 7.3a). The other exception is the augmented irregular cycle interleaved triangle point flexagon type K where these are twofold and onefold (Fig. 7.3k) respectively.

The net for the augmented irregular cycle interleaved triangle point flexagon type A is shown in Fig. 7.28. This was derived by using the net for the irregular cycle interleaved hexagon point flexagon type A (Fig. 7.24) as a precursor and replacing the regular hexagons by equilateral triangles. As assembled, the flexagon is in intermediate position 1 which is, in appearance, a single triangle. The principal6-cycle and the two subsidiary 3-cycles shown in the intermediate position map (Fig. 7.27) can be traversed by using the simple flex. Main positions are triangle vertex pairs connected by pairs of point hinges (Fig. 1.8a). Subsidiary main positions 1(4), 2(5) and 3(6) are absent.

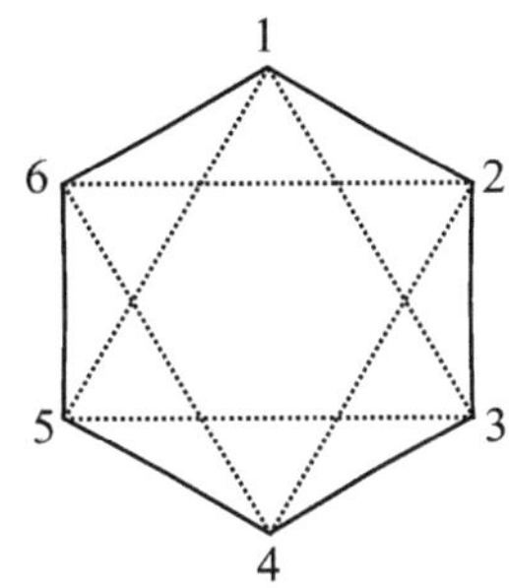

Fig. 7.27 Intermediate position map for the augmented irregular cycle interleaved triangle point flexagons type A

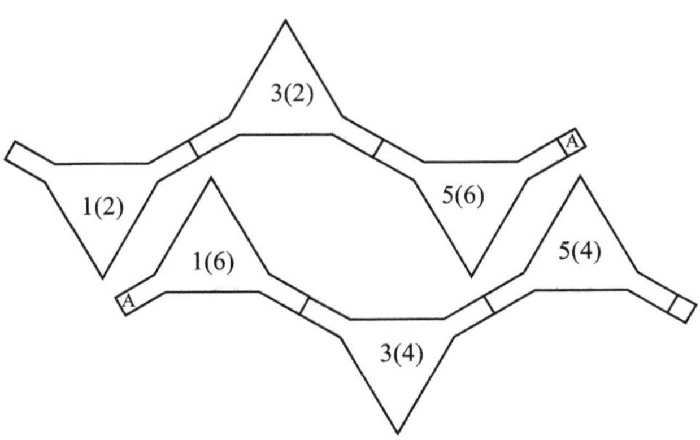

Fig. 7.28 Net for the augmented irregular cycle interleaved triangle point flexagon type A. One copy needed. Join the two parts of the net at A-A. Interleave during assembly, nesting hinges, and with leaves numbered 1 left visible

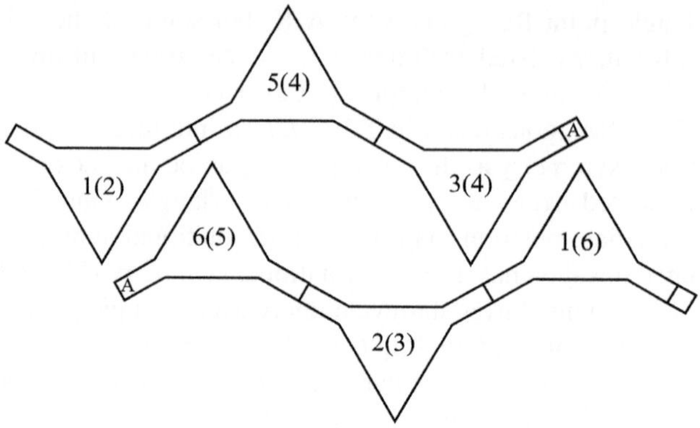

Fig. 7.29 Net for the augmented irregular cycle interleaved triangle point flexagon type C. One copy needed. Join the two parts of the net at A-A. Interleave during assembly, with hinges adjacent, and with leaves numbered 1 left visible

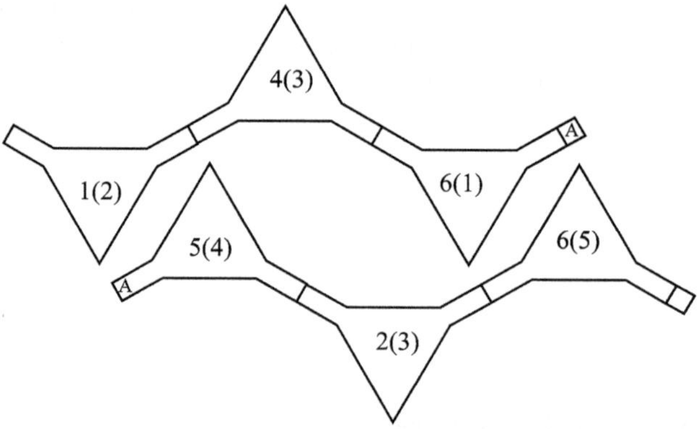

Fig. 7.30 Net for the augmented irregular cycle interleaved triangle point flexagon type J. One copy needed. Join the two parts of the net at A-A. Interleave during assembly, with hinges adjacent, and with leaves numbered 1 left visible

Nets for the augmented irregular cycle interleaved triangle point flexagons types C and J are shown in Figs. 7.29 and 7.30. These were derived by using the nets for the irregular cycle interleaved hexagon point flexagons, types C and J (Figs. 7.25 and 7.26) as precursors and replacing the regular hexagons by equilateral triangles. There are hinges that cannot be nested, so leaves do not overlap exactly, and there is no one correct method of assembly. (Section 5.6.1).

As assembled, the flexagons are in intermediate position 1. These are approximately single triangles (Fig. 5.23a). The intermediate positions maps (Fig. 7.31) can be traversed by using the simple flex. Most of the main positions are approximately triangle vertex pairs connected by pairs of point hinges (Fig. 5.23b). In both flexagons the principal 6-cycle can be traversed. Some subsidiary main positions are absent (Table 7.7). In both flexagons there are two incomplete subsidiary 1/3-cycles. This notation for an incomplete cycle means that one of the three sub-

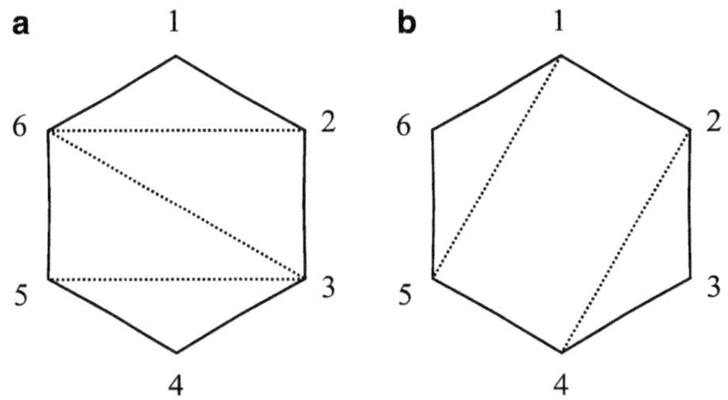

Fig. 7.31 Intermediate position maps for augmented irregular cycle interleaved triangle point flexagons. (**a**) Type C. (**b**) Type J

Table 7.8 Numbers of distinct face numbering sequences for point flexagons made from fundamental vertex nets

Leaf type	Fundamental vertex net	Number of leaves in net	Torsion 0	1	2	3	4	5	6	Total
Triangle	⟨3⟩	3	0	1	0	0	0	0	0	1
Triangle	⟨3⟩	6	27	0	32	0	1	0	0	60
Square	⟨4⟩	4	2	0	1	0	0	0	0	3
Pentagon	⟨5⟩	5	0	11	0	1	0	0	0	12
Hexagon	⟨6⟩	6	33	0	26	0	1	0	0	60
Heptagon	⟨7⟩	7	0	302	0	57	0	1	0	360
Octagon	⟨8⟩	8	1,208	0	1,191	0	120	0	1	2,520

sidiary main positions in a subsidiary 3-cycle can be visited and the other two subsidiary main positions are absent. Type C can also be flexed into subsidiary main position 3(6), which is not in a cycle, and is approximately an equilateral triangle vertex pair connected by a single point hinge.

7.4 Distinct Face Numbering Sequences

General properties of various types of point flexagon made from fundamental vertex nets (Section 3.4), and with point hinges at adjacent vertices, are described in Sections 5.3.1, 5.4.1, 5.5.1, 5.6.1 and 7.3.1. Some of these have more than one possible distinct face numbering sequence. Total numbers of distinct face numbering sequences are given in Table 7.8, These data are taken from Pook (2007) and Sherman (2007).

Transformation between flexagons is not possible for the triangle point flexagons listed in the table. Transformation between square point flexagons with the same number of leaves and torsion, as listed in the table, is always possible. Interleaf flexes are used. Similarly, transformation between flexagons is possible for the pentagon point flexagons and the hexagons point flexagons, and is believed to be possible for the heptagon point flexagons and the octagon point flexagons.

7.5 Irregular Cycle Non Interleaved Point Flexagons

7.5.1 General Properties

The topological invariants for irregular cycle non interleaved point flexagons are the same as those for fundamental point flexagons (Section 5.1). They are made from irregular vertex nets, which are not fundamental vertex nets (Section 3.4). All irregular cycle non interleaved point flexagons can be flexed by using the simple flex and this is their characteristic flex. In intermediate positions, but not in any other position, the dual of an irregular cycle non interleaved point flexagon is an irregular cycle unagon. An irregular cycle unagon is an irregular cycle even edge flexagon (Section 7.2.1) with the number of sectors reduced to one. It can only exist in an intermediate position which is, in appearance, a single polygon. Irregular cycle unagons cannot be flexed, so they are not flexagons (cf. Section 5.3.1).

Hence, irregular cycle non interleaved point flexagons correspond to irregular cycle even edge flexagons, and the irregular polygons shown in Figs. 7.1–7.3 are possible associated polygons for irregular cycle non interleaved point flexagons. All the possible flexagon figures for the irregular cycle non interleaved square point flexagon, irregular cycle non interleaved pentagon point flexagons, and hexagon point flexagons are shown in Figs. 7.32–7.34 (cf. Figs. 7.4–7.6). These were derived by using the irregular polygons shown in Figs. 7.1–7.3 as the inscribed polygons of flexagon figures. To avoid confusion circumscribing polygons are shown as dots representing their vertices (Section 5.2.1).

The net for an irregular cycle non interleaved point flexagon is the dual of the net for the corresponding irregular cycle unagon (Section 5.3.4). Hence, the net for an irregular cycle non interleaved point flexagon may be derived by inscribing polygons with vertices at midpoints of the edges of the polygons of the net for an irregular

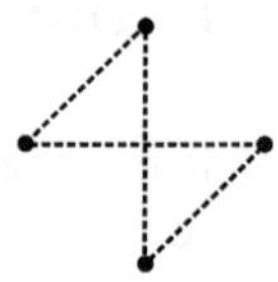

Fig. 7.32 Flexagon figure for the irregular cycle non interleaved square point flexagon

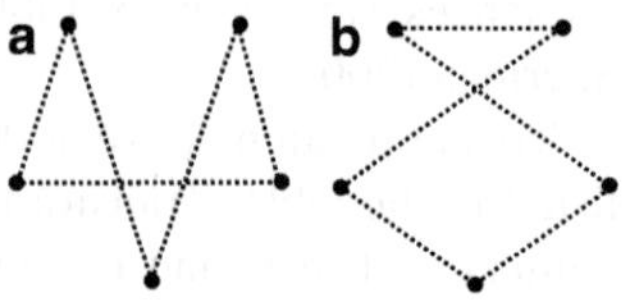

Fig. 7.33 Flexagon figures for irregular cycle non interleaved pentagon point flexagons. (**a**) Type A. (**b**) Type B

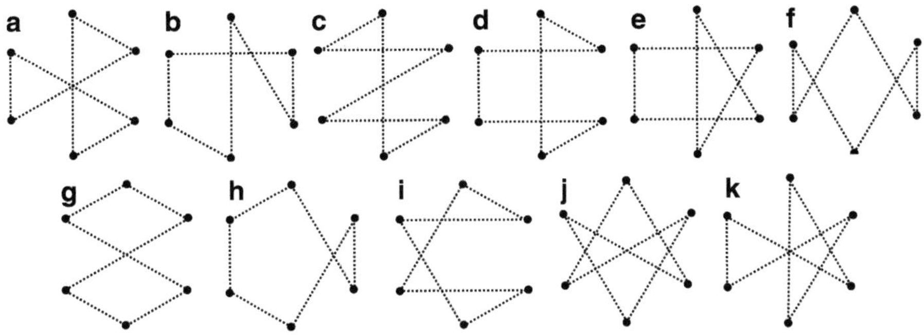

Fig. 7.34 Flexagon figures for irregular cycle non interleaved hexagon point flexagons. (**a**) Type A. (**b**) Type B. (**c**) Type C. (**d**) Type D. (**e**) Type E. (**f**) Type F. (**g**) Type G. (**h**) Type H. (**i**) Type I. (**j**) Type J. (**k**) Type K

cycle unagon. Thus, the edge hinges along edges of the original polygons become point hinges at vertices of the inscribed polygons (Sections 2.4.1 and 3.4). This is shown for fundamental nets in Figs. 3.18–3.20. Two examples are given below.

7.5.2 The Irregular Cycle Non Interleaved Square Point Flexagon

The net for the irregular cycle non interleaved square point flexagon is shown in Fig. 7.35. The torsion is zero. The flexagon figure is shown in Fig. 7.32. As assembled, the flexagon is in intermediate position 1 which is, in appearance, a single square. The principal 4-cycle shown in the intermediate position map can be traversed by using the simple flex. The intermediate position map is the same as that for the fundamental square point flexagon 1⟨4, 4⟩ (Fig. 5.5). The irregular behaviour as the principal 4-cycle is traversed is clearly noticeable. Principal main positions are square vertex pairs linked by pairs of point hinges (Fig. 3.23).

7.5.3 An Irregular Cycle Non Interleaved Pentagon Point Flexagon

The net for the irregular cycle non interleaved pentagon point flexagon type B is shown in Fig. 7.36. The torsion is 1. The flexagon figure is shown in Fig. 7.33b. As assembled, the flexagon is in intermediate position 1 which is, in appearance, a single regular pentagon. The principal 5-cycle shown in the intermediate position map (Fig. 7.37) can be traversed by using the simple flex. The irregular behaviour as the principal 5-cycle is traversed is noticeable. There is also an incomplete

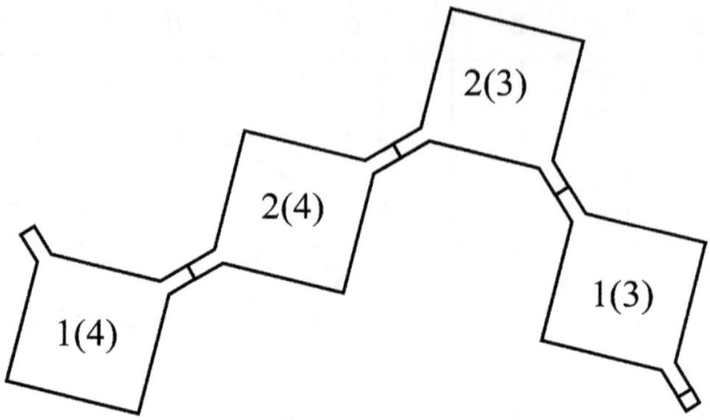

Fig. 7.35 Net for the irregular cycle non interleaved square point flexagon. One copy needed. Fold until leaves numbered 1 are visible

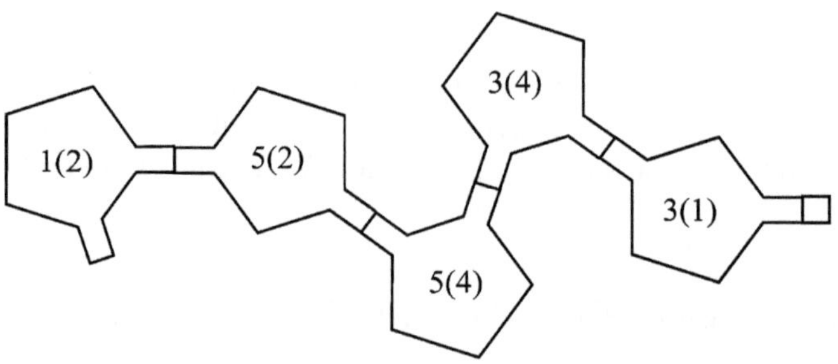

Fig. 7.36 Net for the irregular cycle non interleaved pentagon point flexagon type B. One copy needed. Fold until leaves numbered 1 are visible

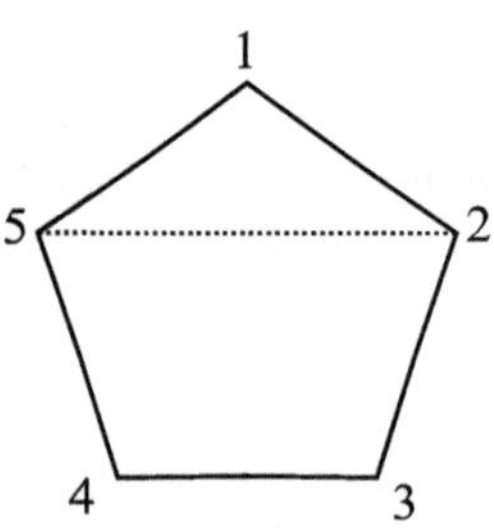

Fig. 7.37 Intermediate position map for the irregular cycle non interleaved pentagon point flexagon type B

subsidiary 1/5-cycle. This notation for an incomplete cycle means that one of the five subsidiary main positions, present in the subsidiary 5-cycle of the two sector irregular cycle pentagon even edge flexagon type B (Section 7.2.4), can be visited, and the other four subsidiary main positions are absent. Subsidiary main position 2(5) can be visited. Main positions are pentagon vertex pairs linked by pairs of point hinges.

References

Conrad AS, Hartline DK (1962) *Flexagons*. RIAS Technical Report 62-11. Baltimore, MD: RIAS
Pook LP (2003) Flexagons Inside Out. Cambridge University Press, Cambridge
Pook LP (2007) Transformations between point flexagons. Flexagon Lovers Group posting 343. http://tech.groups.yahoo.com/group/Flexagon_Lovers/. Accessed 31 May 2007
Pook LP (2009) Enantiomorphism in flexagons. Flexagon Lovers Group posting 559. http://tech.groups.yahoo.com/group/Flexagon_Lovers/. Accessed 11 February 2009
Sherman S (2007) Transformations between point flexagons. Flexagon Lovers Group posting 342. http://tech.groups.yahoo.com/group/Flexagon_Lovers/. Accessed 31 May 2007

Chapter 8
Degenerate Flexagons

8.1 Introduction

In a general sense, a degenerate flexagon is a flexagon derived by deleting one or more faces from a precursor flexagon. Most of the degenerate flexagons described in this chapter are degenerate versions of first order fundamental even edge flexagons, irregular cycle non interleaved point flexagons, irregular cycle interleaved point flexagons, and compound edge flexagons. These were all derived by deleting faces from solitary precursor flexagons so they are solitary flexagons. Topological invariants are the same as those of precursor flexagons.

Nomenclature for degenerate flexagons is difficult because the same degenerate flexagon can have more than one type of precursor, so for some of them arbitrary flexagon type designations are used. A feature of some degenerate flexagons is that they are more stable than the precursor flexagons, and so are easier to handle.

Deleting a face does not change the appearance of main positions and intermediate positions, but it does change the net used to construct the flexagon, the intermediate position map and, where appropriate, the flexagon figure. Intermediate position maps for degenerate flexagons are subsets of intermediate position maps for precursor flexagons. Deleting a face means that the corresponding intermediate position is absent. On an intermediate position map this is indicated by a dot. Main positions that included the deleted face are also absent. At least one cycle that was present in a precursor fundamental flexagon becomes an incomplete cycle in a degenerate version, and so cannot be completely traversed. For example, incomplete principal 3/5-cycle means that three of the principal main positions that are present in the precursor flexagon can be visited but the other two principal main positions are absent. Deleting faces sometimes results in deficient flexagons in which it is not possible to visit all the main positions without disconnecting a hinge, refolding the flexagon, and reconnecting the hinge. It can also result in structures that are not flexagons.

A practical way of deleting a face is to start with a paper model of the net for the precursor flexagon and then glue together pairs of leaves bearing the number of the face to be deleted. This modified net is then used as a template for the net for the desired degenerate flexagon. It is usually convenient to retain the original face numbers. Deletion of faces is straightforward for flexagons in which leaves to be

L. Pook, *Serious Fun with Flexagons*, Solid Mechanics and Its Applications 164, DOI 10.1007/978-90-481-2503-6_8, © Springer Science+Business Media B.V. 2009

glued together are adjacent in nets since the resulting degenerate flexagon is a simple band of hinged polygons. In irregular cycle interleaved fundamental point flexagons, leaves to be glued together may not be adjacent, and deletion of faces sometimes results in flexagons that are not simple bands.

The number of types of degenerate flexagons increases rapidly as the number of edges on the constituent polygons increases. Associated polygons for degenerate versions of first order fundamental even edge flexagons, non interleaved point flexagons and compound edge flexagons are degenerate irregular polygons. Vertices of a degenerate irregular polygon are the same as some of the vertices of a regular polygon. Possible degenerate flexagons can be enumerated by counting possible degenerate irregular polygons. Possible degenerate irregular polygons that have some of the vertices of squares, regular pentagons and regular hexagons are shown in Figs. 8.1–8.5. Rotations and reflections are not regarded as distinct. The arbitrary type letters shown in Figs. 8.2–8.5 are used in descriptions of corresponding flexagons.

Enumeration of possible degenerate versions of irregular cycle interleaved point flexagons is difficult because of the occurrence of flexagons that are not simple bands of hinged polygons. Flexagons that are not simple bands do not have associated polygons, they have associated double polygons or associated semi double polygons, and enumeration of these is not straightforward.

Fig. 8.1 Triangle with three of the vertices of a square

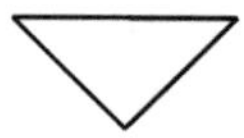

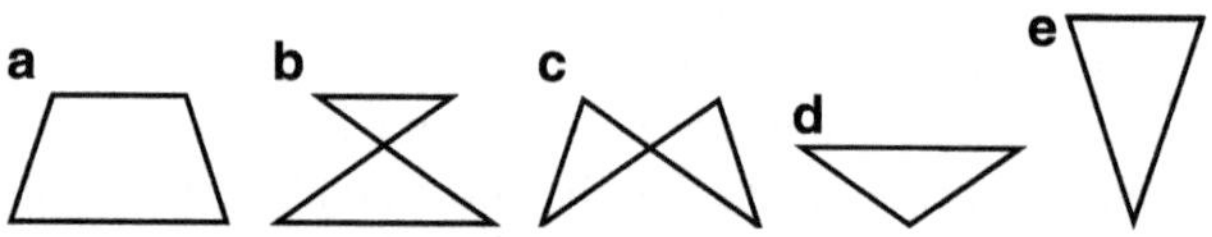

Fig. 8.2 Polygons that have some of the vertices of a regular pentagon. (**a**) Type A. (**b**) Type B. (**c**) Type C. (**d**) Type D. (**e**) Type E

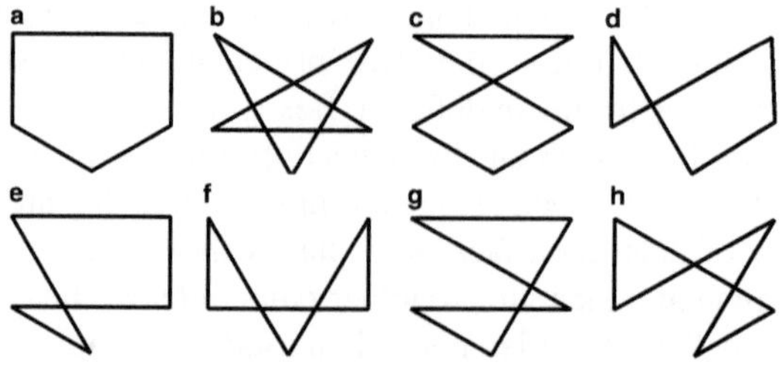

Fig. 8.3 Pentagons that have five of the vertices of a regular hexagon. (**a**) Type A. (**b**) Type B. (**c**) Type C. (**d**) Type D. (**e**) Type E. (**f**) Type F. (**g**) Type G. (**h**) Type H

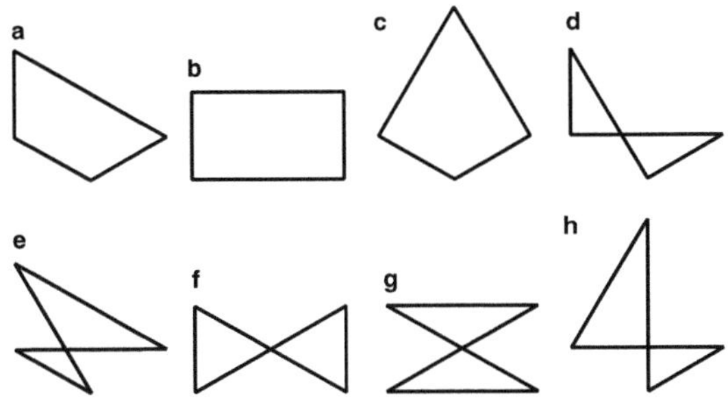

Fig. 8.4 Quadrilaterals that have four of the vertices of a regular hexagon. (**a**) Type I. (**b**) Type J. (**c**) Type K. (**d**) Type L. (**e**) Type M. (**f**) Type N. (**g**) Type O. (**h**) Type P

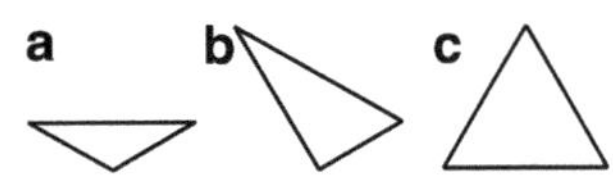

Fig. 8.5 Triangles that have three of the vertices of a regular hexagon. (**a**) Type Q. (**b**) Type R. (**c**) Type S

8.2 Degenerate Even Edge Flexagons

8.2.1 General Properties

Degenerate versions of first order fundamental even edge flexagons are called degenerate even edge flexagons. In the flexagon figure for a degenerate even edge flexagon, the inscribed polygon is a degenerate irregular polygon with its vertices on midpoints on some of the edges of the circumscribing polygon. Possible flexagon figures for degenerate square even edge flexagons, degenerate pentagon even edge flexagons and degenerate hexagon even edge flexagons are shown in Figs. 8.6–8.10. These were derived by using the degenerate irregular polygons shown in Figs. 8.1–8.5 as the inscribed polygons of flexagon figures. Type letters in the figures are used in descriptions of corresponding flexagons, together with an indication of the number of sectors. For a given number of sectors, there are no degenerate triangle even edge flexagons, one degenerate square even edge flexagon, five degenerate pentagon even edge flexagons, and 19 degenerate hexagon even edge flexagons.

Most of the degenerate even edge flexagons described below have two sectors. They illustrate the wide range of possibilities. Their nets were derived by deletion of faces (Section 8.1), either from first order fundamental even edge flexagons or from degenerate even edge flexagons. The dynamic properties of a degenerate flexagon are a simplification of those of the precursor flexagon used in its derivation. Alternatively, nets could have been derived from flexagon figures by using the procedure given in Section 7.2.2.

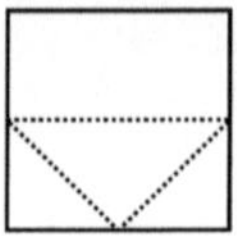

Fig. 8.6 Flexagon figure for a degenerate square even edge flexagon (Les Pook, Flexagons inside out, 2003, © Cambridge University Press 2003, reprinted with permission)

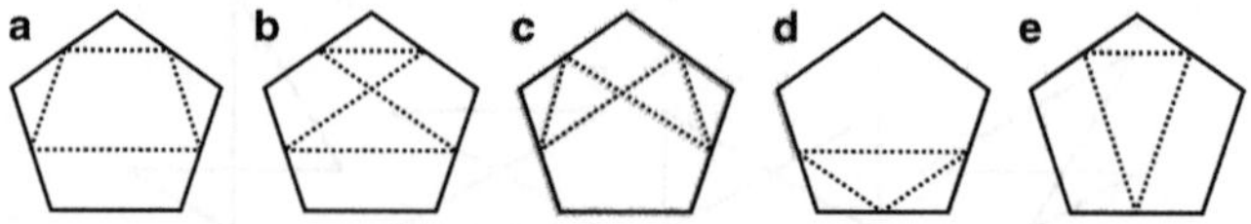

Fig. 8.7 Flexagon figures for degenerate pentagon even edge flexagons. (**a**) Type A. (**b**) Type B. (**c**) Type C. (**d**) Type D. (**e**) Type E (Figs. 8.7a–d: Les Pook, Flexagons inside out, 2003, © Cambridge University Press 2003, reprinted with permission)

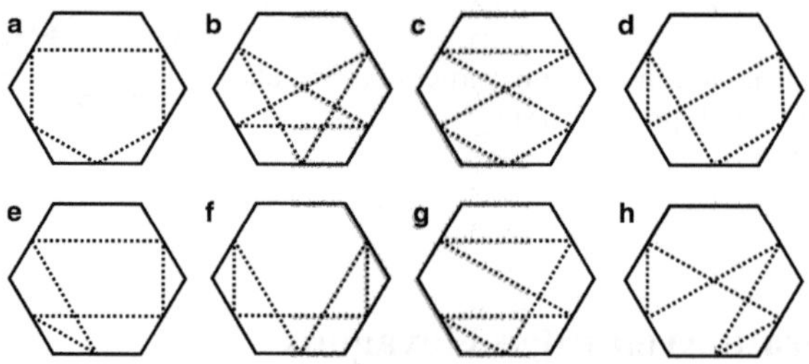

Fig. 8.8 Flexagon figures for degenerate hexagon even edge flexagons. (**a**) Type A. (**b**) Type B. (**c**) Type C. (**d**) Type D. (**e**) Type E. (**f**) Type F. (**g**) Type G. (**h**) Type H (Les Pook, Flexagons inside out, 2003, © Cambridge University Press 2003, reprinted with permission)

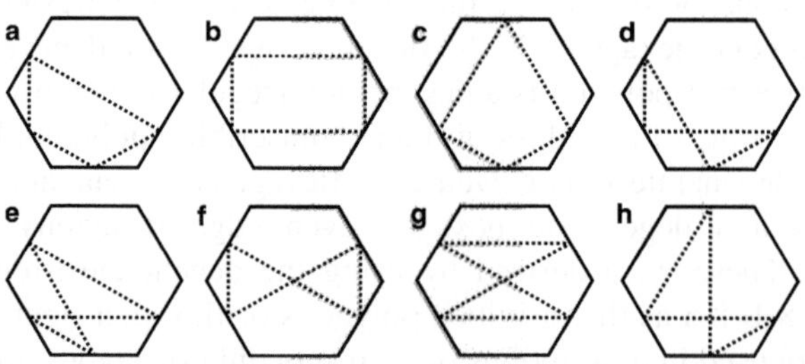

Fig. 8.9 Flexagon figures for degenerate hexagon even edge flexagons. (**a**) Type I. (**b**) Type J. (**c**) Type K. (**d**) Type L. (**e**) Type M. (**f**) Type N. (**g**) Type O. (**h**) Type P (Figs. 8.9a, d and e: Les Pook, Flexagons inside out, 2003, © Cambridge University Press 2003, reprinted with permission)

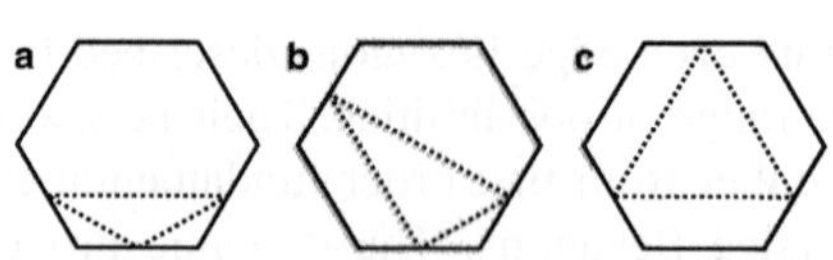

Fig. 8.10 Flexagon figures for degenerate hexagon even edge flexagons. (**a**) Type Q. (**b**) Type R. (**c**) Type S (Fig. 8.10a: Les Pook, Flexagons inside out, 2003, © Cambridge University Press 2003, reprinted with permission)

The simplest possible flexagon of each variety of even edge flexagon is a basic even edge flexagon. A basic even edge flexagon is always a solitary even edge flexagon that has three leaves per sector. There are always three main positions that can be visited, usually of more than one type, and there are always three intermediate positions on an intermediate position map. First order fundamental triangle even edge flexagons (Section 4.2.3) and triangular fundamental compound edge flexagons (Section 6.3.1) are basic even edge flexagons. The lines on their intermediate position map (Fig. 4.7) are an equilateral triangle. All other basic even edge flexagons are degenerate flexagons and the lines on their intermediate position maps are irregular triangles.

A degenerate unagon is a degenerate even edge flexagon with the number of sectors reduced to one. It can exist only in an intermediate position which is, in appearance, a single polygon. Degenerate unagons cannot be flexed, so they are not flexagons (cf. Section 5.3.1).

8.2.2 A Degenerate Square Even Edge Flexagon

There is one two sector degenerate *square even edge-flexagon*. It is one of the simplest possible edge flexagons, and it is a basic even edge flexagon. It is unusual in that all the hinges that operate during flexing are parallel. Some of its properties are given in Table 8.1 (cf. Table 4.4). The torsion per sector is 1. The flexagon figure is shown in Fig. 8.6.

The net for the two sector degenerate square even edge flexagon is shown in Fig. 8.11. This was derived by using the net for the fundamental square even edge

Table 8.1 Properties of the two sector degenerate square even edge flexagon. The incomplete principal cycle is in bold

Typical main position	Cycle type	Number of cycles	Main position type	Ring symbol	Curvature
1(2)	**2/4-cycle**	**1**	**Flat**	**4(90°)**	**0°**
1(3)	None	–	Box	4(0°)	360°

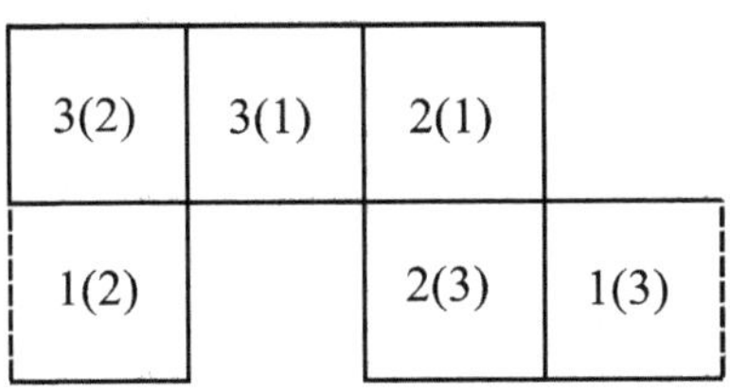

Fig. 8.11 Net for the two sector degenerate fundamental square even edge flexagon. One copy needed

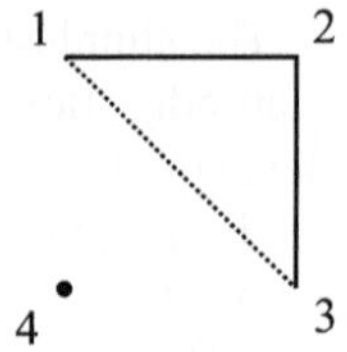

Fig. 8.12 Intermediate position map for the two sector degenerate square even edge flexagon

flexagon 2⟨4, 4⟩ (Figs. 1.2 and 4.16a) as a precursor, and deleting face 4. Except for face numbering, nets obtained by deleting any of the other three faces are identical. Alternatively, the net for the two sector irregular square even edge flexagon (Fig. 7.8) could be used as a precursor, but two different results are obtained depending on which face is deleted. Deleting face 1 or face 4 results in one enantiomorph, whereas deleting face 2 or face 3 results in the other enantiomorph.

As assembled, the flexagon is in principal main position 2(1), which is, in appearance, a flat regular even edge ring of four squares (Fig. 1.1b). The incomplete principal 2/4-cycle shown in the intermediate position map (Fig. 8.12) can be traversed by using the twofold pinch flex. From intermediate position 1 the flexagon can be opened into subsidiary main position 1(3). which is a box edge ring of four squares (Fig. 2.5). Intermediate position 4 is absent.

8.2.3 Degenerate Pentagon Even Edge Flexagons

There are five two sector degenerate pentagon even edge flexagons, types A–E. Types D and E are the simplest possible pentagon even edge flexagons, and they are basic even edge flexagons. Types A and D are described in this section. Some of their properties are given in Table 8.2 (cf. Table 4.5). The torsion per sector is 2 for type A and 1 for type D. Flexagon figures are shown in Fig. 8.7a and d.

The net for the two sector degenerate pentagon even edge flexagons type A is shown in Fig. 8.13. This was derived by using the net for the first order fundamental pentagon even edge flexagon 2⟨5, 5⟩ (Fig. 4.23a) as a precursor and deleting face 5. Deleting different faces leads to the same result, apart from face numbering, and alternative precursors can be used.

As assembled, the flexagon is in intermediate position 1, which is, in appearance, a pentagon edge pair. From here it can be opened into principal main position 1(2), and the incomplete principal 3/5-cycle shown by the solid lines in the intermediate position map (Fig. 8.14) can be traversed by using the twofold pinch flex. Intermediate position 5 is absent (cf. Fig. 4.9). Principal main positions are skew regular even edge rings of four regular pentagons (Fig. 2.7). Intermediate position 1 can also be opened into either subsidiary main position 1(3) or subsidiary main position 1(4), which are slant regular even edge rings of four regular pentagons (Fig. 4.24). The high curvature (216°) means that these cannot be turned inside out, so the incomplete subsidiary 3/5-cycle shown by the dotted lines in the intermediate position map cannot be traversed directly. All three subsidiary main positions can be visited by flexing via principal main positions.

Table 8.2 Properties of the two sector degenerate pentagon even edge flexagons types A and D. Incomplete principal cycles are in bold

Type	Typical main position	Cycle type	Number of cycles	Main position type	Ring symbol	Curvature
A	**1(2)**	**3/5-cycle**	**1**	**Skew**	**4(108°)**	**–72°**
A	1(3)	3/5 cycle	1	Slant	4(36°)	216°
D	**1(2)**	**2/5-cycle**	**1**	**Skew**	**4(108°)**	**–72°**
D	1(3)	1/5 cycle	1	Slant	4(36°)	216°

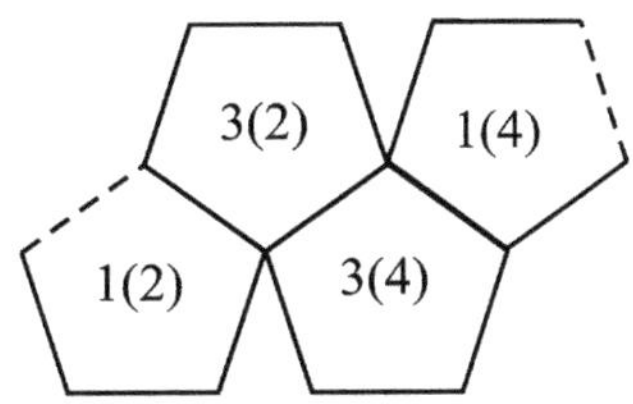

Fig. 8.13 Net for the two sector degenerate pentagon even edge flexagon type A. Two copies needed. Fold until leaves numbered 1 are visible

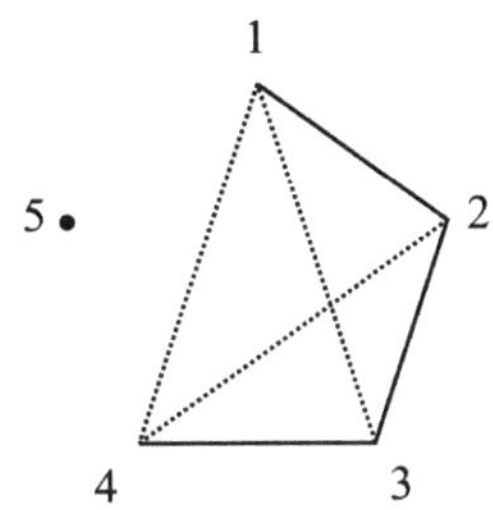

Fig. 8.14 Intermediate position map for the two sector degenerate pentagon even edge flexagon type A

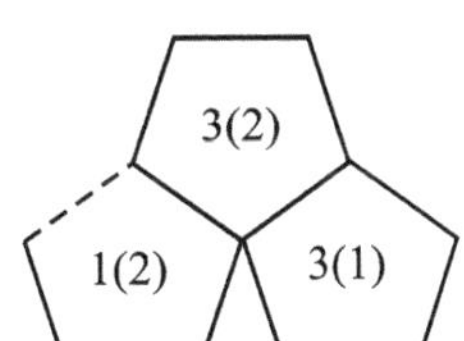

Figure 8.15 Net for the two sector degenerate fundamental pentagon even edge flexagon type D. Two copies needed. Fold until leaves numbered 1 are visible

The net for the degenerate pentagon even edge flexagon type D is shown in Fig. 8.15. This was derived by using the net for the degenerate pentagon even edge flexagon type A as a precursor and deleting face 4. The same result, but with different face numbering, can be derived by deleting face 1. Other precursors can also be used. As assembled, the flexagon is in intermediate position 1. The incomplete principal 2/5-cycle shown in the intermediate positions map (Fig. 8.16) can be traversed by using the twofold pinch flex, and subsidiary main position 1(3), of the incomplete subsidiary 1/5 cycle, can be visited.

Fig. 8.16 Intermediate position map for the two sector degenerate fundamental pentagon even edge flexagon type D

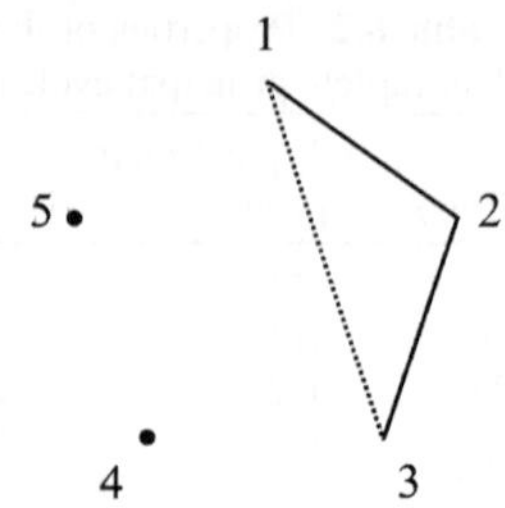

8.2.4 Degenerate Hexagon Even Edge Flexagons

There are 17 two sector degenerate hexagon even edge flexagons, types A–S. Types Q, R and S are the simplest possible pentagon even edge flexagons, and they are basic even edge flexagons. Two sector versions of types K and Q, and the three sector version of type S, are described in this section. Some of their properties are given in Table 8.3 (cf. Table 4.6). The torsion per sector is 2 for type K and 1 for types Q and S. Flexagon figures are shown in Figs. 8.9c and 8.10a and c.

The net for the two sector degenerate hexagon even edge flexagon type K is shown in Fig. 8.17. This was derived by using the net for the first order fundamental hexagon even edge flexagon 2⟨6, 6⟩ (Fig. 4.29a) as a precursor and deleting faces 2 and 6. Other derivations are possible.

As assembled, the flexagon is in intermediate position 3, which is, in appearance, a hexagon edge pair. From here it can be opened into principal main position 3(4), and the incomplete principal 2/6-cycle, shown by the solid lines in the intermediate position map (Fig. 8.18), can be traversed by using the twofold pinch flex. Principal main positions are skew regular even edge rings of four regular hexagons (Fig. 1.5). Intermediate position 3 can be opened into subsidiary main positions 3(1) and 3(5) also, which are slant regular even edge rings of four regular hexagons. The high curvature (120°) means that these cannot be turned inside out so the subsidiary 3-cycle shown by the dotted equilateral triangle in the intermediate position map cannot be traversed directly. It can be traversed indirectly via principal main

Table 8.3 Properties of the two sector degenerate hexagon even edge flexagons types K and Q, and the three sector degenerate hexagon even edge flexagon type S. Incomplete principal cycles are in bold

Type	Typical main position	Cycle type	Number of cycles	Main position type	Ring symbol	Curvature
K	**3(4)**	**2/6 cycle**	**1**	**Skew**	**4(120°)**	**−120°**
K	1(3)	3-cycle	1	Slant	4(60°)	120°
K	1(4)	None	–	Box	4(0°)	360°
Q	**1(2)**	**2/6-cycle**	**1**	**Skew**	**4(120°)**	**−120°**
Q	1(3)	1/3 cycle	1	Slant	4(60°)	120°
S	1(3)	3 cycle	1	Flat	6(60°)	0°

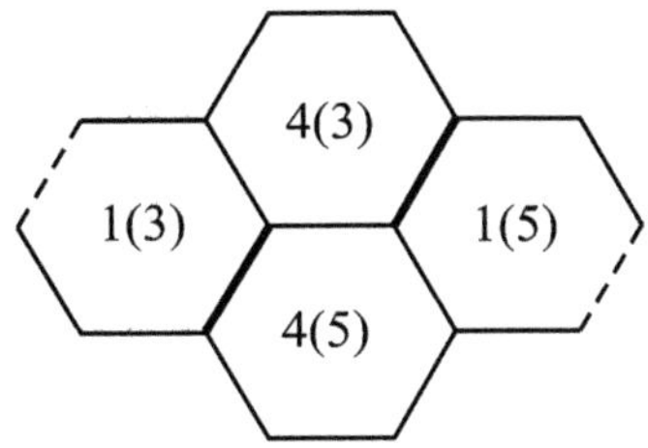

Fig. 8.17 Net for the two sector degenerate hexagon even edge flexagon type K. Two copies needed. Fold until leaves numbered 3 are visible

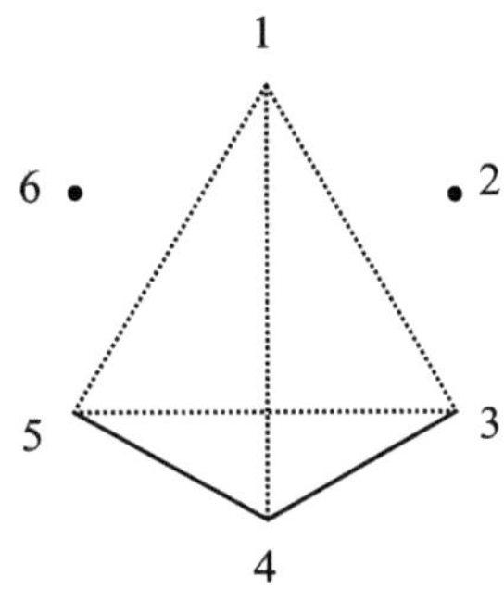

Fig. 8.18 Intermediate position map for the two sector degenerate hexagon even edge flexagon type K

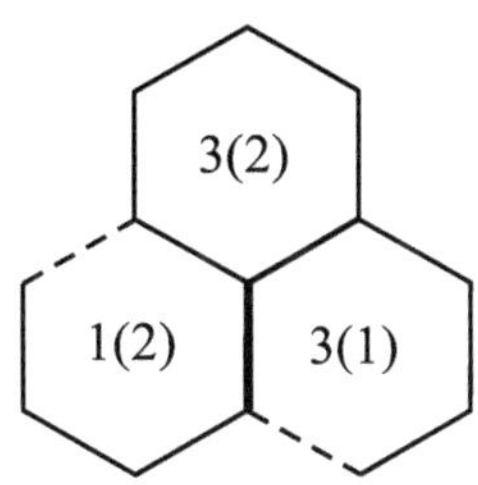

Fig. 8.19 Net for the two sector degenerate fundamental hexagon even edge flexagon type Q. Two copies needed. Fold until leaves numbered 1 are visible

positions. Intermediate position 4 can be opened into subsidiary main position 4(1), which is a box edge ring of four regular hexagons (Fig. 4.31).

The net for the two sector degenerate hexagon even edge flexagon types Q is shown in Fig. 8.19. This was derived by using the net for the first order fundamental hexagon even edge flexagon 2⟨6, 6⟩ (Fig. 4.29a) as a precursor and deleting faces 4, 5 and 6. Other derivations are possible.

As assembled, the flexagon is in intermediate position 1, which is a hexagon edge pair. From here it can be opened into principal main position 1(2), and the incomplete principal 2/6-cycle shown by the solid lines in the intermediate position map (Fig. 8.20a) can be traversed by using the twofold pinch flex. Principal main positions are skew regular even edge rings of four regular hexagons (Fig. 1.5). Intermediate position 1 can also be opened into subsidiary main position 1(3), which is a slant regular even edge ring of four regular hexagons. The high curvature (120°) means that this cannot be turned inside out directly but it can be turned inside out indirectly by flexing via intermediate position 3.

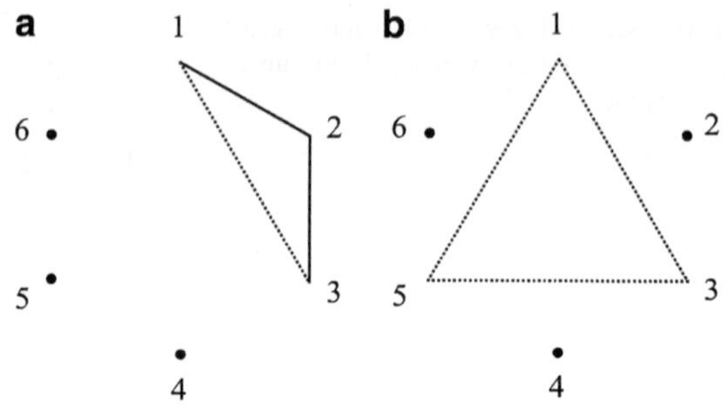

Fig. 8.20 Intermediate position maps. (**a**) The two sector degenerate hexagon even edge flexagon type Q. (**b**) The three sector degenerate hexagon even edge flexagon type S

Fig. 8.21 Twisted-in-between position of the two sector degenerate hexagon even edge flexagon type Q

Fig. 8.22 Net for the three sector degenerate fundamental hexagon even edge flexagon type S. Three copies needed. Fold together pairs of leaves numbered 5

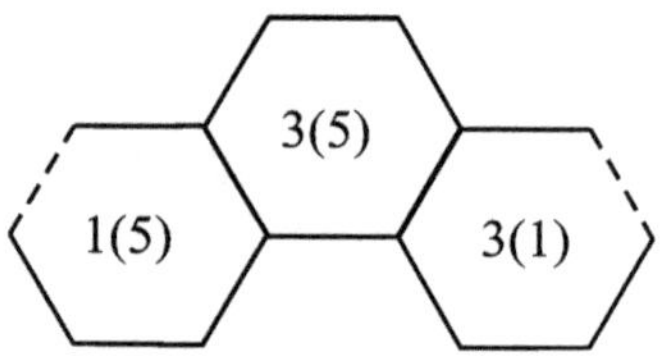

A twist flex can be used to traverse between intermediate positions 1 and 3. Starting from intermediate position 1, hold the pair of leaves numbered 1(3), and pull gently apart, allowing them to twist to reach the twisted-in-between position shown in Fig. 8.21. Continue twisting, allowing the leaves to fold together into intermediate position 3. The continuous movement between intermediate positions 1 and 3 is aesthetically satisfying. Theoretically, this twist flex is possible in some more complicated hexagon even edge flexagons, but in practice the additional degrees of freedom make the twisted-in-between position difficult to find.

The net for the three sector degenerate hexagon even edge flexagons type S is shown in Fig. 8.22. This was derived by using the net for the first order fundamental hexagon even edge flexagon 2⟨6, 6⟩ (Fig. 4.29b) as a precursor and deleting faces 2, 4 and 6. As assembled the flexagon is in subsidiary main position 1(3) which is a flat regular even edge ring of six regular hexagons (Fig. 4.31b). The subsidiary3-cycle shown in the intermediate position map (Fig. 8.20b) can be traversed by using the threefold pinch flex. Intermediate positions are hexagon edge triples.

The net for the three sector degenerate hexagon even edge flexagons type S could also be derived also by using the net for the trihexaflexagon (Fig. 4.14) as a precursor and truncating the equilateral triangles to regular hexagons to produce a truncated flexagon. Retaining the precursor face numbers leads to the net for a truncated trihexaflexagon, which is shown in Fig. 8.23. This is an example of a flexagon that belongs to more than one family of flexagons. As assembled, as a truncated trihexaflexagon it is in principal main position 2(1), which is a flat regular even edge ring of six regular hexagons (Fig. 4.31b). The intermediate position map is the same as that for the trihexaflexagon (Fig. 4.7); the principal 3-cycle shown can be traversed by using the threefold pinch flex.

8.2.5 Degenerate Octagon Even Edge Flexagons

There are numerous two sector degenerate octagon even edge flexagons. Two and four sector versions of one of the simplest possible degenerate octagon even edge flexagons are described in this section. They are basic even edge flexagons, and have been chosen because they are precursors to silver even edge flexagons (Section 10.2.4.1). Some of their properties are given in Table 8.4 (cf. Table 4.7). The torsion per sector is 1. The flexagon figure is shown in Fig. 8.24.

The net for a two sector degenerate octagon even edge flexagon is shown in Fig. 8.25a. This was derived by using the net for the first order fundamental octagon (irregular hexagon) even edge flexagon 2⟨8, 8⟩ (Fig. 4.32) as a precursor, deleting faces 2, 4, 5, 7 and 8 and replacing the irregular hexagons by regular octagons. Other derivations are possible.

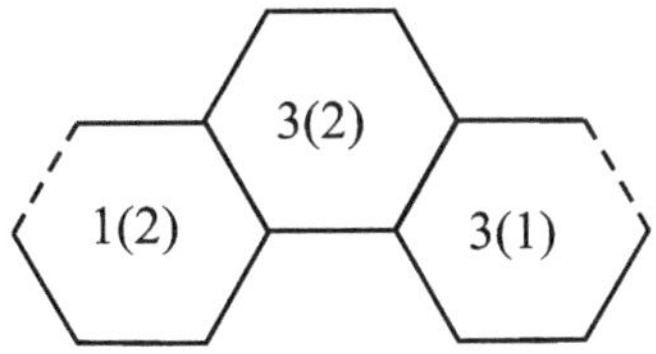

Fig. 8.23 Net for a truncated trihexaflexagon. Three copies needed

Table 8.4 Properties of a degenerate octagon even edge flexagon

Number of sectors	Typical main position	Cycle type	Number of cycles	Main position type	Ring symbol	Curvature
2	1(3)	1/4 cycle	1	Flat	4(90°)	0°
2	1(6)	2/8 cycle	1	Slant	4(45°)	180°
4	1(3)	1/4 cycle	1	Skew	8(90°)	–360°
4	1(6)	2/8 cycle	1	Flat	8(45°)	

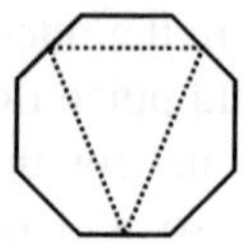

Fig. 8.24 Flexagon figure for degenerate octagon even edge flexagons

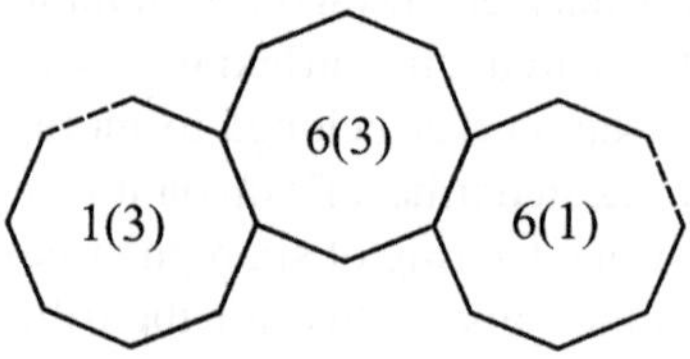

Fig. 8.25 Net for a degenerate octagon even edge flexagon. (a) Two sectors. Two copies needed. Fold together pairs of leaves numbered 6. (b) Four sectors. Four copies needed. Fold together pairs of leaves numbered 3

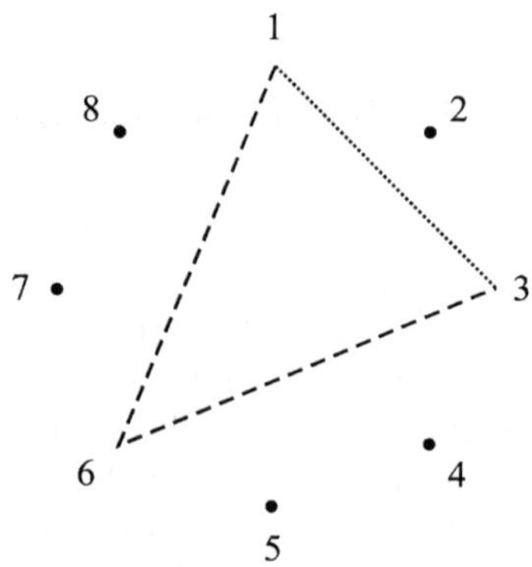

Fig. 8.26 Intermediate position map for degenerate octagon even edge flexagons

As assembled, the flexagon is in subsidiary main position 1(3), which is, in appearance, a flat regular even edge ring of four regular octagons (Fig. 4.33c). This is the only subsidiary main position of the incomplete subsidiary 1/4 cycle shown by the dotted line in the intermediate position map (Fig. 8.26). From here it can be flexed into subsidiary main positions 1(6) and 3(6) by using the twofold pinch flex and hence it can traverse the incomplete subsidiary 2/8-cycle shown by the dashed lines in the intermediate position map. These subsidiary main positions are slant regular even edge rings of four regular octagons Fig. 4.33b. The high curvature (180°) means that they cannot be turned inside out, so intermediate position 6 cannot be visited without disconnecting a hinge, refolding the flexagon, and reconnecting the hinge. All the main positions can be visited so, as assembled, it is not a deficient flexagon. However, if it is assembled by folding together pairs of leaves numbered 1 or 3, with leaves numbered 6 on the outside of the ring, then it becomes a deficient flexagon because subsidiary main position 1(3) cannot be visited. This is an example of a flexagon that is difficult to classify.

The net for a four sector degenerate octagon even edge flexagon is shown in Fig. 8.25b. This was derived by using the net for the first order fundamental octagon (irregular hexagon) even edge flexagon 2⟨8, 8⟩ (Fig. 4.32) as a precursor, increasing the number of sectors from two to four, deleting faces 2, 4, 5, 7 and 8 and replacing the irregular hexagons by regular octagons. Other derivations are possible. As assembled, the flexagon is in subsidiary main position 6(1), which is a flat regular even edge ring of eight regular octagons (Fig. 2.9). The incomplete subsidiary 1/4 cycle (dotted line) and the incomplete subsidiary 2/8-cycle (dashed lines) shown in the intermediate position map (Fig. 8.26) can be traversed by using the fourfold pinch flex. Subsidiary main position 3(6) is a flat regular even edge ring of eight regular octagons (Fig. 2.9), and subsidiary main position 1(3) is a skew regular even edge ring of eight regular octagons. Intermediate positions are octagon edge quadruples. Pocket flexes are theoretically possible but the short hinges make these difficult.

8.2.6 A Degenerate Dodecagon Even Edge Flexagon

There are numerous three sector degenerate dodecagon even edge flexagons; one of the simplest possible is described in is this section. It a basic even edge flexagon, and has been chosen because it is a precursor to a bronze even edge flexagon (Section 10.2.6.1). Some of its properties are given in Table 8.5 (cf. Table 4.8). The torsion per sector is 1. The flexagon figure is shown in Fig. 8.27. Two and six sector versions are precursors to other bronze even edge flexagons.

The net for a three sector degenerate dodecagon even edge flexagon is shown in Fig. 8.28. This was derived by using the net for the first order fundamental dodecagon even edge flexagon 2⟨12, 12/5⟩ (Fig. 4.38) as a precursor, increasing the number of sectors from two to three and deleting faces 2, 3, and 5–11. Other derivations are possible. The table and the intermediate position map (Fig. 8.29) show that there are three main positions which belong to three different incomplete cycles. These are an incomplete principal 1/8-cycle (solid line), an incomplete subsidiary 1/4-cycle (dotted line), and an incomplete subsidiary 1/3-cycle (dashed line).

Table 8.5 Properties of a three sector degenerate dodecagon flexagon. The incomplete principal cycle is in bold

Main position	Cycle type	Number of cycles	Main position type	Ring symbol	Curvature
1(4)	1/4-cycle	1	Skew	6(90°)	−180°
4(12)	1/3-cycle	1	Flat	6(60°)	0°
12(1)	**1/12-cycle**	**1**	**Slant**	**6(30°)**	**180°**

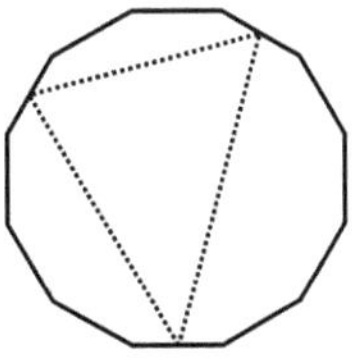

Fig. 8.27 Flexagon figure for a three sector degenerate dodecagon even edge flexagon

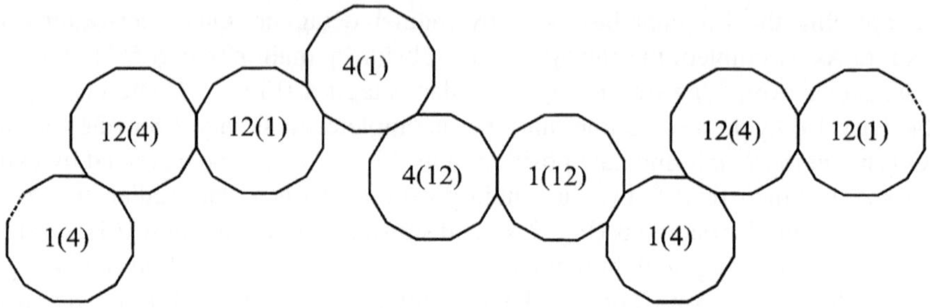

Fig. 8.28 Net for a three sector degenerate dodecagon even edge flexagon. One copy needed. Fold together pairs of leaves numbered 1

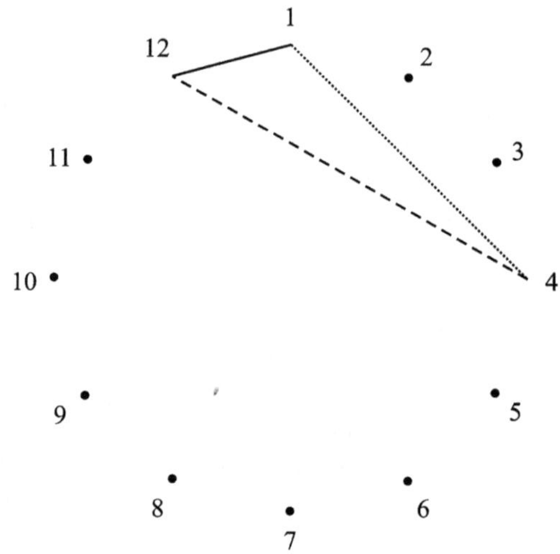

Fig. 8.29 Intermediate position map for a three sector degenerate dodecagon even edge flexagon

As assembled, the flexagon is in subsidiary main position 4(12), which is, in appearance, a flat regular even edge ring of six regular dodecagons (Fig. 8.30). From here, by using the threefold pinch flex, it can be flexed into subsidiary main position 1(4), which is a is a skew regular even edge ring of six regular dodecagons (Fig. 8.31a) or into principal main position 12(1). This is a slant regular even edge ring of six regular dodecagons (Fig. 8.31b) that, because of the high curvature (180°), cannot be turned inside out directly. It can be turned inside out indirectly by flexing via subsidiary main position 1(4). A snap flex is needed but leaves do not have to be bent. Intermediate positions are dodecagon edge triples.

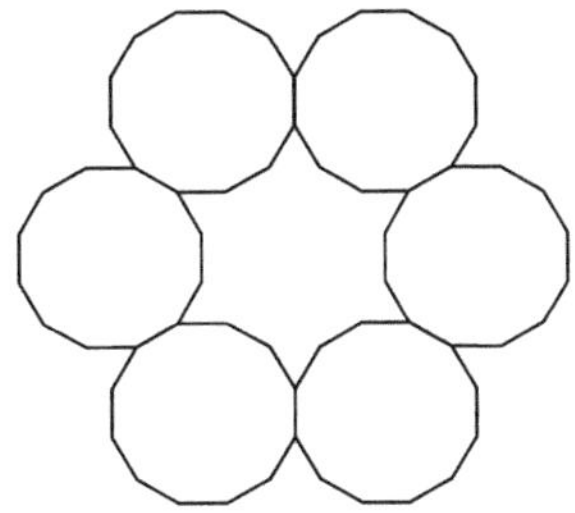

Fig. 8.30 Flat regular even edge ring of six regular dodecagons

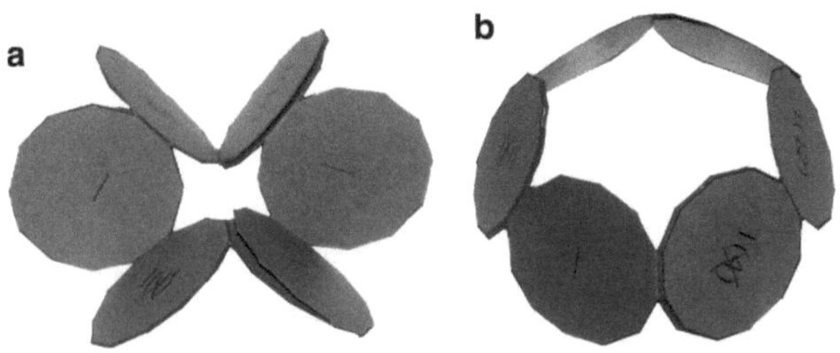

Fig. 8.31 A flexagon as even edge rings of six regular dodecagons. (**a**) Skew. (**b**) Slant

8.3 Degenerate Non Interleaved Point Flexagons

8.3.1 General Properties

Irregular cycle non interleaved point flexagons correspond to irregular cycle even edge flexagons, and the irregular polygons shown in Figs. 7.1–7.3 are possible associated polygons for irregular cycle non interleaved point flexagons (Section 7.5.1). Similarly, degenerate non interleaved point flexagons, which are solitary flexagons, correspond to degenerate even edge flexagons, and the degenerate irregular polygons shown in Figs. 8.1–8.5 are possible associated polygons for degenerate non interleaved point flexagons.

All the possible flexagon figures for the degenerate non interleaved square point flexagon, degenerate non interleaved pentagon point flexagons, and degenerate non interleaved hexagon point flexagons are shown in Figs. 8.32–8.34 (cf. Figs. 8.6–8.10). These were derived by using irregular polygons shown in Figs. 8.1–8.5 as the inscribed polygons of flexagon figures. To avoid confusion, circumscribing polygons are shown as dots representing their vertices (Section 5.2.1).

For a flexagon figure for a degenerate non interleaved point flexagon to be valid, the vertices of the inscribed polygon must coincide with a group of adjacent vertices of the circumscribing polygon. This condition is needed to avoid deficient flexagons with faces that cannot be visited without disconnecting a hinge, refolding the flexagon, and reconnecting the hinge. The condition means that there is a degenerate

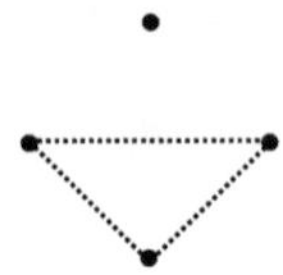

Fig. 8.32 Flexagon figure for the degenerate non interleaved square point flexagon

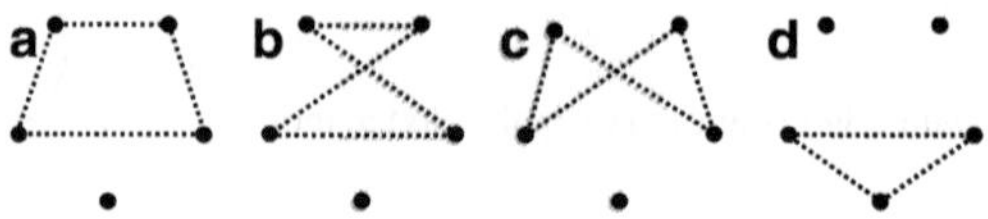

Fig. 8.33 Flexagon figures for degenerate non interleaved pentagon point flexagons. (a) Type A. (b) Type B. (c) Type C. (d) Type D

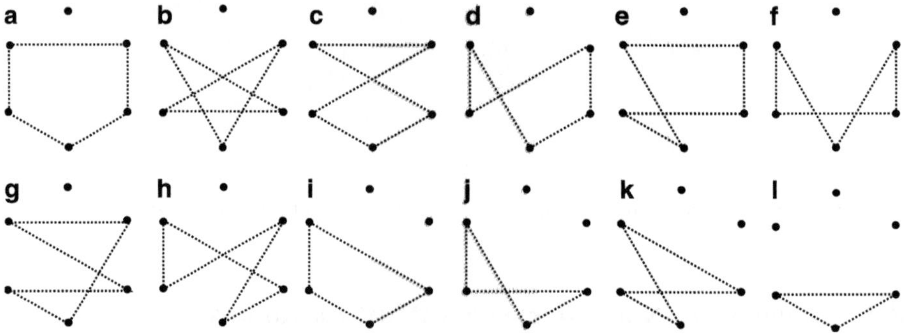

Fig. 8.34 Flexagon figures for degenerate non interleaved hexagon point flexagons. (a) Type A. (b) Type B. (c) Type C. (d) Type D. (e) Type E. (f) Type F. (g) Type G. (h) Type H. (i) Type I. (j) Type L. (k) Type M. (l) Type Q

even edge flexagon that corresponds to every degenerate non interleaved point flexagon, but the converse is not true. For example, there is no degenerate non interleaved pentagon point flexagon that corresponds to the degenerate pentagon even edge flexagons type E (Fig. 8.7e) and what would be a corresponding flexagon figure is absent from Fig. 8.33. Similarly, flexagon figures for types J, K, N, O, P, R and S are absent from Fig. 8.34. There are no degenerate non interleaved triangle point flexagons, one degenerate non interleaved square point flexagon, four degenerate non interleaved pentagon point flexagons and 12 degenerate non interleaved hexagon point flexagons. Type letters in Figs. 8.33 and 8.34 are the same as in Figs. 8.7–8.10 and are used in descriptions of corresponding flexagons.

Nets for degenerate non interleaved point flexagons described below were derived by deleting faces from appropriate precursors. Some deletions result in structures which are not flexagons, and an example is given. In intermediate positions, a degenerate non interleaved point flexagon is the dual of the corresponding degenerate unagon (Section 8.2.1), so nets are duals of nets for degenerate unagons and can be constructed as described in Section 7.5.1. Theoretically, nets could be derived from flexagon figures by using the procedure given in Section 7.2.1, with face numbering sequences derived by the procedure given in Section 5.4.1, but this is not straightforward.

The simplest possible flexagon of each variety of non interleaved point flexagon is a basic point flexagon (cf. Section 8.2.1). A basic point flexagon is a solitary point flexagon that consists of three leaves. There are always three main positions that can be visited, and there are always three intermediate positions on an intermediate position map. The fundamental triangle point flexagon 1⟨3, 3⟩ (Section 5.3.2) is a basic point flexagon. The lines on its intermediate position map, which is the same as that for the first order fundamental triangle even edge flexagons *S*⟨3, 3⟩ (Fig. 4.7) are an equilateral triangle. All other basic point flexagons are degenerate flexagons and the lines on their intermediate position maps are two edges of irregular triangles.

8.3.2 The Degenerate Non Interleaved Square Point Flexagon

There is one degenerate non interleaved square point flexagon. It is a basic point flexagon. Some of its properties are given in Table 8.6 (cf. Tables 4.4 and 8.1). The torsion is 1. The flexagon figure is shown in Fig. 8.32.

The net for the degenerate non interleaved square point flexagon is shown in Fig. 8.35. This was derived by using the net for the fundamental square point flexagon 1⟨4, 4⟩ (Fig. 5.4b) as a precursor and deleting face 4. Except for face numbering, nets obtained by deleting any of the other three faces are identical. The net for the irregular cycle interleaved square point flexagon (Fig. 7.17) is an alternative precursor. Different results are obtained from this precursor depending on which face is deleted. Deleting face 2 results in one enantiomorph, whereas deleting face 4 results in the other enantiomorph. Deleting either face 1 or face 3 results in a straight strip of

Table 8.6 Properties of the degenerate non interleaved square point flexagon. The incomplete principal cycle is in bold

Typical main position	Cycle type	Number of cycles
1(2)	**2/4-cycle**	1

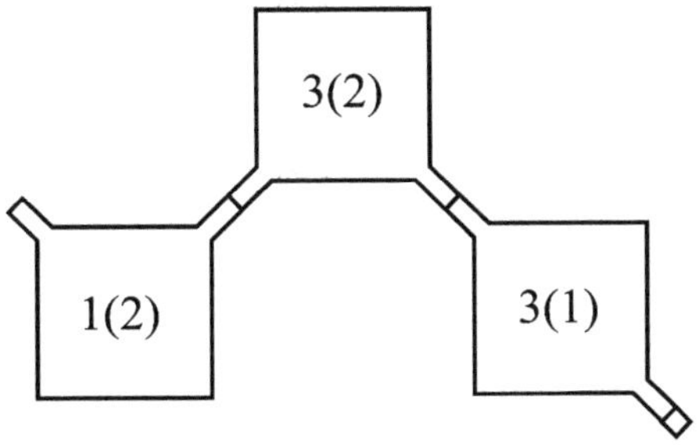

Fig. 8.35 Net for the degenerate non interleaved square point flexagon. One copy needed. Fold until leaves numbered 1 are visible

three squares linked by pairs of point hinges, which is not a flexagon. As assembled, the flexagon is in intermediate position 1, which is, in appearance, a single square. The incomplete principal 2/4-cycle shown in the intermediate position map (Fig. 8.36) can be traversed by using the simple flex. In principal main positions the flexagon is a square vertex pair linked by a pair of point hinges (Fig. 3.23).

8.3.3 *Degenerate Non Interleaved Pentagon Point Flexagons*

There are four degenerate non interleaved pentagon point flexagons, types A–D. Type D is the simplest possible degenerate non interleaved pentagon point flexagon, and is a basic point flexagon. Types A and D are described in this section. Some of their properties are given in Table 8.7 (cf. Tables 4.5 and 8.2). The torsion is 2 for type A and 1 for type D. Flexagon figures are shown in Fig. 8.33a and d.

The net for the degenerate non interleaved pentagon point flexagon type A is shown in Fig. 8.37. This was derived by using the net for the fundamental pentagon point flexagon 1⟨5, 5⟩ (Fig. 5.6) as a precursor and deleting face 5. Other derivations are possible. As assembled, the flexagon is in intermediate position 1, which is, in appearance, a single regular pentagon. The incomplete principal 3/5-cycle

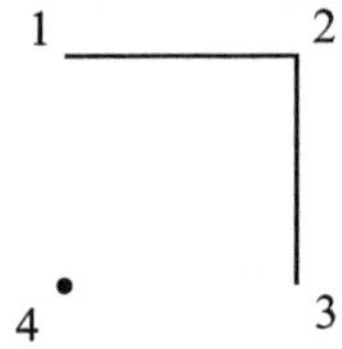

Fig. 8.36 Intermediate position map for the degenerate non interleaved square point flexagon

Table 8.7 Properties of the degenerate non interleaved pentagon point flexagons types A and D. Incomplete principal cycles are in bold

Type	Typical main position	Cycle type	Number of cycles
A	**1(2)**	**3/5-cycle**	**1**
D	**1(2)**	**2/5-cycle**	**1**

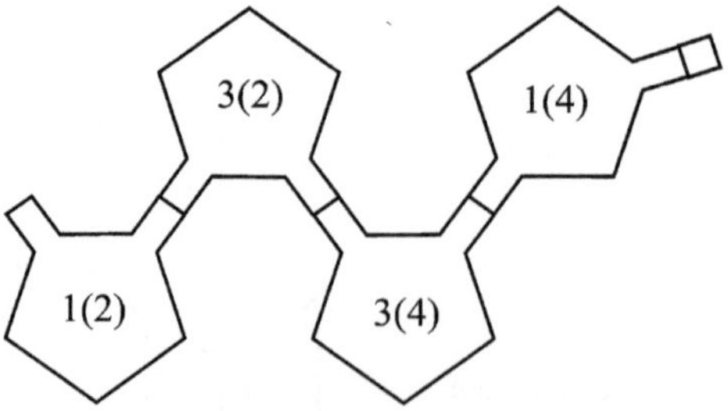

Fig. 8.37 Net for the degenerate non interleaved pentagon point flexagon type A. One copy needed. Fold until leaves numbered 1 are visible

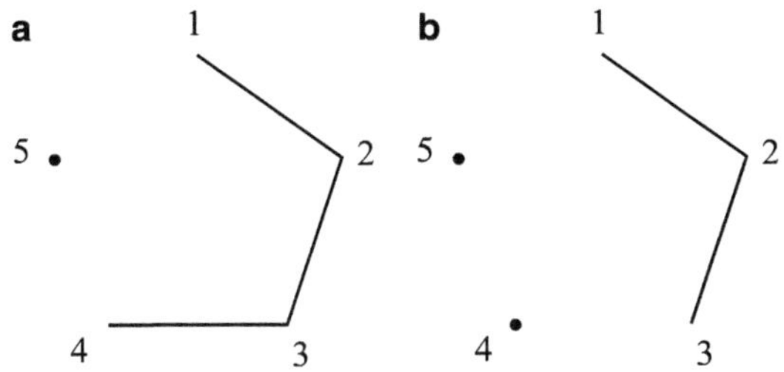

Fig. 8.38 Intermediate position maps for degenerate non interleaved pentagon point flexagons. (**a**) Type A. (**b**) Type D

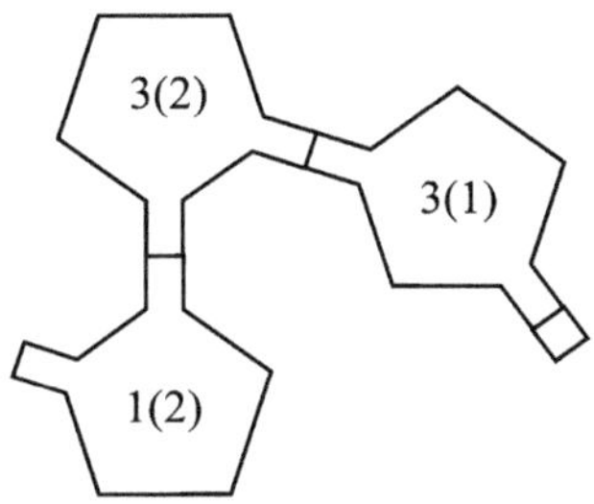

Fig. 8.39 Net for the degenerate non interleaved pentagon point flexagon, type D. One copy needed. Fold until leaves numbered 1 are visible

shown in the intermediate position map (Fig. 8.38a) can be traversed by using the simple flex. Principal main positions are regular pentagon vertex pairs linked by pairs of point hinges.

The net for the degenerate non interleaved pentagon point flexagon type D is shown in Fig. 8.39. This was derived by using the net for type A (Fig. 8.37) as a precursor and deleting face 4. As assembled, the flexagon is in intermediate position 1. The incomplete principal 2/5-cycle shown in the intermediate position map (Fig. 8.38b) can be traversed by using the simple flex. Intermediate positions and principal main positions have the same appearances as those of type A.

8.4 Degenerate Irregular Cycle Interleaved Point Flexagons

8.4.1 General Properties

Irregular cycle interleaved point flexagons (Section 7.3.1) are solitary flexagons, so degenerate versions are also solitary flexagons. A non interleaved point flexagon (Section 7.5.1) is always a simple band, that is the band is a single loop of polygons, and degenerate point flexagons derived by deletion of one or more faces are also simple bands. An irregular cycle interleaved point flexagon is always a simple band. However, there are three possibilities if a degenerate point flexagon is derived by deletion of a face from an interleaved point flexagon. In the first possibility, the two leaves to be glued together to delete a face (Section 8.1) are adjacent in the net and the degenerate interleaved point flexagon derived is a simple band. The associated polygon of the precursor interleaved point flexagon, for example Fig. 7.3a becomes an associated polygon with one fewer vertices, for example Fig. 8.3h.

In the second possibility, the two leaves to be glued together are separated by two or more leaves in the net. The resulting degenerate interleaved point flexagon is a figure-of-eight band, and can also be called a double point flexagon. It is an example of a flexagon that belongs to more than one family of flexagons. A simple definition of the torsion is the algebraic sum of the torsion for the two conjoined bands. For a full definition of the torsion of the flexagon, the torsion of each of the two conjoined bands has to be defined individually. The associated polygon of the precursor flexagon becomes an associated double polygon. This consists of two polygons with a common vertex, for example Fig. 8.40. The common vertex is indicated by a dot.

In the third possibility, the two leaves to be glued together are separated by one leaf in the net. The resulting degenerate point flexagon is a simple band with a single leaf attached by a pair of point hinges, and can also be called a semi double point flexagon. This single leaf is ignored in the calculation of torsion. The single leaf appears as a loose flap during flexing. Whether such loose flaps are acceptable is a matter of taste. The associated polygon of the precursor flexagon becomes an associated semi double polygon. This consists of a polygon with a line segment attached to a vertex, for example Fig. 8.41.

Enumeration of possible degenerate point flexagons that are not simple bands is difficult because this requires enumeration of possible double polygons and possible semi double polygons, which is not straightforward. Some possibilities are illustrated by the simple examples described below. Deleting two or more faces can result in more complicated bands.

8.4.2 Degenerate Interleaved Triangle Point Flexagons

Three of the degenerate interleaved triangle point flexagons, which can be derived by deleting faces from the net for the augmented irregular cycle triangle point flexagon type A (Fig. 7.28), are types AA–AC. Type AD is non interleaved, and is a

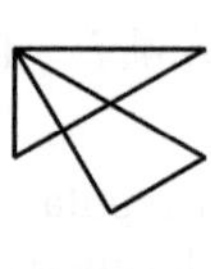

Fig. 8.40 Associated double polygon for the degenerate interleaved triangle point flexagon type AB. It has five of the vertices of a regular hexagon

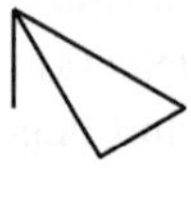

Fig. 8.41 Associated semi double polygon for the degenerate interleaved triangle point flexagon type AC. It has four of the vertices of a regular hexagon

basic point flexagon. Some of their properties are given in Table 8.8 (cf. Tables 5.6 and 8.3). The torsion is 3 for type AA, 2 for type AB and 1 for types AC and AD. Intermediate positions are, in appearance, single equilateral triangles and main positions are equilateral triangle vertex pairs connected by pairs of point hinges (Fig. 1.8a).

The net for the degenerate interleaved triangle point flexagon type AA is shown in Fig. 8.42. This was derived by using the net for the augmented irregular cycle triangle point flexagon type A (Fig. 7.28) as a precursor and deleting face 6. The flexagon is a simple band of equilateral triangles. The associated polygon is the pentagon that has five of the vertices of a regular hexagon type H shown in Fig. 8.3h. Deleting either face 2 or face 4 leads to the same result, apart from face numbering. As assembled, the flexagon is in intermediate position 1. The incomplete principal 4/6 cycle shown in the intermediate position map (Fig. 8.43a) can be traversed by using the simple flex. A subsidiary 3-cycle and an incomplete subsidiary 1/3-cycle can also be traversed by using the simple flex.

Table 8.8 Properties of degenerate interleaved triangle point flexagons types AA–AC, and the degenerate interleaved triangle point flexagon AD. Incomplete principal cycles are in bold

Type	Typical main position	Cycle type	Number of cycles
AA	**1(2)**	**4/6-cycle**	**1**
AA	1(3)	3-cycle	1
AA	2(4)	1/3-cycle	1
AB	**1(2)**	**4/6-cycle**	**1**
AB	1(3)	3-cycle	1
AB	2(4)	1/3-cycle	1
AC	**1(2)**	**3/6-cycle**	**1**
AC	1(3)	1/3	2
AD	**1(2)**	**2/6-cycle**	**1**
AD	1(3)	1/3-cycle	1

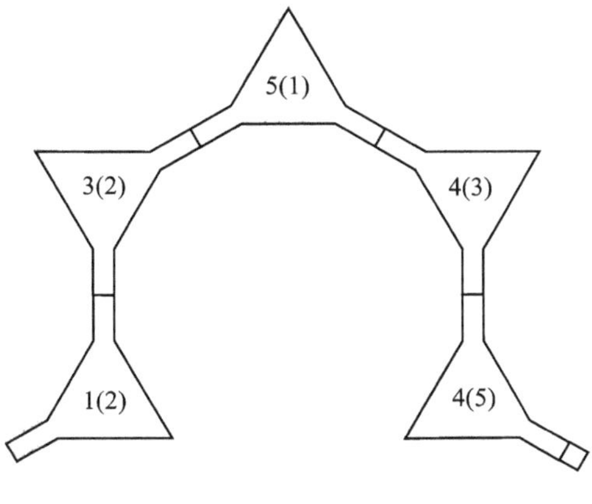

Fig. 8.42 Net for the degenerate interleaved triangle point flexagon type AA. One copy needed. Interleave during assembly, nesting hinges, with leaves numbered 1 left visible

The net for type AB is shown in Fig. 8.44. This was derived by using the net for the augmented irregular cycle triangle point flexagon type A as a precursor, deleting face 5 and renumbering so that the intermediate position map is the same as that for type AA (Fig. 8.43a). The flexagon is a figure-of-eight band of equilateral triangles. The torsion per loop is 1 so the total torsion is 2. The associated double polygon is shown in Fig. 8.40. Deleting either face 1 or face 3 leads, after renumbering, to the same result. As assembled, the flexagon is in intermediate position 1. Apart from pat structure, the dynamic properties are the same as those of type AA, and the intermediate position map can be traversed by using the simple flex.

The net for type AC is shown in Fig. 8.45. This was derived by using the net for type AA as a precursor and deleting face 5. Most of the flexagon is a band of three

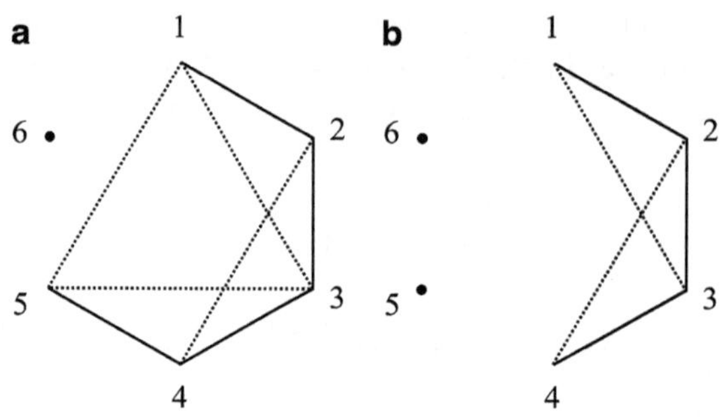

Fig. 8.43 Intermediate position maps for degenerate interleaved triangle point flexagons. (**a**) Types AA and AB. (**b**) Type AC

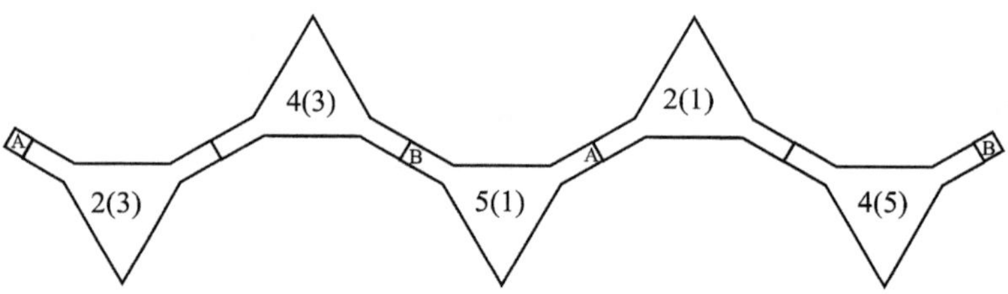

Fig. 8.44 Net for the degenerate interleaved triangle point flexagon type AB. One copy needed. Interleave during assembly, nesting hinges, with leaves numbered 1 left visible. Join at A-A and B-B

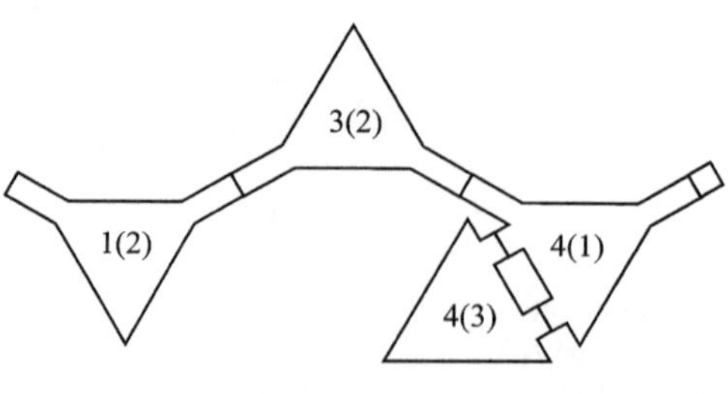

Fig. 8.45 Net for the degenerate interleaved triangle point flexagon type AC. One copy needed.. Fold until leaves numbered 1 are visible

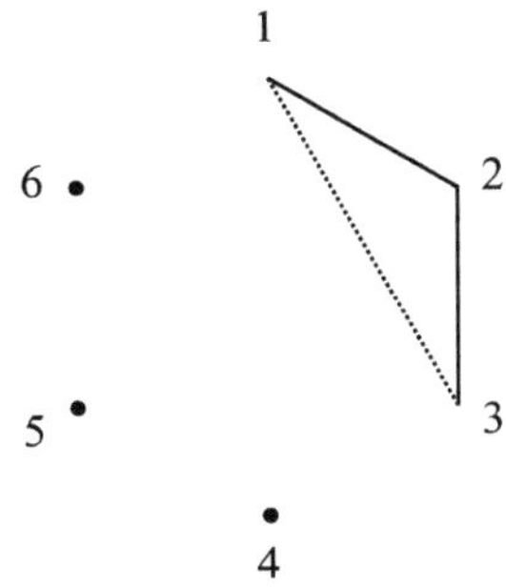

Fig. 8.46 Intermediate position map for the degenerate triangle point flexagons type AD

equilateral triangles. A fourth equilateral triangle is attached, as a loose flap, by a pair of point hinges, so the flexagon is a semi double point flexagon. The associated semi double polygon is shown in Fig. 8.41. The net can also be derived by using the net for the augmented irregular cycle triangle point flexagon type A as a precursor and deleting any two faces. As assembled, the flexagon is in intermediate position 1. Despite the loose flap, the incomplete principal 4/6-cycle shown in the intermediate position map (Fig. 8.43b) can be traversed by using the simple flex. Two incomplete subsidiary 1/3-cycles can also be traversed by using the simple flex.

The net for the degenerate triangle point flexagon type AD is shown in Fig. 5.2d. This was derived by using the net for type AC as a precursor and deleting face 4. The associated polygon is the triangle that has three of the vertices of a regular hexagon type R shown in Fig. 8.5b. Deleting any of the other three faces results in a strip of three triangles linked by pairs of point hinges. This is not a flexagon. The incomplete principal 2/6-cycle and the incomplete subsidiary 1/3-cycle, shown in the intermediate position map (Fig. 8.46) can be visited by using the simple flex. The flexagon is identical to the fundamental triangle point flexagon 1⟨3, 3⟩ (Section 5.3.2). It is an example of a flexagon that belongs to more than one family of flexagons.

8.5 Degenerate Compound Edge Flexagons

8.5.1 General Properties

Fundamental compound edge flexagons are solitary flexagons, so degenerate compound edge flexagons are also solitary flexagons. Flexagon figures, as described in Section 4.2.1, are not appropriate for the characterisation of fundamental compound edge flexagons This is because, in attempted flexagon figures, the inscribed part is either a line which is not a complete polygon, or is an overlapping line, as illustrated in Fig. 8.47 for the fundamental hexagon compound edge flexagons 3⟨6, 6, 4⟩, 2⟨6, 6, 5⟩, 2⟨6, 6, 7⟩ and 3⟨6, 6, 8⟩ (Section 6.2).

The equivalent flexagon figure for a fundamental compound edge flexagon is the same as the flexagon figure for the first order fundamental even edge flexagons

(Section 4.2.1) that have the same associated polygon. Thus, the equivalent flexagon figure for the square-like fundamental hexagon compound edge flexagon 3⟨6, 6, 4⟩ (Section 6.4.2) is the same as the flexagon figure for the first order fundamental square even edge flexagons *S*⟨4, 4⟩ (Fig. 4.2), and is shown in Fig. 8.48a. Other parts of Fig. 8.48 show equivalent flexagon figures for the fundamental hexagon compound edge flexagons 2⟨6, 6, 5⟩, 2⟨6, 6, 7⟩ and 3⟨6, 6, 8⟩. Similarly, equivalent flexagon figures for degenerate compound edge flexagons are the same as flexagon figures for degenerate even edge flexagons (Section 8.2.1). Equivalent flexagon figures cannot be used to derive nets as described for flexagon figures in Section 7.2.2.

Possible degenerate compound edge flexagons that may be derived by deletion of faces from a particular fundamental compound edge flexagon can be enumerated by considering possible equivalent flexagon figures. By definition, the associated polygon for a fundamental compound edge flexagon is a convex polygon, so associated polygons for degenerate compound edge flexagons are also convex polygons. Hence, the inscribed portions of the corresponding equivalent flexagon figures are convex polygons. In addition, for an equivalent flexagon figure for a degenerate compound edge flexagon to be valid the vertices of the inscribed polygon must coincide with a group of adjacent vertices of the circumscribing polygon. This condition is needed to avoid deficient flexagons with faces that cannot be visited without disconnecting a hinge, refolding the flexagon, and reconnecting the hinge. Therefore, from Figs. 8.6–8.10, there is one type of degenerate compound edge flexagon if the associated polygon for the precursor flexagon is a square. There are two types if it is a regular pentagon, and three types if it is a regular hexagon. Some examples are given below.

The simplest possible flexagon of each variety of compound edge flexagon is a basic compound edge flexagon. A basic compound edge flexagon is always a solitary

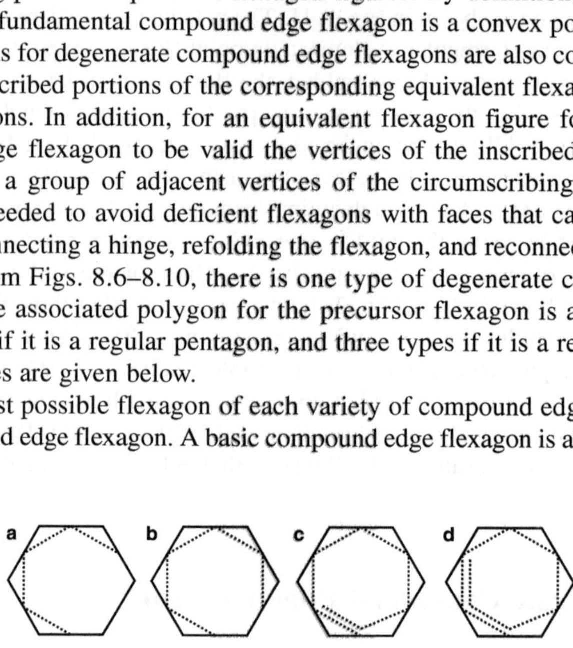

Fig. 8.47 Attempted flexagon figures for fundamental hexagon compound edge flexagons. (**a**) 3⟨6, 6, 4⟩. (**b**) 2⟨6, 6, 5⟩. (**c**) 2⟨6, 6, 7⟩. (**d**) 3⟨6, 6, 8⟩

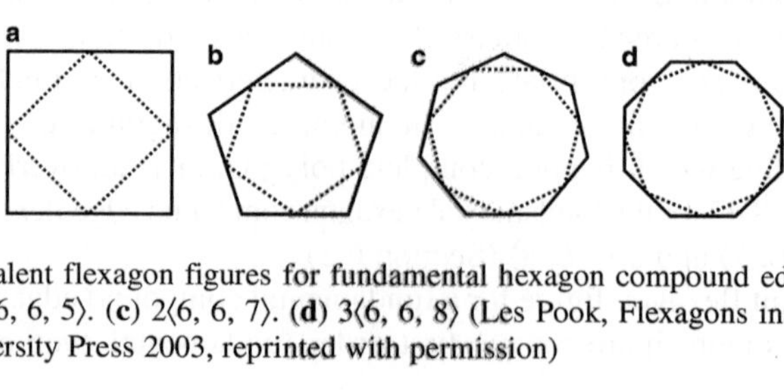

Fig. 8.48 Equivalent flexagon figures for fundamental hexagon compound edge flexagons. (**a**) 3⟨6, 6, 4⟩. (**b**) 2⟨6, 6, 5⟩. (**c**) 2⟨6, 6, 7⟩. (**d**) 3⟨6, 6, 8⟩ (Les Pook, Flexagons inside out, 2003, © Cambridge University Press 2003, reprinted with permission)

even edge flexagon that has three leaves per sector. There are always three intermediate positions on an intermediate position map. Triangular compound edge flexagons (Section 6.3.1) are basic compound edge flexagons. The lines on their intermediate position map are the edges of an equilateral triangle. All other basic compound edge flexagons are degenerate flexagons, and the lines on their intermediate position maps are either the edges of irregular triangles or two of the edges of an irregular triangle.

8.5.2 *The Degenerate Square-Like Hexagon Compound Edge Flexagon*

There is one three sector degenerate square-like hexagon compound edge flexagon. It is one of the simplest possible compound edge flexagons, and it is a basic compound edge flexagon. Some of its properties are given in Table 8.9 (cf. Table 6.4). The torsion per sector is 1. The equivalent flexagon figure is the same as the flexagon figure for a degenerate square even edge flexagon shown in Fig. 8.6.

The net for the three sector degenerate square-like hexagon compound edge flexagon is shown in Fig. 8.49. This was derived by using the net for the fundamental hexagon compound edge flexagon 3⟨6, 6, 4⟩ (Fig. 6.8) as a precursor and deleting face 4. As assembled, the flexagon is in principal main position 2(1). This is, in appearance, a flat compound edge ring of six regular hexagons (Fig. 6.9a). The incomplete principal 2/4 cycle shown in the intermediate position map can be traversed by using a twist flex. The intermediate position map is the same as that

Table 8.9 Properties of the degenerate square-like hexagon compound edge flexagon. The principal cycle is in bold

Typical main position	Cycle type	Number of cycles	Main position type	Ring symbol	Curvature
1(2)	**2/4-cycle**	**1**	**Flat**	**3(0°, 120°).**	**0°**
1(3)	None	–	Antibox	3(−60°, 60°).	360°

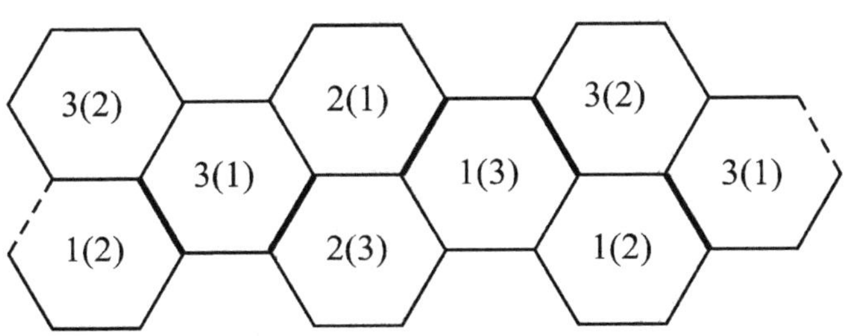

Fig. 8.49 Net for the three sector degenerate square-like hexagon compound edge flexagon. One copy needed

for the two sector degenerate square even edge flexagon (Fig. 8.12). This is a basic compound edge flexagon in which the lines on the intermediate positions map are the edges of an irregular triangle.

Starting from principal main position 2(1), to flex to principal main position 3(2) hold leaves numbered 2(3) and pull them gently outwards. The flexagon can also be flexed into subsidiary main position 1(3), which is an antibox edge ring of six regular hexagons (Fig. 2.13), by using another type of twist flex. Starting from principal main position 2(1), work round the flexagon by folding leaves numbered 2 together in pairs to reach subsidiary main position 1(3), which cannot be turned inside out directly because of the high curvature (360°). This can be done indirectly by starting from principal main position 3(2) to reach subsidiary main position 3(1). Each part of this type of twist flex is a local flex because some pats are left unchanged.

8.5.3 Degenerate Pentagonal Compound Edge Flexagons

There are four pentagonal compound edge flexagons with flat main positions listed in Table 6.2. Two of these are the pentagonal fundamental square compound edge flexagon 4⟨4, 4, 5⟩ and the pentagonal fundamental hexagon compound edge flexagon 2⟨6, 6, 5⟩. There are two degenerate versions of each of these, types A and D. The equivalent flexagon figures are the same as the flexagon figures for degenerate pentagon even edge flexagons types A and D respectively. These are shown in Fig. 8.7a and d. Types D are basic compound edge flexagons. Some of the properties of the four sector degenerate pentagonal square compound edge flexagon type D, and the two sector degenerate pentagonal hexagon compound edge flexagon type A, are given in Table 8.10 (cf. Table 6.5). The torsion per sector is 1 for type D and 2 for type A.

8.5.3.1 A Degenerate Pentagonal Square Compound Edge Flexagon

The net for the four sector degenerate pentagonal square compound edge flexagon type D is shown in Fig. 8.50. This was derived by using the net for the pentagonal fundamental square compound edge flexagon 4⟨4, 4, 5⟩ (Fig. 6.10) as a precursor and deleting faces 2 and 4. As assembled, the flexagon is in subsidiary main position

Table 8.10 Properties of pentagonal compound edge flexagons. The principal cycle is in bold

Leaf type	Flexagon type	Typical main position	Cycle type	Number of cycles	Main position type	Ring symbol	Curvature
Square	D	1(3)	2(5)	1	Flat	4(0°, 90°)	0°
Hexagon	**A**	**1(2)**	**3/5-cycle**	**1**	**Flat**	**2(60°, 120°)**	**0°**
Hexagon	A	1(3)	3/5-cycle	1	Slant	2(0°, 60°)	240°

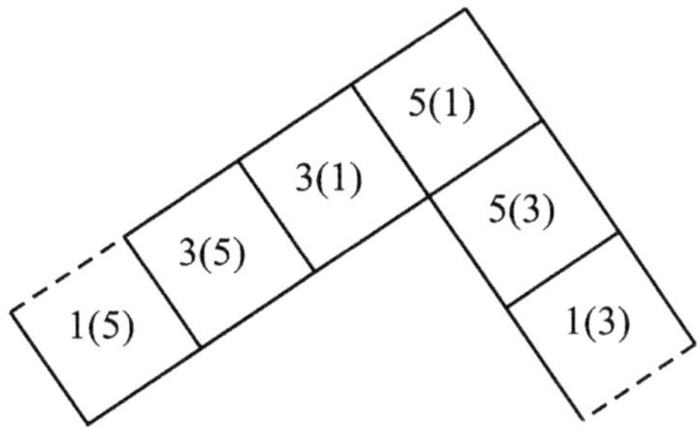

Fig. 8.50 Net for the four sector degenerate pentagonal square compound edge flexagon type D. Two copies needed. Fold together pairs of leaves numbered 5

Fig. 8.51 Intermediate position map for the four sector degenerate pentagonal square compound edge flexagon type D

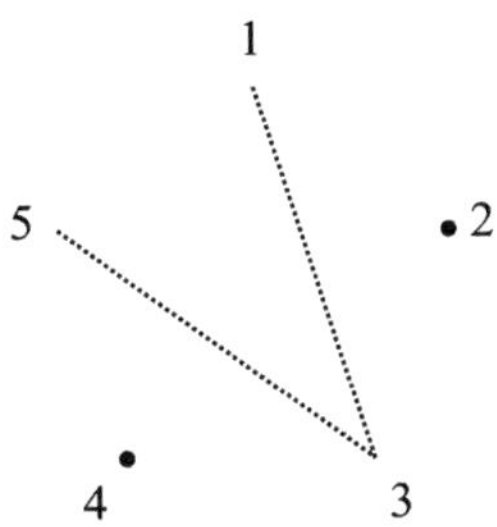

1(3) which is, in appearance, a flat compound edge ring of eight squares (Fig. 1.11). The incomplete subsidiary 2/5-cycle shown in the intermediate position map (Fig. 8.51) can be traversed by using a twist flex. Starting at subsidiary main position 1(3), open the pats at diagonally opposite corners at leaves numbered 5, and fold the pairs of leaves numbered 1 together to reach intermediate position 3. This is clearly defined, and is an irregular box edge ring of eight squares (Fig. 6.11), with leaves numbered 3 on the outside of the ring. To complete the twist flex to subsidiary main position 3(5), open the pats that were at the other pair of corners at leaves numbered 5, and fold the pairs of leaves numbered 1 together.

This is a basic compound edge flexagon for which the lines on the intermediate positions map are two of the edges of an irregular triangle. Principal main position 5(1) is absent because he principal 5-cycle, and corresponding main positions are absent from the precursor flexagon (Section 6.5.2).

8.5.3.2 A Degenerate Pentagonal Hexagon Compound Edge Flexagon

The net for the two sector degenerate pentagonal hexagon compound edge flexagon type A is shown in Fig. 8.52. This was derived by using the net for the fundamental compound hexagon flexagon 2⟨6, 6, 5⟩ (Fig. 6.12) as a precursor and deleting face 5. Other derivations are possible. As assembled the flexagon is in principal main position 1(2). This is, in appearance, a flat compound ring of four regular hexagons

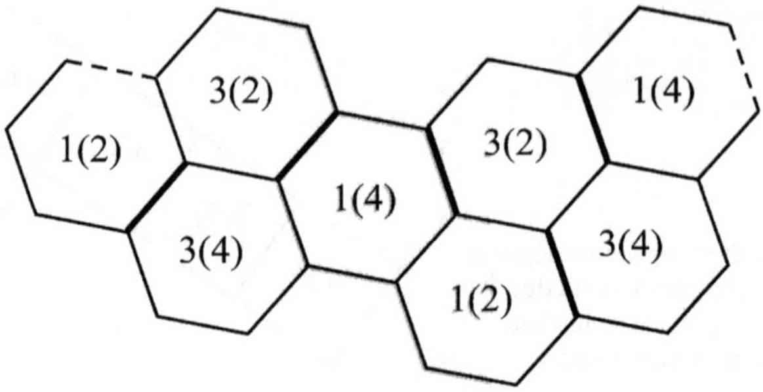

Fig. 8.52 Net for the two sector degenerate pentagonal hexagon compound edge flexagon type A. One copy needed

(Fig. 6.9b). The incomplete principal 3/5-cycle shown in the intermediate position map can be traversed by using the twofold pinch flex. The intermediate position map is the same as that for the two sector degenerate pentagon even edge flexagon type A (Fig. 8.14). A twist flex is easier, as follows. Starting from principal main position 1(2), hold the pairs of leaves numbered 3(2). Then turn them over and twist so that the two pairs of leaves numbered 1 fold together to reach principal main position 4(2). The high curvature (240°) means that the incomplete subsidiary 3/5-cycle shown in the intermediate position map cannot be traversed directly, but all three subsidiary main positions can be reached from principal main positions. For example, to flex to subsidiary main position 1(3) from principal main position 2(1) fold together pairs of leaves numbered 2 in another type of twist flex. Subsidiary main positions are slant compound edge rings of four regular hexagons (Fig. 2.14).

Chapter 9
Irregular Ring Even Edge Flexagons

9.1 Introduction

Irregular ring even edge flexagons are even edge flexagons with main positions that are, in appearance, irregular even edge rings of regular convex polygons (Section 2.2.5). An even edge ring consists of an even number of edge hinged polygons (Section 1.1). An irregular even edge ring is an even edge ring that is neither regular (Section 2.2.2) nor compound (Section 2.2.4). Identical sectors in irregular even edge rings contain at least three polygons. For example, the flat irregular even edge ring of six regular pentagons (Fig. 2.17) has two sectors with three pentagons in each. The flat irregular ring of eight squares type A (Fig. 9.1a) has two sectors with four squares in each, but type B (Fig. 9.1b) cannot be divided so it has one sector containing eight squares. Descriptions in this chapter are largely restricted to flat non overlapping irregular even edge rings and corresponding flexagons.

Fundamental irregular ring even edge flexagons are made from first order fundamental edge nets (Section 3.2), and a standard face numbering sequence is used (Section 4.1.1). They are regular cycle flexagons in which all the main positions of a cycle have the same appearance and the same pat structure. Fundamental irregular ring even edge flexagons can be regarded as generalisations of fundamental compound edge flexagons, so they are solitary flexagons, and the topological invariants are the same (Sections 4.2.1 and 6.2). Various flexes are used but there is no characteristic flex that can be used for all irregular ring even edge flexagons. Most of the fundamental irregular ring even edge flexagons described in this chapter have a cycle in which main positions are flat non overlapping irregular even edge rings. There are numerous possible fundamental irregular ring even edge flexagons, most of which are difficult to handle, so only a selection is described. Most of these are reasonably easy to handle. Degenerate versions are usually easier to handle and examples are included.

9.1.1 Fundamental Irregular Ring Even Edge Flexagons

Possible fundamental irregular ring even edge flexagons can be enumerated by fan folding an appropriate first order fundamental edge net (Section 3.2) onto a sector

L. Pook, *Serious Fun with Flexagons*, Solid Mechanics and Its Applications 164,
DOI 10.1007/978-90-481-2503-6_9, © Springer Science+Business Media B.V. 2009

of an irregular even edge ring. In general, neglecting enantiomorphs, there are two distinct solutions, but for some rings there is only one distinct solution. Each distinct solution for an irregular even edge ring leads to a distinct fundamental irregular ring even edge flexagon. In a fundamental irregular ring even edge flexagon there is at least one cycle that can be traversed that is regular in that all the main positions have the same pat structure and are, in appearance, irregular even edge rings. These main positions can be characterised either by sector diagrams or by flexagon diagrams. The associated polygon for a fundamental irregular ring even edge flexagon is a regular convex polygon with the same number of edges as there are leaves per sector. Arbitrary flexagon type designations are used.

It is not possible to devise a face numbering sequence for fundamental irregular ring even edge flexagons which is consistent in the sense that face numbers for main positions do not become mixed up as a cycle is traversed. However, if a standard numbering sequence (Section 4.1.1) is used then, as a cycle is traversed, consistent patterns of numbers appear in cyclic order on pairs of faces. A key pat can be selected in each sector, and the numbers on the two visible leaves of a pat used in a main position code. Where possible, it is better to use a pat that is a single leaf as a key pat. As examples, sector diagrams for fundamental irregular ring triangle even edge flexagons, with key pats identified, are shown in Fig. 9.2. The key pats are single leaves.

Main position codes based on key pats can be used in Tuckerman diagrams. Tuckerman diagrams are sometimes modified to main position sequence diagrams in order to show sequences of main positions. No distinction is made between principal main positions and subsidiary main positions. In some fundamental irregular ring even edge flexagons more than one main position has the same main position code at a key pat. To avoid ambiguity a full main position code, which includes all the numbers visible on faces, can be used.

Mixed up face numbers mean that deletion of faces to derive degenerate irregular ring even edge flexagons requires deletion of face numbers. A practical way of

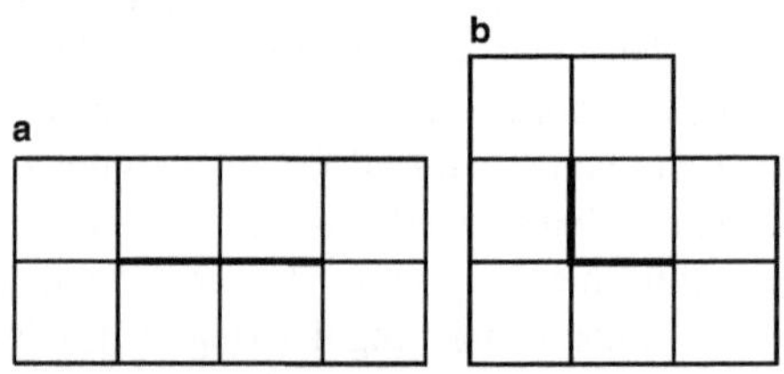

Fig. 9.1 Flat irregular even edge rings of eight squares. (**a**) Type A. (**b**) Type B

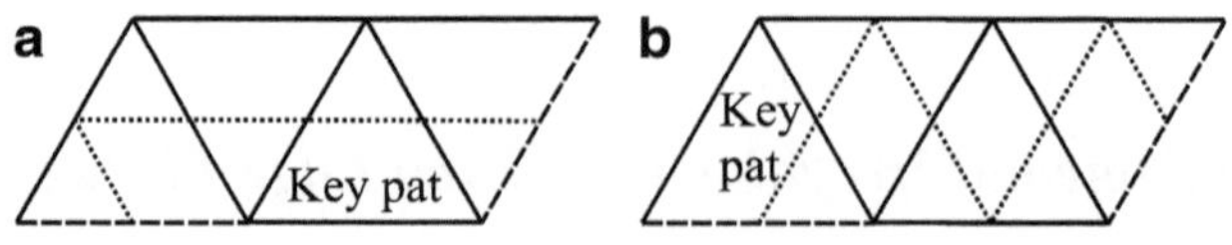

Fig. 9.2 Sector diagrams for fundamental irregular ring triangle even edge flexagons. (**a**) Type A. (**b**) Type B

deleting a face number is to start with a paper model of the net for the precursor flexagon and then glue together pairs of leaves bearing the face number to be deleted. This modified net is then used as a template for the net for the desired degenerate irregular ring even edge flexagon. It is usually convenient to retain the original face numbers (Section 8.1).

Odd edge rings of regular convex polygons, for example the flat irregular odd edge ring of five regular hexagons (Fig. 2.18), cannot be matched by first order fundamental edge nets, so there are no fundamental irregular ring odd edge flexagons.

9.2 Irregular Ring Triangle Even Edge Flexagons

9.2.1 General Properties

The flat irregular even edge ring of ten equilateral triangles shown in Fig. 2.15 is the simplest possible flat non overlapping irregular ring of equilateral triangles. Paper models of corresponding ten triangle irregular ring even edge flexagons are not satisfactory. The flat irregular even edge ring of 12 equilateral triangles shown in Fig. 1.13 is the next simplest. It can be divided into three identical sectors. The first order fundamental triangle edge net ⟨3⟩ (Fig. 3.2) can be fan folded onto this ring in two distinct ways. These two solutions, types A and B, are shown in Fig. 9.2 as sector diagrams for fundamental irregular ring 12 triangle even edge flexagons. Key pats are identified. They are single leaves.

Reducing the number of sectors in the flat irregular even edge ring of 12 equilateral triangles to two, leads to a slant irregular even edge ring of eight equilateral triangles (curvature 120°), and increasing the number of sectors to four leads to a skew irregular even edge ring of 16 equilateral triangles (curvature −120°). Sector diagrams for corresponding flexagons are unchanged.

9.2.2 An Irregular Ring 12 Triangle Even Edge Flexagon

The net for the fundamental irregular ring 12 triangle even edge flexagon type A is shown in Fig. 9.3a. There are three sectors and the torsion per sector is 1. The sector diagram is shown in Fig. 9.2a. The associated polygon is a pentagon. A transformation between flexagons is possible with the five sector first order fundamental triangle even edge flexagon 5⟨3, 3⟩ (Section 4.2.3).

As assembled, the flexagon is in main position 2(1) at the key pats. This is, in appearance, a flat irregular even edge ring of 12 equilateral triangles (Fig. 1.13). The 5-cycle shown in the Tuckerman diagram (Fig. 9.4) can be traversed by using a version of the threefold pinch flex. Starting from main position 2(1), fold together pairs of pats so that pairs of leaves numbered 5 are folded together to reach an intermediate position. This is a combination of an antibox edge ring of six equilateral

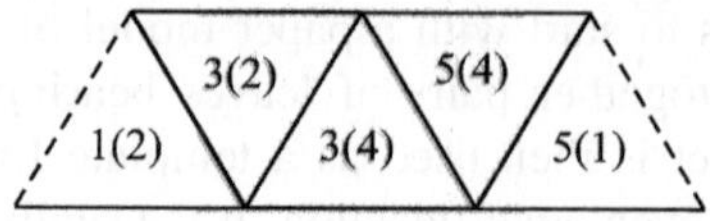

Fig. 9.3 Net for fundamental triangle irregular ring even edge flexagons type A. Fold together pairs of leaves numbered 4. (**a**) Twelve triangle version. Three copies needed. (**b**) Eight triangle version. Two copies needed. Fold so that leaves numbered 5 are on the inside of the ring. (**c**) Sixteen triangle version. Four copies needed

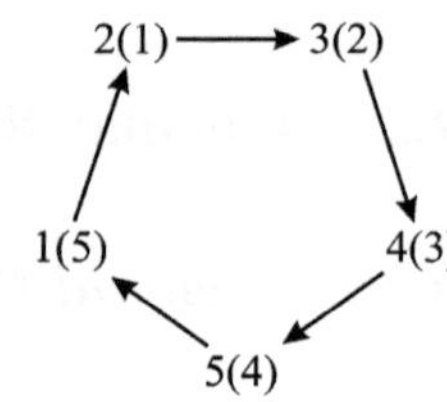

Fig. 9.4 Tuckerman diagram for the 5-cycle fundamental irregular ring triangle even edge flexagons type A

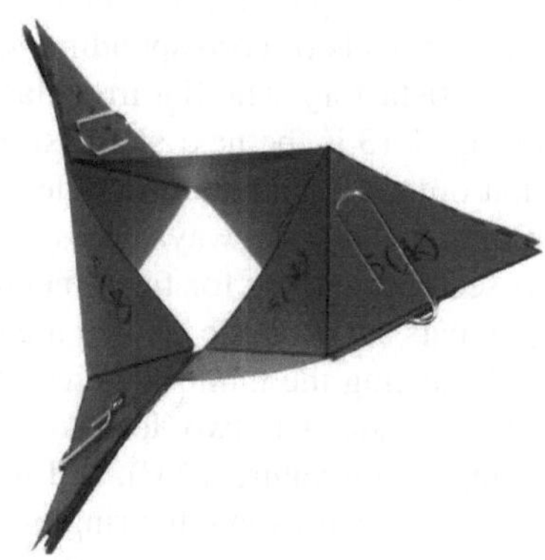

Fig. 9.5 A flexagon as a combination of an antibox edge ring of six equilateral triangles and three equilateral triangle edge triples

triangles and three equilateral triangle edge triples (Fig. 9.5). It is intermediate position 2, with leaves numbered 2 visible on the inside of the antibox edge ring. Then open the flexagon into main position 3(2).

9.2.3 An Irregular Ring Eight Triangle Even Edge Flexagon

The net for the fundamental irregular ring eight triangle even edge flexagon type A is shown in Fig. 9.3b. There are two sectors and the torsion per sector is 1. The sector diagram is shown in Fig. 9.2a. The associated polygon is a pentagon. As assembled, the flexagon is in main position 2(1) at the key pats. This is, in appearance, a slant irregular even edge ring of eight equilateral triangles. Starting from main position 2(1), to flex to main position 3(2) use a version of the twofold pinch flex. Fold together pairs of pats, so that pairs of leaves numbered 5 are folded together, to reach an intermediate position. This is a combination of an antibox edge ring of four equilateral triangles and two equilateral triangle edge triples. It is intermediate position 2, with leaves numbered 2 on the inside of the antibox edge ring. Then open the flexagon into main position 3(2). The high curvature (120°) means that main positions cannot be turned inside out. It is a deficient flexagon (Section 4.2.1) because the 5-cycle shown in the Tuckerman diagram (Fig. 9.4) cannot be traversed without disconnecting a hinge, refolding the flexagon, and reconnecting the hinge.

9.2.4 An Irregular Ring 16 Triangle Even Edge Flexagon

The net for the fundamental irregular ring 16 triangle even edge flexagon type A is shown in Fig. 9.3c. There are four sectors and the torsion per sector is 1. The sector diagram is shown in Fig. 9.2a. The associated polygon is a pentagon. As assembled, the flexagon is in main position 2(1) at the key pats. This is, in appearance, a skew irregular even edge ring of 16 equilateral triangles. The 5-cycle shown in the Tuckerman diagram (Fig. 9.4) can be traversed by using a version of the fourfold pinch flex. Starting from main position 2(1), fold together pairs of pats, so that pairs of leaves numbered 5 are folded together, to reach an intermediate position. This is a combination of an antibox edge ring of eight equilateral triangles and four equilateral triangle edge triples. It is intermediate position 2, with leaves numbered 2 visible on the inside of the antibox edge ring. Then open the flexagon into main position 3(2). Paper clips are needed to keep the flexagon under control.

9.3 Irregular Ring Square Even Edge Flexagons

9.3.1 General Properties

The flat irregular even edge ring of six squares shown Fig. 2.16 is the simplest possible flat non overlapping irregular even edge ring of squares. It can be divided into two identical sectors. The first order fundamental square edge net ⟨4⟩ (Fig. 3.3) can be fan folded onto this ring in one distinct way. This solution is shown in Fig. 9.6 as the flexagon diagram for the fundamental irregular ring six square even edge flexagon. A key pat is identified. This is a folded pile of two leaves with the hinge on an outside edge of the flexagon.

The two flat non overlapping irregular even edge rings of eight squares shown in Fig. 9.1 are the next simplest. Type A (Fig. 9.1a) can be divided into two identical sectors. The first order fundamental square edge net ⟨4⟩ can be fan folded onto this ring in one distinct way. This solution is shown in Fig. 9.7 as a sector diagram for the fundamental irregular ring eight square even edge flexagon type A. A key pat is identified. It is a single leaf. Type B (Fig. 9.1b) consists of one sector. The first

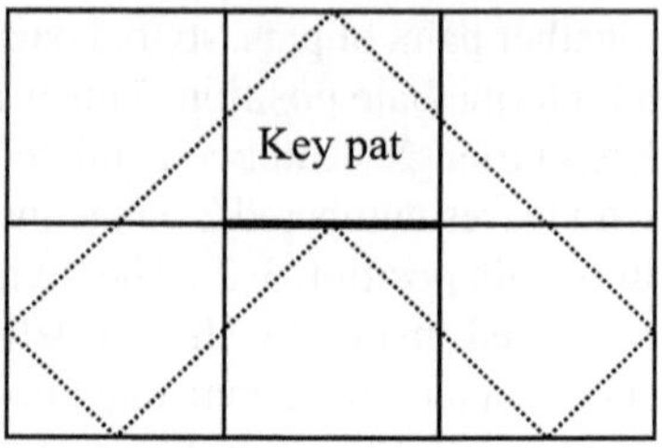

Fig. 9.6 Flexagon diagram for the fundamental irregular ring six square even edge flexagon

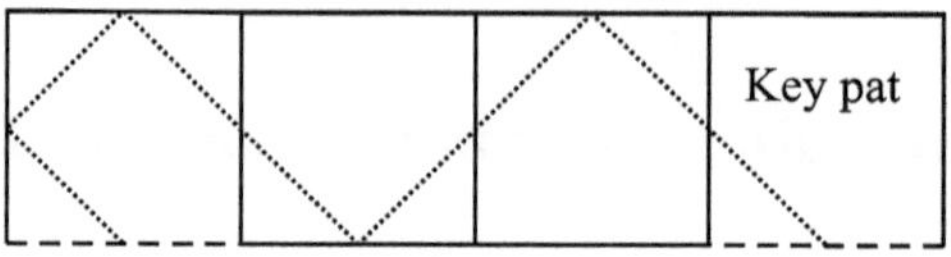

Fig. 9.7 Sector diagram for the fundamental irregular ring eight square even edge flexagon type A

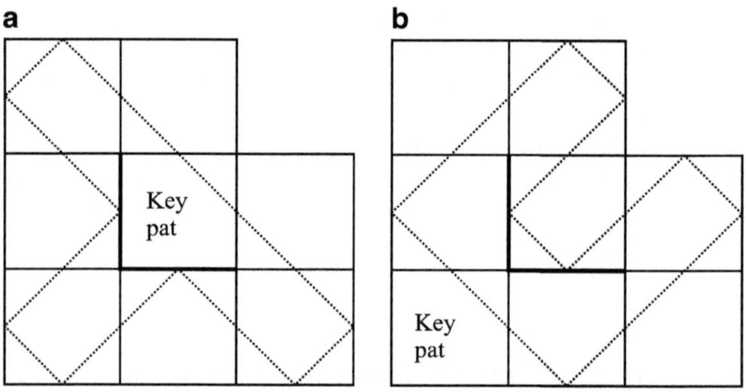

Fig. 9.8 Flexagon diagrams for fundamental irregular ring eight square even edge flexagons. (a) Type BA. (b) Type BB

order fundamental square edge net ⟨4⟩ can be fan folded onto this ring in two distinct ways. These solutions, types BA and BB, are shown in Fig. 9.8 as flexagon diagrams for the fundamental irregular ring eight square even edge flexagons, types BA and BB. Key pats are identified. They are single leaves. A flat overlapping irregular even edge ring of eight squares is shown in Fig. 6.4b. In the figure, the eighth square is covered by the central square.

9.3.2 *Irregular Ring Six Square Even Edge Flexagons*

The net for the fundamental irregular ring six square even edge flexagon is shown in Fig. 9.9. A standard face numbering sequence (Section 4.1.1) is used. The torsion is 6.

The flexagon diagram is shown in Fig. 9.6. The associated polygon is a dodecagon. A transformation between flexagons is possible with the three sector version of the first order fundamental square even edge flexagon 3⟨4, 4⟩ (Section 4.2.4).

As assembled, the flexagon is in main position 1(3) at the key pat. The main position is, in appearance, a flat irregular even edge ring of six squares (Fig. 2.16). The three 4-cycles shown by the solid lines in the main position sequence diagram (Fig. 9.10) can be traversed by using an asymmetric 2-fold pinch flex. To traverse the 4-cycle shown in the Tuckerman diagram (Fig. 9.11), start from main position 1(3). Fold over a flap consisting of two pats, so that leaves numbered 1 and 9 are folded together, to reach a first intermediate position. This is a flat regular even edge ring of four squares (Fig. 1.1b). Complete the flex by unfolding a flap of two pats, so that leaves numbered 6 and 10 are visible, to reach main position 4(6). The hinges used in the second part of the flex are at 90° to those used in the first part.

A zigzag flex can be used to traverse between the 4-cycles, and also for a transformation between flexagons to the three sector version of the first order fundamental

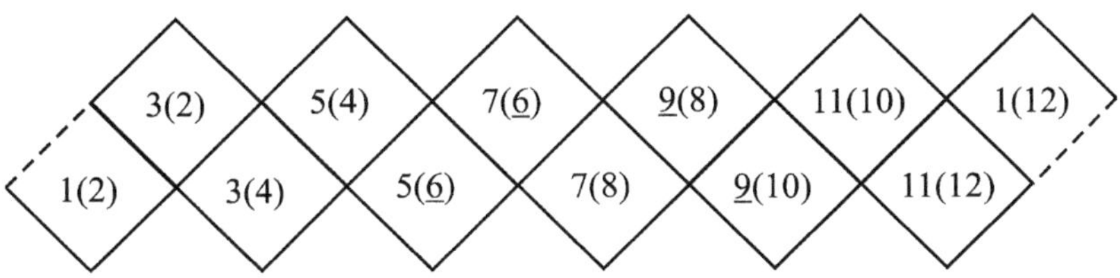

Fig. 9.9 Net for the fundamental irregular ring six square even edge flexagon. One copy needed. Fold together pairs of leaves numbered 2, 5, 6, 8, 10 and 11

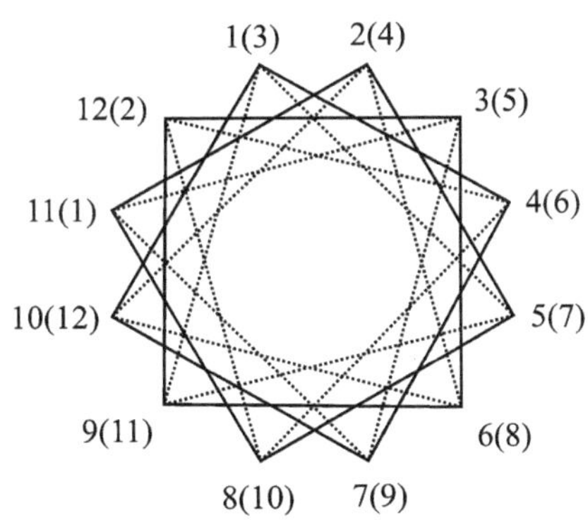

Fig. 9.10 Main position sequence diagram for the fundamental irregular ring six square even edge flexagon

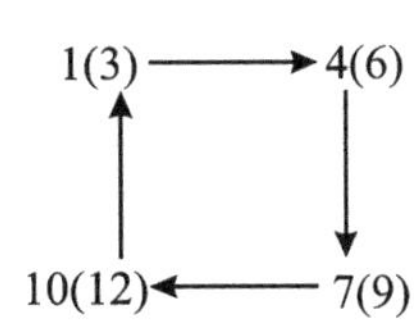

Fig. 9.11 Tuckerman diagram for the a 4-cycle of the fundamental irregular ring six square even edge flexagon

square even edge flexagon 3⟨4, 4⟩. To flex from main position 1(3) to main position 5(7), first fold over a flap consisting of two pats, so that leaves numbered 1 and 9 are folded together. Then fold over another flap so that pairs of leaves numbered 3 and 7 are folded together to reach a second intermediate position. This is a square edge pair (Fig. 1.15) with leaves numbered 4 and 12 visible. It can be opened into a square edge triple. This is an intermediate position of the three sector version of the first order fundamental square even edge flexagon 3⟨4, 4⟩ so the flex sequence so far is a transformation between flexagons to 3⟨4, 4⟩. Close the square edge triple to another second intermediate position with leaves numbered 4 and 8 visible. Then unfold the flexagon in the second part of the zigzag flex, so that leaves numbered 1, 5, 7 and 11 are visible to reach main position 5(7). Possible traverses using the zigzag flex are shown by dotted lines in the main position sequence diagram. Other flexes are possible.

The sixer flexagon is a degenerate version of the fundamental irregular ring six square even edge flexagon. Its net is shown in Fig. 9.12. This was derived from the net for the fundamental irregular ring six square even edge flexagon (Fig. 9.9) by deleting face numbers 2, 5, 8 and 11. The torsion is 2, and it is a basic even edge flexagon. The flexagon diagram is shown in Fig. 9.13. As assembled the flexagon is in main position 1(3) at the key pat. This is, in appearance, a flat irregular even edge ring of six squares (Fig. 2.16). To traverse the 4-cycle shown in the Tuckerman diagram start from main position 1(3). The Tuckerman diagram is the same as that for a 4-cycle of the fundamental irregular ring six square even edge flexagon (Fig. 9.11). Fold over a flap consisting of two pats, so that leaves numbered 1 and 9 are folded together, to reach an intermediate position. This is a flat regular even edge ring of four squares (Fig. 1.1b). Complete the flex by unfolding a flap of two

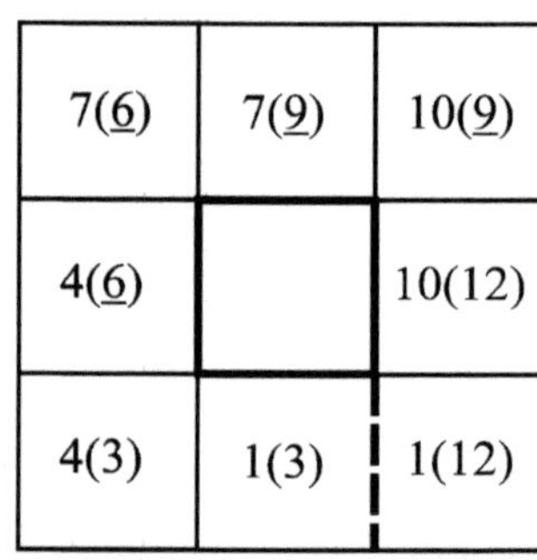

Fig. 9.12 Net for the sixer flexagon. One copy needed. Fold together pairs of leaves numbered 6 and 10

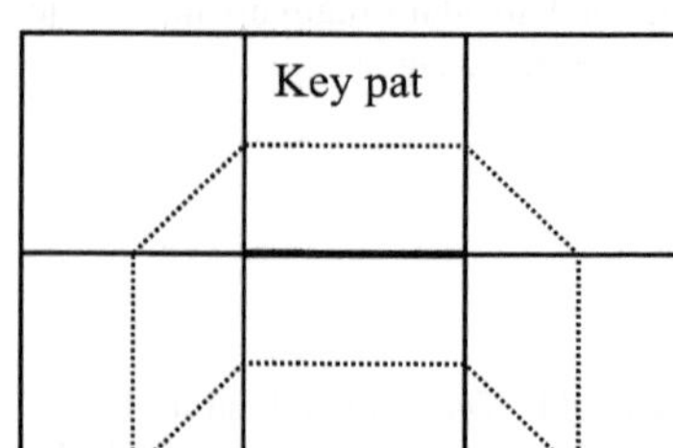

Fig. 9.13 Flexagon diagram for the sixer flexagon

pats, so that leaves numbered 6 and 10 are visible, to reach main position 4(6). The hinges used in the second part of the flex are at 90° to those used in the first part.

9.3.3 Irregular Ring Eight Square Even Edge Flexagons

Fundamental irregular ring eight square even edge flexagons types A, BA and BB are described in this section. Transformations between flexagons are possible between these three types. They also exists in two other forms, the four sector version of the first order fundamental square even edge flexagon 4⟨4, 4⟩ and a square duplex edge flexagon. These are described in Sections 4.2.4 and 12.6.2. Transformations between flexagons with these other forms are possible. The various forms are called, collectively, the octopus flexagon. The net for the fundamental irregular ring eight square even edge flexagon type A is shown in Fig. 9.14. A standard face numbering sequence (Section 4.1.1) is used. There are two sectors and the torsion per sector is 4. The sector diagram is shown in Fig. 9.7. The associated polygon is a 16-gon.

As assembled, the flexagon is in main position 1(2) at key pats. The main position is, in appearance, a flat irregular even edge ring of eight squares (Fig. 9.1a). Two different flexes are needed to traverse the 16-cycle shown by the solid lines in the main position sequence diagram (Fig. 9.15). The two flexes are used alternately, one of each is possible at each main position. Single leaves with the same numbers appear as key pats twice during the 16-cycle, so there are two different versions of each main position, each of which is accompanied by different numbers elsewhere. Pairs of main positions with the same main position code appear symmetrically, as shown by the dashed lines in the main position sequence diagram. To avoid ambiguity, full main position codes can be used (Section 9.1.1). For example, as assembled, the flexagon is in main position 1(2) at the key pats, and using the full main position code it is in main position 1, 5(2, 7). This means that leaves numbered 1 and 5 are visible on the upper face, and leaves numbered 2 and 7 on the lower face. The other version of main position 1(2) at the key pats is, using the full main position code, main position 1, 4(2, 6).

One of the flexes is a version of the twofold pinch flex. Starting from main position 1, 5(2, 7), fold the flexagon in two lengthways to reach a first intermediate position

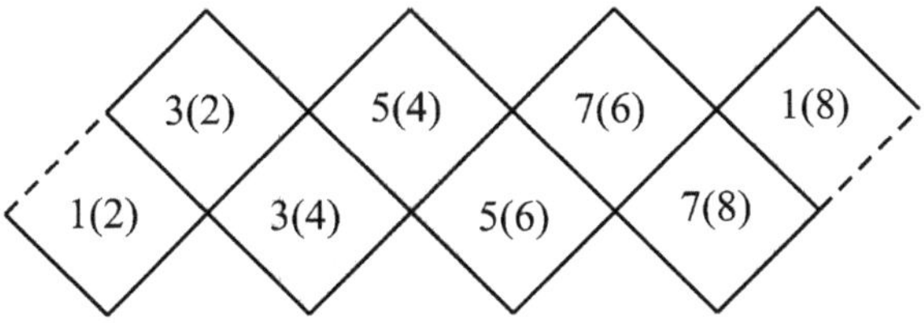

Fig. 9.14 Net for the fundamental irregular ring eight square even edge flexagon type A. Two copies needed. Fold together pairs of leaves numbered 3, 4, 6 and 8

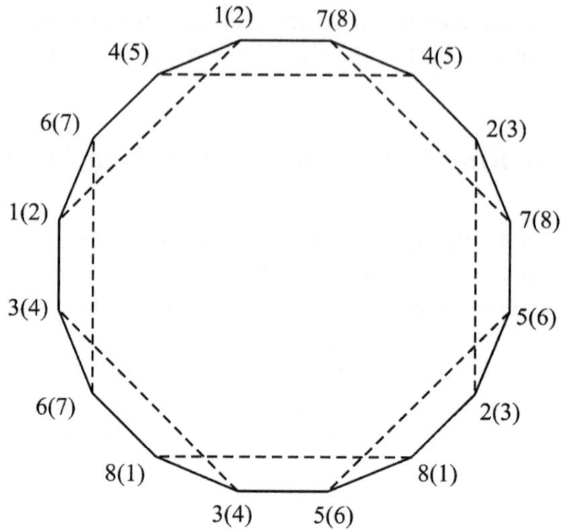

Fig. 9.15 Main position sequence diagram for the fundamental irregular ring eight square even edge flexagon type A

with leaves numbered 1 and 5 visible. This is a straight edge strip of four squares. Then unfold, using the hinges at the opposite edge, to reach main position 4, 7(1, 5). The other flex is the double 2-fold pinch flex. Starting from main position 1, 5(2, 7), at both ends of the flexagon, fold over a flap of two leaves, so that pairs of leaves numbered 1 and 5 are folded together, to reach a second intermediate position. This is a flat regular even edge ring of four squares (Fig. 1.1b) with leaves numbered 2 visible on one face and leaves numbered 7 on the other. Complete the flex by unfolding two flaps, each consisting of two pats, so that leaves numbered 4 and 8 are visible, to reach main position 2, 7(4, 8). The hinges used in the second part of the flex are at 90° to those used in the first part.

To transform to the four sector version of the first order fundamental square even edge flexagon 4⟨4, 4⟩, start from a first intermediate position. Open the centre of the flexagon to an in between position. This is a combination of a box edge ring of four squares and two square edge triples (Fig. 9.16). Then complete this *box flex* by closing the box in the opposite direction to reach an intermediate position of 4⟨4, 4⟩. This is a square edge quadruple.

The net for the fundamental irregular ring eight square edge flexagon type BA is shown in Fig. 9.17a. A standard face numbering sequence (Section 4.1.1) is used The torsion is 8. The flexagon diagram is shown in Fig. 9.8a. The associated polygon is a 16-gon. As assembled, the flexagon is in main position 1(2) at the key pat. This is a flat irregular even edge ring of eight squares (Fig. 9.1b). The dynamic properties of the flexagon are complicated, and various sequences can be used to flex between the main positions shown in the main position sequence diagram (Fig. 9.18). Starting from any main position, the flexagon can be flexed to five different main positions by using one of the two flex sequences described below.

Fig. 9.16 A flexagon as a combination of a box edge ring of four squares and two square edge triples

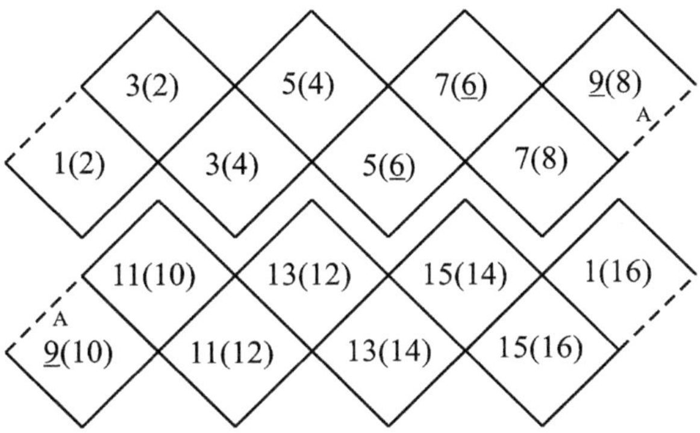

Fig. 9.17 Net for fundamental irregular ring eight square even edge flexagons. One copy needed. Join the two parts of the net at A-A. (**a**) Type BA. Fold together pairs of leaves numbered 4, 5, 7, 9, 10, 12, 14 and 15. (**b**) Type BB. Fold together pairs of leaves numbered 3, 6, 7, 9, 10, 12, 13 and 16

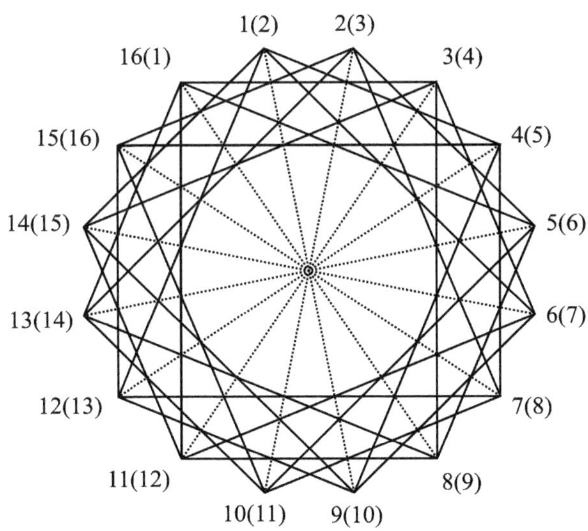

Fig. 9.18 Main position sequence diagram for the fundamental irregular ring square even edge flexagon type BA

In the first sequence, asymmetric twofold pinch flexes are used to flex between main positions that are diametrically opposite in the main position sequence diagram, as shown by dotted lines. Parts of the sequence are transformations between flexagons between types BA and A. Starting from main position 1(2), first fold over a flap consisting of two pats, so that pairs of leaves numbered 2 and 6 are folded together to reach a third intermediate position. This is a flat irregular even edge ring of six squares (Fig. 2.16). Next, unfold so that leaves numbered 5 and 9 are visible, to reach a fourth intermediate position. This is a flat irregular even edge ring of eight squares type A (Fig. 9.1a). It is also a type A main position so the sequence so far is a transformation between flexagons to type A. Then fold over a flap consisting of two pats, so that pairs of leaves numbered 1 and 13 are folded together, to reach another third intermediate position. Finally, to reach main position 9(10), unfold so that leaves numbered 10 and 14 are visible. This second part of the sequence is also a transformation from type A to type BA.

The second sequence uses a combination of asymmetric twofold pinch flexes and the twofold pinch flex. It is used to flex between main positions that are not diametrically opposite, as shown by solid lines in the intermediate position map. Starting from main position 1(2), first flex to a fourth intermediate position as in the first sequence. Next, using a twofold pinch flex, fold the flexagon in two lengthways so that pairs of leaves numbered 8 and 16 are folded together, and the pair of leaves numbered 3 is folded onto the pair of leaves numbered 11. Complete the flex by unfolding about the opposite edge so that leaves numbered 3, 6, 11 and 14 are visible, to reach another fourth intermediate position. Then, using an asymmetric twofold pinch flex fold over a flap consisting of two pats so that leaves numbered 1 and 13 are folded together to reach a third intermediate position. Finally, to reach main position 4(5), unfold so that leaves numbered 4 and 6 are visible. The same flexing sequence, but folding over different flaps, can be used to reach main positions 6(7), 12(13) and 14(15).

The net for the fundamental irregular ring eight square even edge flexagon type BB is shown in Fig. 9.17b. The face numbering sequence is the same as that for type BA. The torsion is 8. The flexagon diagram is shown in Fig. 9.8b. The associated polygon is a 16-gon. As assembled the flexagon is in main position 1(2) at the key pat. This is a flat irregular even edge ring of eight squares (Fig. 9.1b). Flexing sequences are more complicated than for type BA.

9.4 Irregular Ring Pentagon Even Edge Flexagons

9.4.1 General Properties

The flat irregular even edge ring of six regular pentagons shown in Fig. 2.17 is the simplest possible flat non overlapping irregular even edge ring of pentagons. It can be divided into two identical sectors. The first order fundamental pentagon edge net ⟨5⟩ (Fig. 3.4a) can be fan folded onto this ring in one distinct way. This solution is shown in Fig. 9.19a, as the flexagon diagram for the fundamental irregular ring six

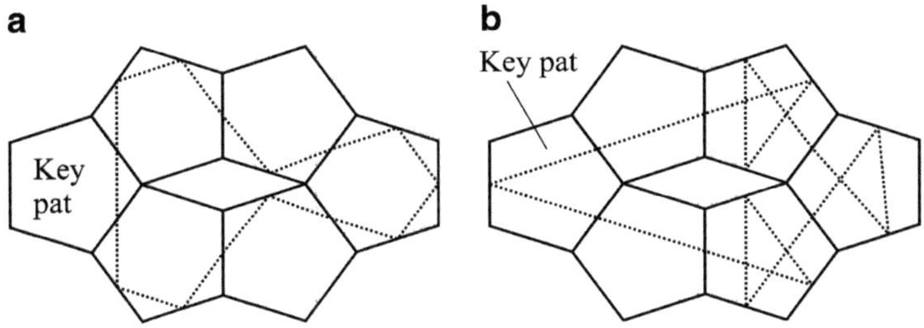

Fig. 9.19 Flexagon diagrams for fundamental irregular ring six pentagon even edge flexagons. (**a**) Type A. (**b**) Type B

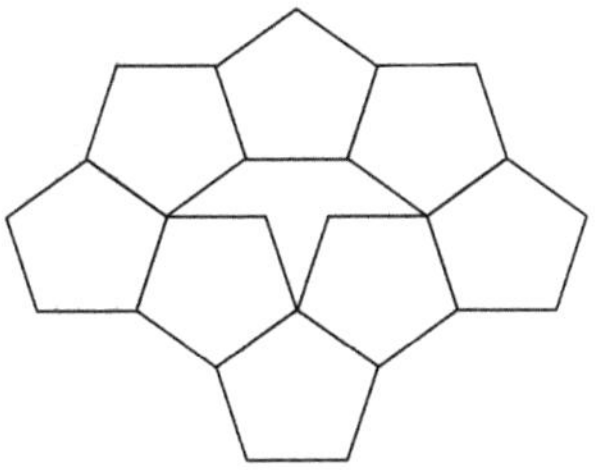

Fig. 9.20 Flat irregular even edge ring of eight regular pentagons

pentagon even edge flexagon type A, with a key pat identified. This is a single leaf. The first order fundamental pentagon edge net ⟨5/2⟩ (Fig. 3.4b) can also be fan folded onto the ring in one distinct way. This solution is shown in Fig. 9.19b, as the flexagon diagram for the fundamental irregular ring six pentagon even edge flexagon type B, with a key pat identified. This is a folded pile of two leaves.

The flat irregular even edge ring of eight regular pentagons shown in Fig. 9.20 is the next simplest possible flat irregular even edge ring of pentagons. The first order fundamental pentagon edge net ⟨5⟩ can be fan folded onto this ring in two distinct ways. These solutions, types A and B, are shown in Figs. 9.21a and b as flexagon diagrams for fundamental irregular ring eight pentagon even edge flexagons, types A and B. Key pats are identified in the figures. For type A, the key pat is a single leaf and for type B it is a folded pile of three leaves. Similarly, the first order fundamental pentagon edge net ⟨5/2⟩ can be fan folded onto the ring in two distinct ways. Figure 9.21c and d show flexagon diagrams for fundamental irregular ring eight pentagon even edge flexagons, types C and D, with key pats identified. For type C the key pat is a single leaf and for type D it is a folded pile of two leaves.

9.4.2 An Irregular Ring Six Pentagon Even Edge Flexagon

The net for the fundamental irregular ring six pentagon even edge flexagon type A is shown in Fig. 9.22a. A standard face numbering sequence (Section 4.1.1) is used. The torsion is 9. The flexagon diagram is shown in Fig. 9.19a. The associated polygon is a 15-gon. A transformation between flexagons is possible with the three sector version of the first order fundamental pentagon even edge flexagon 3⟨5, 5⟩ (Section 4.2.6).

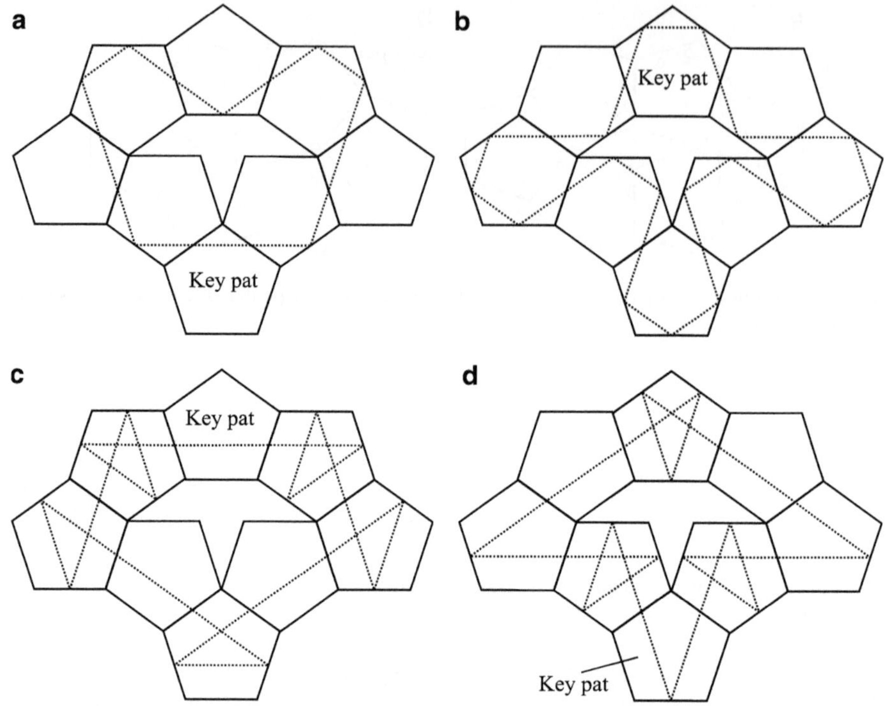

Fig. 9.21 Flexagon diagrams for fundamental irregular ring eight pentagon even edge flexagons. (**a**) Type A. (**b**) Type B. (**c**) Type C. (**d**) Type D

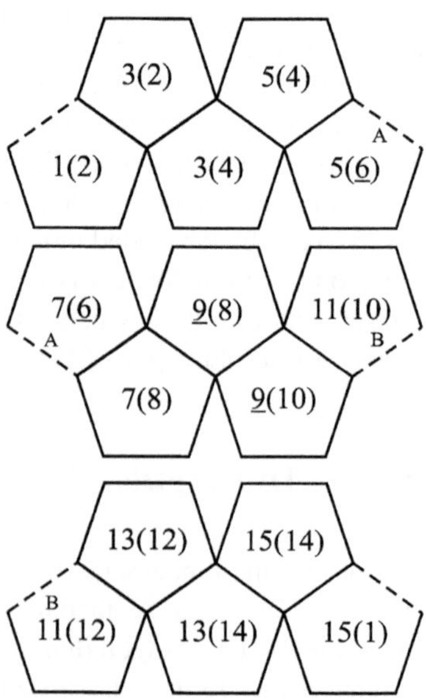

Fig. 9.22 Net for fundamental irregular ring even edge flexagons. One copy needed. Join the three parts of the net at A-A and B-B. (**a**) Six pentagon type A. Fold together pairs of leaves numbered 3, 4, 6, 8, 9, 10, 12, 14 and 15. (**b**) Eight pentagon type A. Fold together pairs of leaves numbered 3, 6, 7, 9, 11, 12 and 15

As assembled, the flexagon is in main position 1(2) at the key pat. The main position is a flat irregular even edge ring of six pentagons (Fig. 2.17). The main position sequence diagram (Fig. 9.23) can be traversed by using an asymmetric threefold pinch flex. To flex from main position 1(2) to main position 2(3), start by folding together pairs of leaves numbered 1, 5 and 11 to reach a first intermediate position. This is a pentagon edge pair with leaves numbered 2 and 7 visible. Then open the flexagon into a second intermediate position. This is a pentagon edge triple (Fig. 9.24). It is also an intermediate position of the three sector version of the first order fundamental pentagon even edge flexagon 3⟨5, 5⟩, so the flex sequence so far is a transformation between flexagons to 3⟨5, 5⟩. Next, close the second intermediate position to another first intermediate with leaves numbered 2 and 12 visible. Finally, complete the asymmetric threefold pinch flex by opening into main position 2(3). In practice this final part of the flex is difficult, and it is easier to break the asymmetric threefold pinch flex into several steps as follows.

To flex from main position 1(2) to main position 2(3), start by folding together leaves numbered 13 to form a pat of five leaves. Then arrange the remaining leaves

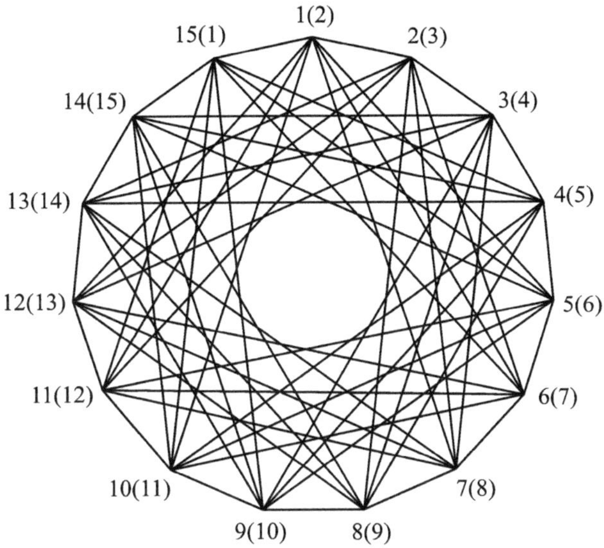

Fig. 9.23 Main position sequence diagram for the fundamental irregular ring six pentagon even edge flexagon type A

Fig. 9.24 A flexagon as a pentagon edge triple

into two more pats of five leaves each to reach an intermediate position. This is a pentagon edge triple (Fig. 9.24). Next, unfold the leaf numbered 2(3) and then piles of three leaves from each of the adjacent pats to form two new pats of three leaves each. At this stage, the remaining pats should pop into position. If not ease them into position, noting that there are two pats of two leaves, and the pat at the end of the flexagon has four leaves. Other main positions can be reached from this intermediate position because there are six single leaves that can be unfolded. Also, one could start by folding leaves numbered 5 together. There is some duplication but, in total, eight different main positions can be reached from main position 1(2), as shown in the main position sequence diagram (Fig. 9.23).

9.4.3 An Irregular Ring Eight Pentagon Even Edge Flexagon

The net for the fundamental irregular ring eight pentagon even edge flexagon type A is shown in Fig. 9.22b. A standard face numbering sequence (Section 4.1.1) is used. The torsion is 5. The flexagon diagram is shown in Fig. 9.21a. The associated polygon is a 15-gon. The net is the same as that for the fundamental irregular ring six pentagon even edge flexagon type A (previous section), but the torsion is different so no transformation between flexagons is possible.

As assembled, the flexagon is in main position 1(2) at the key pat. The main position is a flat irregular even edge ring of eight pentagons (Fig. 9.20). The partial main position sequence diagram (Fig. 9.25) can be traversed by using a twist flex.

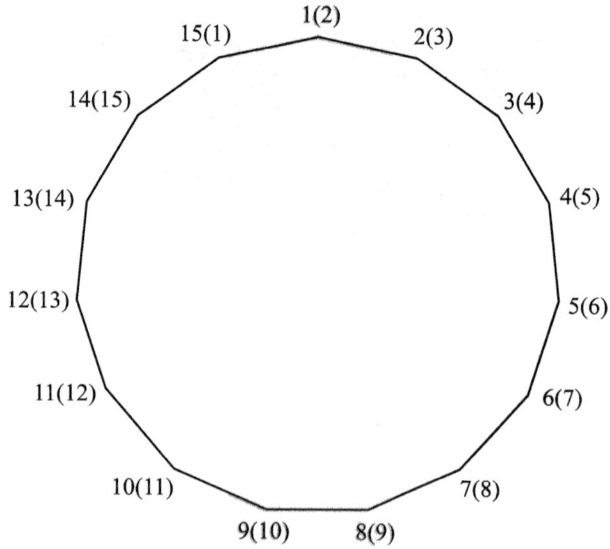

Fig. 9.25 Partial main position sequence diagram for the fundamental irregular ring eight pentagon even edge flexagon type A

Flexing is difficult, and the flexagon tends to pop into unexpected main positions. To flex from main position 1(2) to main position 2(3). First, using paper clips, fasten together the pairs of leaves numbered 7 and 12. Then fold together pairs of leaves numbered 8, 10 and 13. Pairs of leaves numbered 1 and 4 should fold together to complete the flex, if not ease them into position. Remove the paper clips. It is possible, but difficult, to flex between any two main positions by using various types of twist flex.

9.5 Irregular Ring Hexagon Even Edge Flexagons

9.5.1 General Properties

The flat irregular even edge ring of six regular hexagons shown in Fig. 9.26 is one of the two simplest possible flat non overlapping irregular even edge rings of regular hexagons. It can be divided into two identical sectors. The first order fundamental hexagon edge net ⟨6⟩ (Fig. 3.5) can be fan folded onto this ring in one distinct way. This solution is shown in Fig. 9.27 as the flexagon diagram for a fundamental irregular ring six hexagon even edge flexagon, with a key pat identified. This is a single leaf.

The next simplest flat non overlapping irregular even edge rings of regular hexagons contain eight hexagons. Two of these are shown in Fig. 9.28. Type A (Fig. 9.28a) can be divided into two identical sectors. The first order fundamental

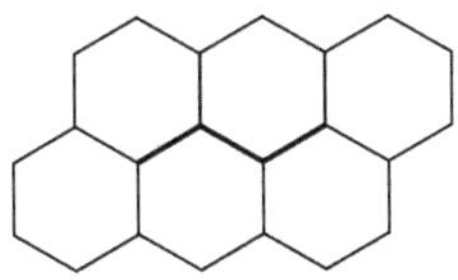

Fig. 9.26 A flat irregular even edge ring of six regular hexagons

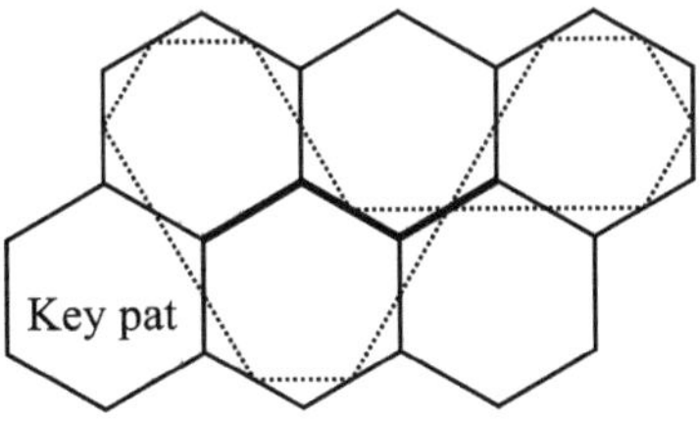

Fig. 9.27 Flexagon diagram for a fundamental irregular ring six hexagon even edge flexagon

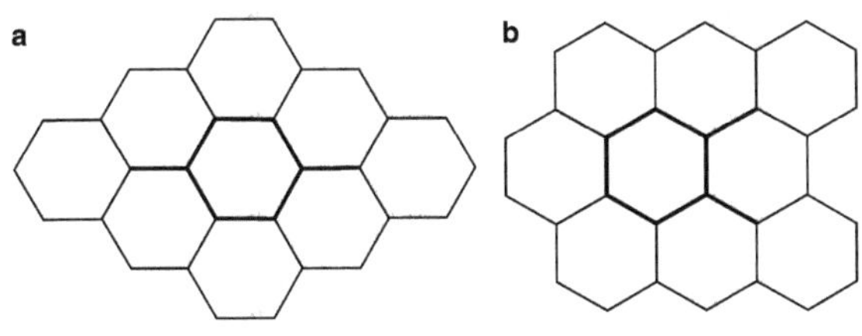

Fig. 9.28 Flat irregular even edge rings of eight regular hexagons. (**a**) Type A. (**b**) Type B

square edge net ⟨6⟩ can be fan folded onto this ring in two distinct ways. These solutions, types AA and AB, are shown in Fig. 9.29 as sector diagrams for the fundamental irregular ring eight hexagon even edge flexagons, types AA and AB. Key pats are identified. The key pat is a single leaf for type AA and a folded pile of four leaves for type AB. Type B (Fig. 9.28b) consists of one sector. The first order fundamental square edge net ⟨6⟩ can be fan folded onto this ring in two distinct ways. These solutions, types BA and BB, are shown in Fig. 9.30 as flexagon diagrams for the fundamental irregular ring eight hexagon even edge flexagons, types BA and BB. Key pats are identified. They are folded piles of two leaves.

9.5.2 An Irregular Ring Six Hexagon Even Edge Flexagon

The net for a fundamental irregular ring six hexagon even edge flexagon is shown in Fig. 9.31. A standard face numbering sequence (Section 4.1.1) is used. The torsion is 9. The flexagon diagram is shown in Fig. 9.27. The associated polygon is an18-gon. A transformation between flexagons is possible with the three sector version of the first order fundamental hexagon even edge flexagon 3⟨6, 6⟩ (Section 4.2.7).

As assembled, the flexagon is in main position 1(2) at the key pat. This is, in appearance a flat irregular even edge ring of six regular hexagons (Fig. 9.26). The main position sequence diagram (Fig. 9.32) can be traversed by using zigzag

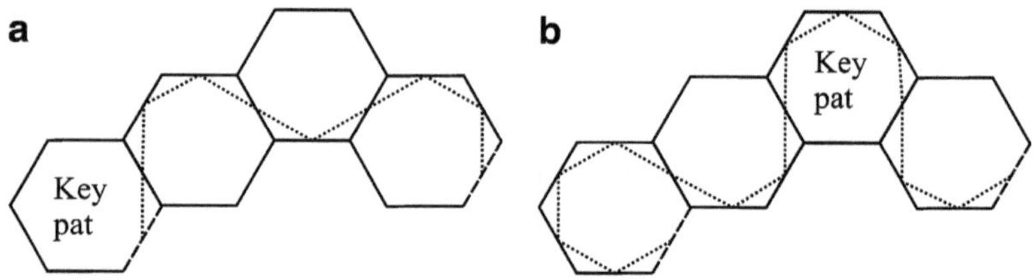

Fig. 9.29 Sector diagrams for fundamental irregular ring eight hexagon even edge flexagons (**a**) Type AA. (**b**) Type AB

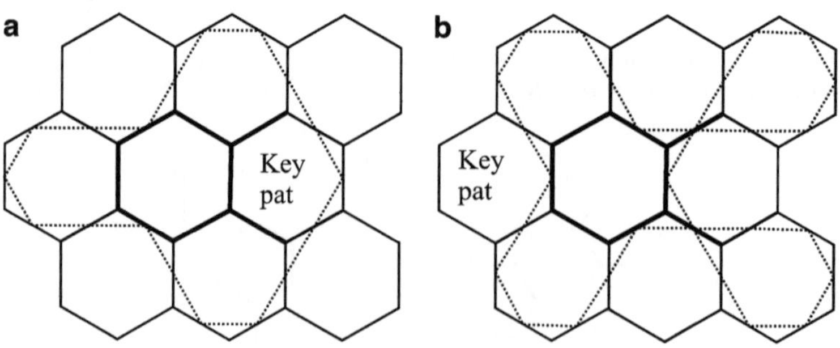

Fig. 9.30 Flexagon diagrams for fundamental irregular ring eight hexagon flexagons. (**a**) Type BA. (**b**) Type BB

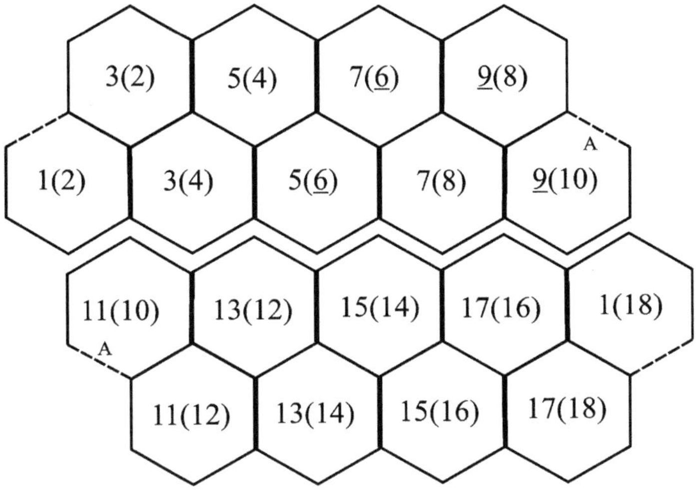

Fig. 9.31 Net for a fundamental irregular ring six hexagon even edge flexagon. One copy needed. Join the two parts of the net at A-A. Fold together pairs of leaves numbered 3, 4, 6, 8, 9, 10, 11, 13, 14, 16, 17 and 18

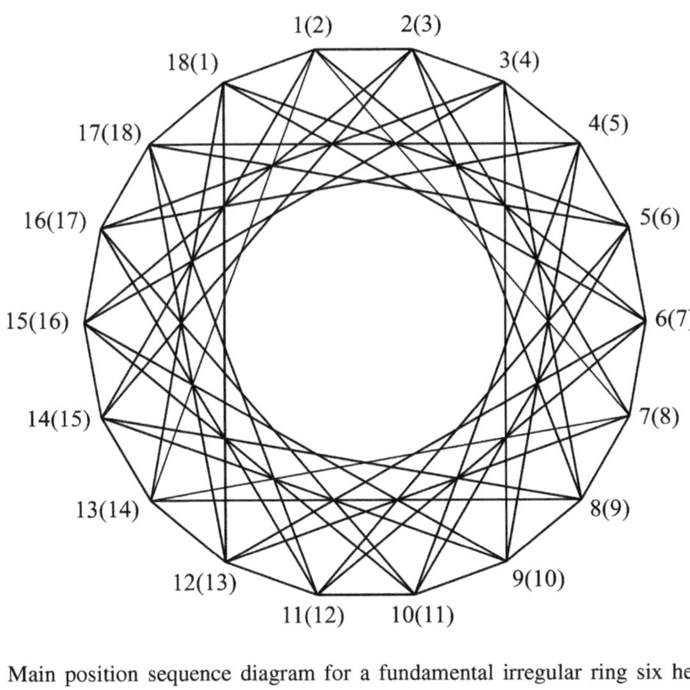

Fig. 9.32 Main position sequence diagram for a fundamental irregular ring six hexagon even edge flexagon

Fig. 9.33 A fundamental irregular ring six hexagon even edge flexagon part way through a zigzag flex

flexes and threefold pinch flexes, but this is difficult, and paper clips may be needed to keep the flexagon under control. Starting from main position 1(2), to flex to main position 6(7) fold together pairs of leaves numbered 2, 5, 12 and 15, in the first part of a zigzag flex, to reach intermediate position 1(7). This is a hexagon edge pair with leaves numbered 1 and 7 visible. Figure 9.33 shows the flexagon part way through a zigzag flex. Unfold so that leaves numbered 3, 6, 11 and 14 are visible, in the second part of the zigzag flex, to reach main position 6(7). The intermediate position can be opened into a hexagon edge triple, with leaves numbered 1, 7 and 13 visible. This is an intermediate position of the first order fundamental hexagon even edge flexagon3⟨6, 6⟩ so this is a transformation between flexagons to 3⟨6, 6⟩.

Six different main positions can be reached from main position 1(2), as shown in the main position sequence diagram. To reach main position 13(14), flex to intermediate position 1(7) and turn over a pat to reach intermediate position 1(13). Then unfold, by using a zigzag flex so that leaves numbered 6, 9, 14 and 17 are visible. To reach some main positions, a threefold pinch flex is needed. For example, to reach main position 7(8) flex to intermediate position 1(7), and open it into a hexagon edge triple with leaves numbered 1, 7 and 13 visible. Apply a threefold pinch flex to reach another hexagon edge triple with leaves numbered 2, 8 and 14 visible. Next, fold together the pair of leaves numbered 2 to reach intermediate position 8(14). Finally, unfold the flexagon using a zigzag flex so that leaves numbered, 2, 4, 12 and 15 are visible.

9.5.3 Irregular Ring Eight Hexagon Even Edge Flexagons

A dual marked net for the fundamental irregular ring eight hexagon even edge flexagons, types AA and BA, is shown in Fig. 9.34. Standard face numbering sequences (Section 4.1.1) are used. Numbers are used for type A. There are two sectors and the torsion per sector is 4. The sector diagram is shown in Fig. 9.29a. The associated polygon is an enneagon. Letters are used for type BA. The torsion is 8. The flexagon diagram is shown in Fig. 9.30a. The associated polygon is an 18-gon. Transformation between flexagons is possible between the two types.

As assembled, the flexagon is type AA, and is in main position 1(2) at the key pats. This is, in appearance, a flat irregular even edge rings of eight regular hexagons type A (Fig. 9.28a). The main position sequence diagram (Fig. 9.35) can be traversed

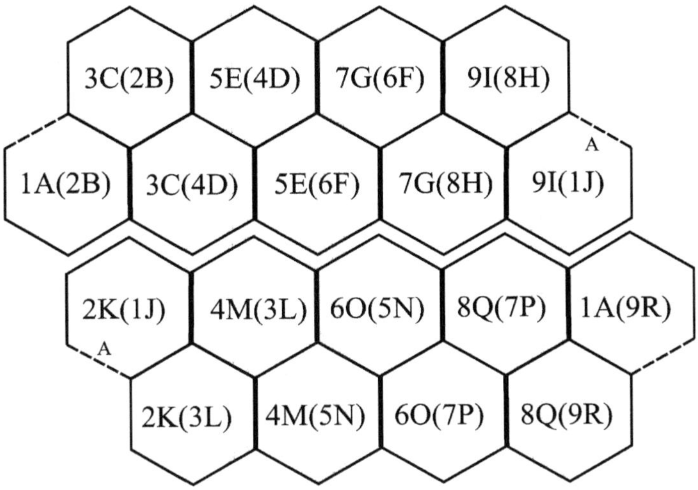

Fig. 9.34 Dual marked net for fundamental irregular ring eight hexagon even edge flexagons, types AA and BA. One copy needed. Fold together pairs of leaves numbered 3, 4, 6, 8 and 9

Fig. 9.35 Main position sequence diagram for the fundamental irregular ring eight hexagon even edge flexagon type AA

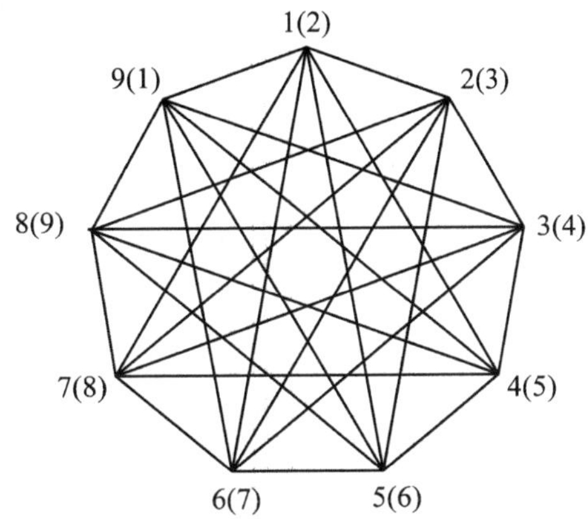

by using three different twist flexes. The hinges must be well creased for these to work smoothly. Starting from main position 1(2), to flex to main position 5(6), hold the pair of leaves marked 5(6) and pull gently so that the remaining leaves twist into position. To flex to main position 2(3) from main position 1(2) hold the leaves marked 2(3), turn them over and pull gently apart. To flex to main position 4(5) from main position 1(2) hold the pairs of leaves with leaves numbered 1 and 8 visible, turn them over and pull gently apart.

Main positions of type BA, which are flat irregular even edge rings of eight regular hexagons type B (Fig. 9.28b), can only be reached by transformation between flexagons from type AA. To transform to a main position of type BA from type AA turn over a pat of two leaves at one side of the flexagon. This can be done in four

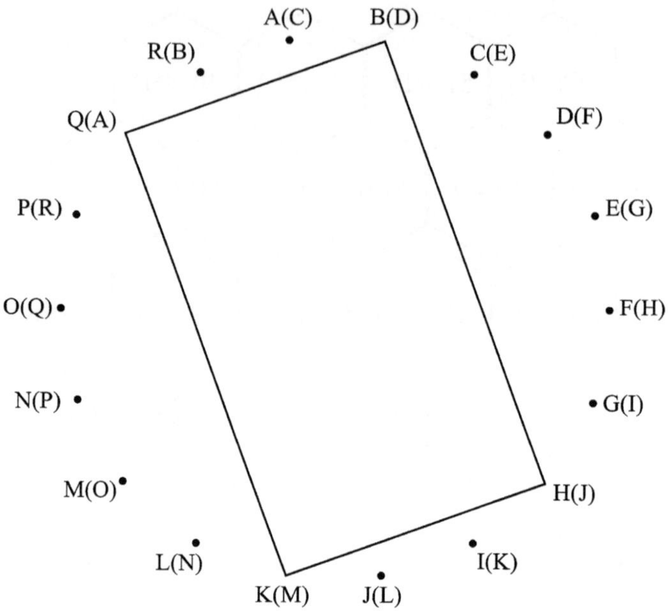

Fig. 9.36 Partial main position sequence diagram for the fundamental irregular ring eight hexagon even edge flexagon type BA

different ways. The four main positions of type BA, at the key pat, that can be reached from main position 1(2) of type AA are shown in the partial main position sequence diagram (Fig. 9.36). Since it is not possible to flex directly between main positions of type BA it could be regarded as a muddled version of type AA. Other flexes and transformations are possible.

The net for the minitwist flexagon, which is a degenerate version of the irregular ring eight hexagon even edge flexagon type AA, is shown in Fig. 9.37. This was derived by using the dual marked net for the fundamental irregular ring eight hexagon even edge flexagons, types AA and BA (Fig. 9.34) as a precursor, deleting face numbers 3, 6 and 9, and renumbering. There are two sectors and the torsion per sector is 2. it is a basic even edge flexagon. The sector diagram is shown in Fig. 9.38 with a key pat identified. This is a single leaf.

As assembled, the flexagon is in main position 1(2). This is a flat irregular ring of eight regular hexagons type A (Fig. 9.28a). The 3-cycle shown in the Tuckerman diagram (Fig. 9.39) can be traversed by using twist flexes. Face numbers do not become mixed up. Starting from main position 1(2), to flex to main position 2(3) hold one of the pair of diametrically opposite pair of leaves marked 2(3) and turn them over, allowing the other leaves to twist into position.. Using the other pair of leaves marked 2(3) gives the same result, but the twist flex is different.

Transformation between flexagons to a degenerate version of the fundamental irregular ring eight hexagon even edge flexagon is possible. To do this, turn over a single leaf at one side of the flexagon. If this does not work, turn the leaf over in

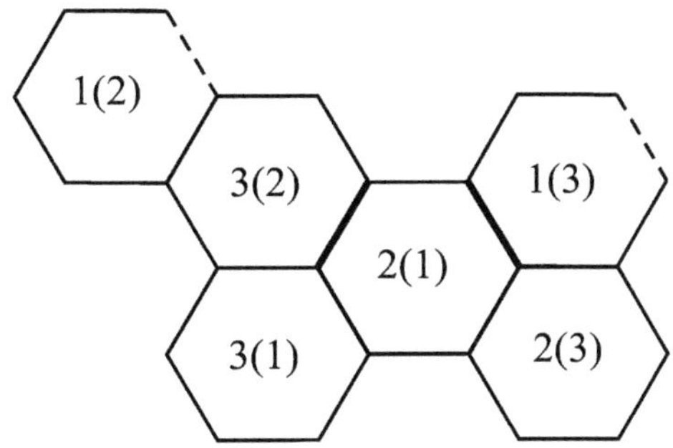

Fig. 9.37 Net for the minitwist flexagon. Two copies needed

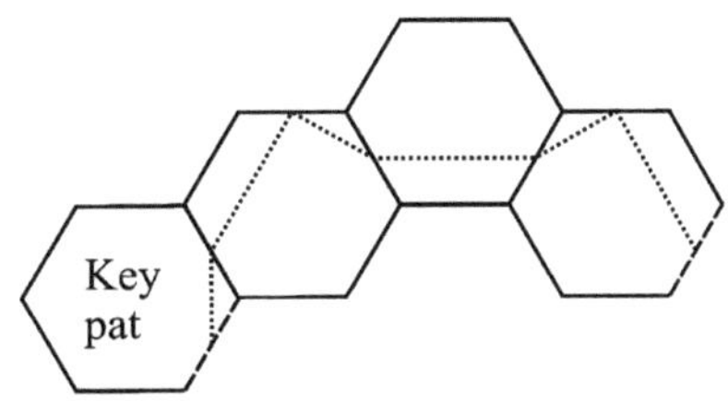

Fig. 9.38 Sector diagram for the minitwist flexagon

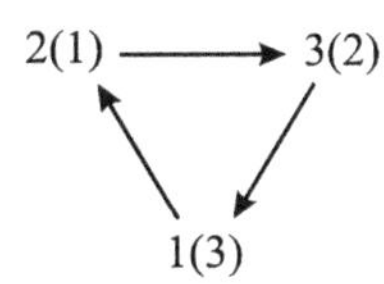

Fig. 9.39 Tuckerman diagram for the 3-cycle of the minitwist flexagon

the other direction. Main positions are flat irregular rings of eight hexagons type B (Fig. 9.28b) and face numbers are mixed up.

9.6 Irregular Ring Dodecagon Even Edge Flexagons

9.6.1 General Properties

One of the possible flat irregular even edge rings of eight regular dodecagons is shown in Fig. 9.40. It can be divided into two identical sectors. The first order fundamental dodecagon edge net ⟨12⟩. (Fig. 3.11a) can be fan folded onto this ring in one distinct way. This solutions is shown in Fig. 9.41 as a sector diagram for a

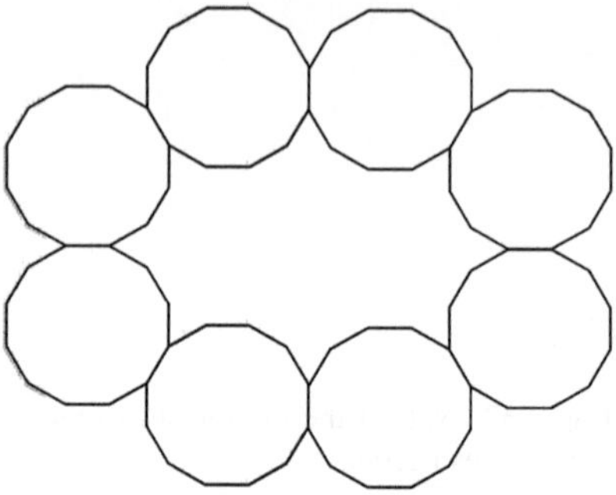

Fig. 9.40 A flat irregular even edge ring of eight regular dodecagons

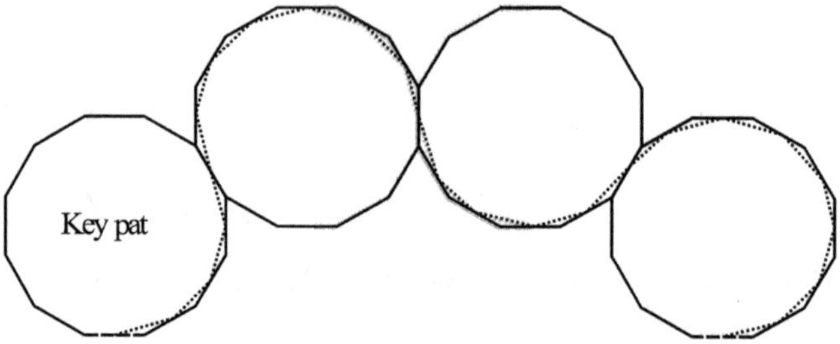

Fig. 9.41 Sector diagram for a fundamental irregular ring dodecagon even edge flexagon

fundamental irregular ring eight dodecagon even edge flexagon, with a key pat identified. This is a folded pile of five leaves.

9.6.2 Irregular Ring Eight Dodecagon Even Edge Flexagons

The net for a fundamental irregular ring dodecagon even edge flexagon is shown in Fig. 9.42. The leaves have been truncated to irregular hexagons to simplify assembly. The resulting truncated flexagon is a partial overlap flexagon and could also be called an irregular hexagon even edge flexagon. It is an example of a flexagon that belongs to more than one family of flexagons. The flexagon is very unstable and it is unsatisfactory as a paper model. It is included because it is a precursor to a degenerate version (below), that is much easier to handle, and demonstrates some of the dynamic properties. There are two sectors and the torsion per sector is 20. The sector diagram is shown in Fig. 9.41. The associated polygon is a 24-gon.

As assembled, the flexagon is in main position 1(5) at the key pats. This is, in appearance, approximately a flat irregular even edge ring of eight regular dodecagons type A (Fig. 9.40). The truncated leaves overlap reasonably well. The only feasible

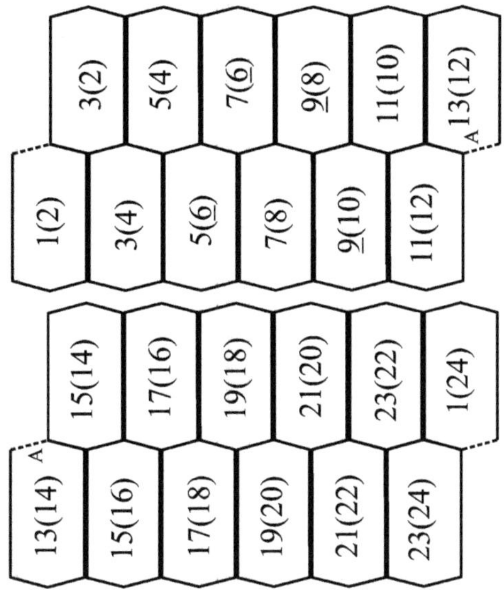

Fig. 9.42 Net for a fundamental irregular ring eight dodecagon (irregular hexagon) even edge flexagon. Two copies needed. Join the two parts of the net at A-A. Fold together pairs of leaves numbered 2, 3, 4, 6, 7, 8, 9, 10, 11, 13, 14, 15, 16, 18, 19, 20, 21, 22 and 23

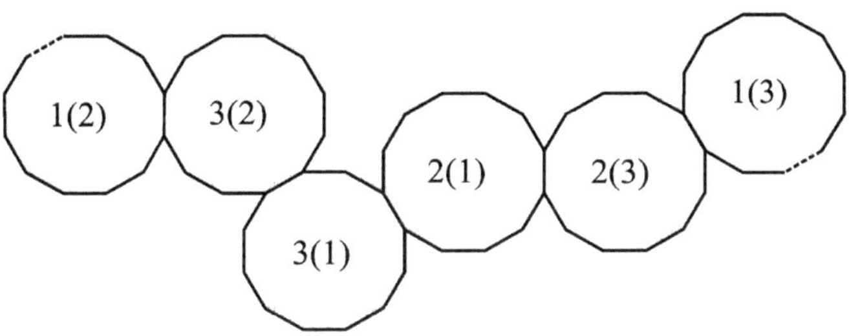

Fig. 9.43 Net for a degenerate irregular ring eight dodecagon even edge flexagon. Two copies needed

flex is a band flex in which the flexagon is deliberately collapsed into an open band, and then reassembled into the desired main position. This is tedious and paper clips are needed to keep the flexagon under control.

The net for a degenerate irregular ring eight dodecagon even edge flexagon is shown in Fig. 9.43. This was derived by using the net for an irregular ring dodecagon even edge flexagon (Fig. 9.42) as a precursor, deleting face numbers 2, 3, 4, 6, 7, 8, 10, 11, 13, 14, 15, 16, 18, 19, 20, 21, 23 and 24, replacing the irregular hexagons by regular dodecagons, and renumbering so that face numbers do not become mixed up. There are two sectors, and the torsion per sector is 2. It is a basic even edge flexagon. The 3-cycle shown in the Tuckerman diagram can be traversed by using the fourfold pinch flex. The Tuckerman diagram is the same as that for the minitwist flexagon (Fig. 9.39). Main positions 1(2) and 2(3) are flat irregular even

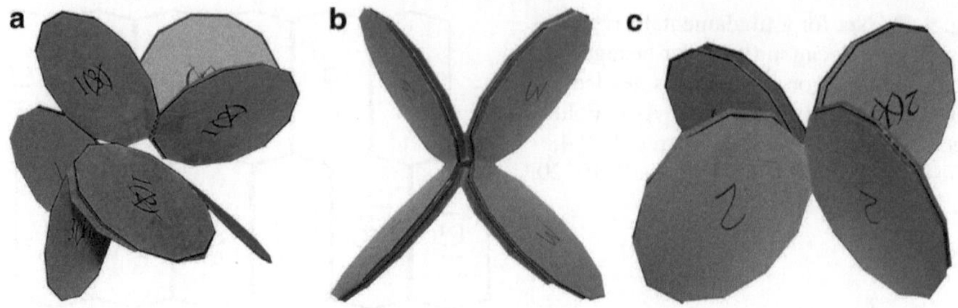

Fig. 9.44 A flexagon as the following. (**a**) A skew irregular even edge ring of eight regular dodecagons. (**b**) A skew regular even edge ring of four regular dodecagons. (**c**) A dodecagon edge quadruple

edge rings of eight regular dodecagons type A (Fig. 9.40). Main position 3(1) is a skew irregular even edge ring of eight regular dodecagons, which cannot be arranged symmetrically (Fig. 9.44a). Intermediate positions 1 and 3 are skew regular even edge rings of four regular dodecagons (Fig. 9.44b). Intermediate position 2 is a dodecagon edge quadruple (Fig. 9.44c).

Chapter 10
Irregular Polygon Edge Flexagons

10.1 Introduction

Most of the edge flexagons described in Chapters 4 and 6–9 are regular polygon edge flexagons that are made from regular convex polygons. An irregular polygon edge flexagon is an edge flexagon made from irregular convex polygons. Only a limited range of irregular polygons leads to irregular polygon edge flexagons whose paper models are reasonably easy to handle (Section 2.3.1). Irregular polygon edge flexagons include some of the more interesting flexagons. Irregular polygon even edge flexagons with at least some flat main positions were briefly described by Conrad and Hartline (1962) and in more detail by Pook (2003), who called them distorted polygon flexagons. The chapter has been deferred to this point because material in earlier chapters is needed for the discussion of irregular polygon edge flexagons. Similarly, the point flexagons described in Chapters 5, 7 and 8 are regular polygon point flexagons, and an irregular polygon point flexagon is a point flexagon made from irregular convex polygons. The dynamic properties of irregular polygon point flexagons do not differ significantly from their precursors so they are not discussed.

Any irregular polygon edge flexagon can be derived by replacing the regular convex polygons in a precursor regular polygon edge flexagon by appropriate irregular convex polygons. The characteristic flex for an irregular polygon edge flexagon is the same as that for the precursor flexagon. Most of the irregular polygon edge flexagons described in this chapter have at least some main positions which are, in appearance, flat regular even edge rings. Some have main positions that are irregular even edge rings. All are solitary flexagons because their precursors are solitary flexagons. In an ideal flexagon (Section 1.2), all the leaves in a flexagon are identical polygons that overlap exactly in main positions. However, there are some irregular polygon even edge flexagons that are partial overlap flexagons in which the leaves do not always overlap exactly in main positions.

In a transformation of polygons, a regular convex polygons is transformed into an irregular convex polygon. An irregular polygon edge flexagon inherits some, but not all, the dynamic properties of its precursor regular polygon edge flexagon. Hence, an understanding of the transformation of polygons is needed for an

L. Pook, *Serious Fun with Flexagons*, Solid Mechanics and Its Applications 164,
DOI 10.1007/978-90-481-2503-6_10, © Springer Science+Business Media B.V. 2009

understanding of the dynamic properties of irregular polygon edge flexagons. There are some irregular polygon edge flexagons which do not have precursors that are edge flexagons made from regular convex polygons. These are stretch flexagons in which a flat main position of a flexagon, is stretched uniformly in one direction.

The irregular polygon edge flexagons described in this chapter are irregular triangle even edge flexagons, including the special cases of silver even edge flexagons and bronze even edge flexagons, isosceles triangle odd edge flexagons, irregular quadrilateral even edge flexagons, and irregular pentagon even edge flexagons.

10.1.1 Transformation of Polygons

A regular convex polygon can be transformed into an irregular convex polygon by using a transformation of polygons. There are four main methods of transformation. They are not mutually exclusive because different methods can sometimes be used to achieve the same transformation.

In the first method, the relative edge lengths of a regular convex polygon are altered without changing the vertex angles. This leads to an irregular convex polygon that is an equiangular polygon in which the vertex angles are identical. For example, a square can be transformed into a rectangle. The range of possibilities increases as the number of edges on a polygon increases (Ball 2002). Replacing regular convex polygon leaves by irregular convex polygons with altered relative edge lengths, but with unaltered vertex angles, does not, in general, change the dynamic properties of a flexagon. However, increasing the relative lengths of the hinged edges of leaves sometimes improves the stability of an edge flexagon. It is not possible to alter relative edge lengths of an equilateral triangle without also altering vertex angles.

In the second method, the vertex angles of a regular convex polygon are altered, without changing relative edge lengths. For example, a square can be transformed into a rhombus. The range of possibilities increases as the number of edges on a polygon increases. Replacing regular convex polygon leaves by irregular convex polygons with altered vertex angles only makes a significant difference to the dynamic properties of a flexagon if hinge angles are altered. It is not possible to alter vertex angles of an equilateral triangle without also altering relative edge lengths.

In the third method, a regular convex polygon is truncated at one or more vertices. This usually alters the number of edges, and always removes one or more vertices. It can introduce new vertices and alter the number of edges. For example, a square can be truncated to a 45°–45°–90° (silver) triangle. This reduces the number of edges from four to three, removes one vertex, and alters two vertex angles from 90° to 45. An equilateral triangle can be truncated to a 60°–120° trapezium. This increases the number of edges from three to four, removes one of the original 60° vertices, and introduces two 120° vertices. Truncating regular convex polygons to irregular convex polygons does not make a significant difference to the dynamic properties of a flexagon if hinge angles are left unaltered. However, such truncated flexagons are partial overlap flexagons in which leaves do not overlap exactly during flexing.

In the fourth method, a regular convex polygon is partially stellated to an irregular convex polygon. This reduces the number of edges. For example, if a regular hexagon is partially stellated to a 60°–120° rhombus then the number of edges is reduced from six to four. Replacing regular convex polygons by partially stellated irregular convex polygons does not make a significant difference to the dynamic properties of a flexagon provided that hinge angles are left unaltered. Partial stellation that increases hinge lengths improves the stability of flexagons.

10.1.2 Stretch Flexagons and Stretch Polygon Rings

In a stretch flexagon, a flat main position of a precursor edge flexagon is stretched uniformly in one direction. Thus, the flexagon as a whole is transformed, rather than individual leaves. The flat main position becomes, in appearance, a stretch polygon ring.

For a precursor even edge flexagon in which the leaves are identical and overlap exactly during flexing, there are three possible outcomes. In the first outcome, the leaves of the stretch flexagon are all identical and overlap exactly during flexing. In the second outcome the leaves of the stretch flexagon are all identical, but do not always overlap exactly during flexing, so it is a partial overlap flexagon. In the third outcome the leaves of the stretch flexagon are not all identical so cannot always overlap exactly during flexing and it is a partial overlap flexagon.

For a flat ring of identical convex polygons which is stretched uniformly in one direction. There are two possible outcomes. In the first outcome the polygons in the stretch polygon ring are all identical. In the second outcome the polygons in the stretch polygon ring are not all identical.

10.2 Irregular Triangle Edge Flexagons

10.2.1 Irregular Triangles

An irregular triangle is either a scalene triangle, in which all three vertex angles are different, or an isosceles triangle. In an isosceles triangle two of the vertex angles, the base angles, are the same, and the third vertex angle, the apex angle, is different. Any irregular triangle can be regarded as a transformation from an equilateral triangle (previous section). If ratios between the vertex angles are rational, then an irregular triangle can also be regarded as a partially stellated regular convex polygon. For example, a 30°–60°–90° (bronze) triangle (Section 2.3.1) is a scalene triangle that can be regarded as a partially stellated regular convex dodecagon (Fig. 10.1). Similarly, a 45°–45°–90° (silver) triangle is an isosceles triangle that can be regarded as a partially stellated regular convex octagon (Fig. 2.20).

10.2.2 Isosceles Triangle Edge Rings

There are two series of flat regular isosceles triangle edge rings. In the cart wheel edge ring series there are three or more triangles in the ring, so rings can be either even or odd. Apex angles of isosceles triangles are at the centre of the ring. In the star edge ring series the number of triangles in the ring is even, and there are six or more triangles in the ring. One of the base angles of each isosceles triangle is at the centre of the ring. The two series include all the possible flat regular edge rings of isosceles triangles. The flat regular even edge ring of six equilateral triangles (Fig. 1.1a) is a special case that belongs to both series.

The first six members of the cart wheel isosceles triangle edge ring series are given in Table 10.1. Outlines of cart wheel isosceles triangle ring series members are regular convex polygons. The first member of the series is the flat regular odd edge ring of three 30°–30°–120° isosceles triangles (Fig. 10.2). The second member is the flat regular even edge ring of four 45°–45°–90° (silver) triangles (Fig. 2.21a). This can be regarded as a flat regular even edge ring of four regular convex octagons (Fig. 4.33c) with the octagons partially stellated to silver triangles. The third, fifth and sixth members are the cart wheel edge rings of five 54°–54° 72° isosceles triangles (Fig. 10.3a), seven 64.3°–64.3°–51.4° isosceles triangles

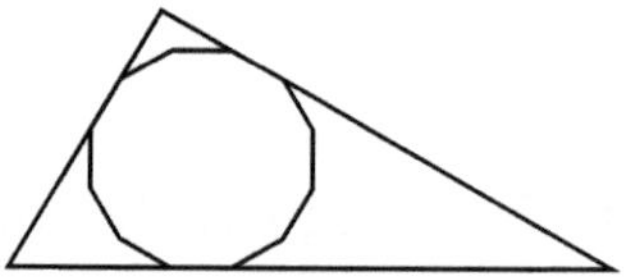

Fig. 10.1 A 30°–60°–90° (bronze) triangle as a partially stellated regular convex dodecagon

Table 10.1 The cart wheel isosceles triangle edge ring series

Number of triangles	Triangle type	Ring symbol
3	30°–30°–120°	3(120°)
4	45°–45°–90° (silver)	4(90°)
5	54°–54° 72°	5(72°)
6	Equilateral	6(60)°
7	64.3°–64.3°–51.4°	7(51.4°)
8	67.5°–67.5°–45°	8(45°)

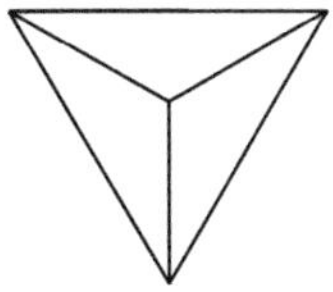

Fig. 10.2 A flat regular odd edge ring of three 30°–30°–120° isosceles triangles

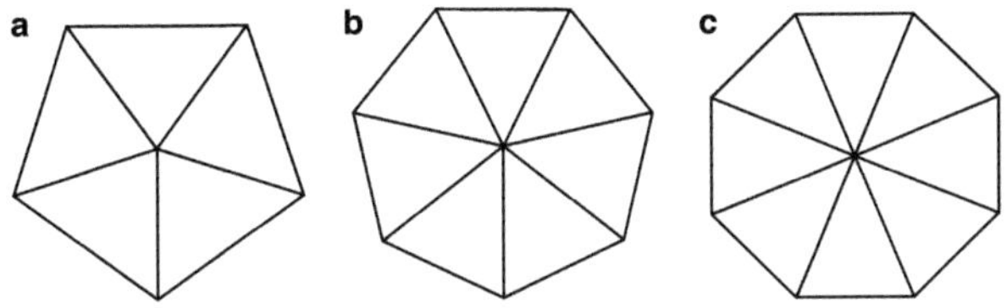

Fig. 10.3 Cart wheel edge rings. (**a**) Five 54°–54°–72° isosceles triangles. (**b**) Seven 64.3°–64.3°–51.4° isosceles triangles. (**c**) Eight 67.5°–67.5°–45° isosceles triangles

Table 10.2 The star isosceles triangle edge ring series

Number of triangles	Triangle type	Ring symbol
6	Equilateral	6(60°)
8	45°–45°–90° (silver)	8(45°)
10	36°–36°–108°	10(36°)
12	30°–30°–120°	12(30°)

(Fig. 10.3b), and eight 67.5°–67.5°–45° isosceles triangles (Fig. 10.3c). The fourth member is the flat regular even edge ring of six equilateral triangles (Fig. 1.1a). Edge rings of five or more isosceles triangles, in the cart wheel isosceles triangle edge ring series, are called cart wheel edge rings.

The first four members of the star isosceles triangle edge ring series are listed in Table 10.2. The first member is the flat regular even edge ring of six equilateral triangles (Fig. 1.1a). The second member is the flat regular even edge ring of eight 45°–45°–90° (silver) triangles (Fig. 2.21b). This can be regarded as a flat regular even edge ring of eight regular convex octagons (Fig. 2.9) with the octagons partially stellated to silver triangles. The third and fourth members are the star edge rings of ten 36°–36°–108° isosceles triangles (Fig. 10.4a) and 12 30°–30°–120° isosceles triangles (Fig. 10.4b). Rings of ten or more isosceles triangles, in the star isosceles triangle edge ring series, have a star outline and are called star edge rings.

There are numerous possible flat irregular edge rings of isosceles triangles. The simplest possible flat irregular even edge ring of silver triangles consists of ten triangles (Fig. 10.5).

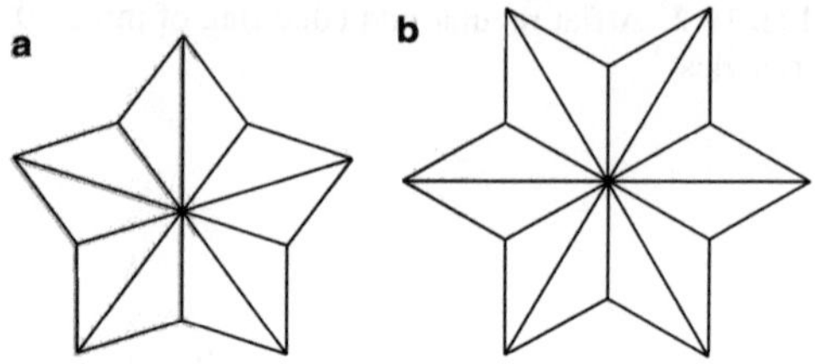

Fig. 10.4 Star edge rings. (a) Ten 36°–36°–108° isosceles triangles. (b) Twelve 30°–30°–120° isosceles triangles

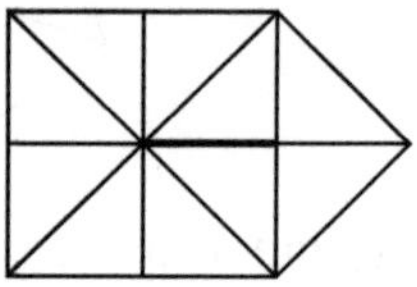

Fig. 10.5 A flat irregular even edge ring of ten 45°–45°–90° isosceles (silver) triangles

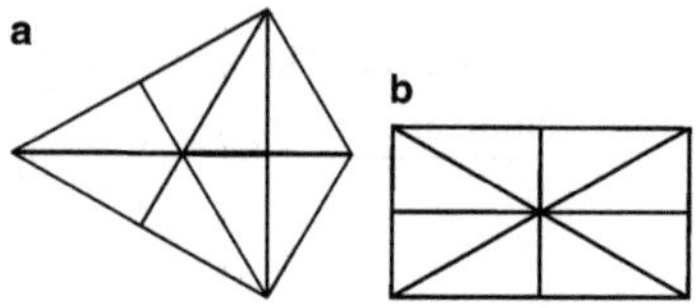

Fig. 10.6 Flat irregular even edge rings of eight 30°–60°–90° (bronze) triangles. (a) Type A. (b) Type B

10.2.3 Bronze Even Edge Rings

There is only one series of flat regular scalene triangle edge rings. This is the star ring series of flat regular scalene triangle even edge rings. Numbers of triangles in a ring and ring symbols are the same as for members of the star isosceles triangle edge ring series (Table 10.1), but each member has an infinity of forms because an infinity of different scalene triangles can be used. Regular scalene triangle odd edge rings are not possible.

Flat regular bronze even edge rings are a special case of flat regular scalene triangle even edge rings. The 30°–60°–90° (bronze) triangle is the only scalene triangle where three types of flat regular even edge rings are possible. There is one ring of four bronze triangles (Fig. 2.22a), one ring of six bronze triangles (Fig. 2.22b), and one ring of 12 bronze triangles (Fig. 2.22c). The ring of six bronze triangles can be regarded as a flat regular even edge ring of eight regular dodecagons (Fig. 8.30) with the dodecagons partially stellated to bronze triangles. For other scalene triangles, either one or two types of flat regular even edge rings are possible. There are numerous possible flat irregular edge rings of scalene triangles. The two simplest possible flat irregular even edge rings of bronze triangles consist of eight triangles (Fig. 10.6).

10.2.4 Isosceles Triangle Even Edge Flexagons

A fundamental isosceles triangle even edge flexagon can be derived by using a first order fundamental triangle even edge flexagon (Section 4.2.3) as a precursor and replacing the equilateral triangles by isosceles triangles. The torsion per sector is 1 and they are basic even edge flexagons. There are no degenerate versions (Section 8.2.1). *Cart* wheel even edge flexagons and star ring even edge flexagons are two varieties of isosceles triangle even edge flexagons. Silver flexagons are isosceles triangle flexagons made from 45°–45°–90° isosceles (silver) triangles, and are often regarded as a distinct variety (Mitchell 2002).

If *S*-fold pinch flexes are used, where *S* is the number of sectors, then the intermediate position map is the same for all fundamental isosceles triangle even edge flexagons, and is the same as the intermediate position map for the fundamental triangle even edge flexagons *S*⟨3, 3⟩ (Fig. 4.7). A 3-cycle can be traversed, so there are three main positions that can be visited. There are two types of main position appearance, one of which appears twice. This is because, due to rotation of leaves (Section 4.2.5.3) the sum of the hinge angles at the centre of the flexagon has two different values, one of which appears twice. Other flexes are usually possible. General features of isosceles triangle even edge flexagons are illustrated by the silver even edge flexagons described below.

10.2.4.1 Fundamental Silver Even Edge Flexagons

A 45°–45°–90° (silver) triangle is an isosceles triangle, so fundamental silver even edge flexagons could also be called fundamental isosceles triangle even edge flexagons. Fundamental silver even edge flexagons are made from the fundamental silver edge net ⟨silver⟩ (Fig. 3.24), and a standard face numbering sequence (Section 4.1.1) is used. A convenient notation is *S*⟨silver⟩ where *S* is the number of sectors. There are two flat regular even edge rings of silver triangles (Fig. 2.21). These are main position appearances of the two sector fundamental silver even edge flexagon 2⟨silver⟩, and the four sector fundamental silver even edge flexagon 4⟨silver⟩. Some of the properties of these flexagons are given in Table 10.3

Table 10.3 Properties of fundamental silver even edge flexagons regarded as partially stellated degenerate octagon even edge flexagons

Flexagon symbol	Typical main position	Cycle type	Number of cycles	Main position type	Ring symbol	Curvature
2⟨silver⟩	1(2)	1/4 cycle	1	Flat	4(90°)	0°
2⟨silver⟩	2(3)	2/8 cycle	1	Slant	4(45°)	180°
4⟨silver⟩	1(2)	1/4 cycle	1	Skew	8(90°)	–360°
4⟨silver⟩	2(3)	2/8 cycle	1	Flat	8(45°)	0°
5⟨silver⟩	1(2)	1/4 cycle	1	Skew	10(90°)	–540°
5⟨silver⟩	2(3)	2/8 cycle	1	Skew	10(45°)	–90°

(cf. Tables 4.3, 4.7 and 8.4). Some of the properties of the five sector fundamental silver even edge flexagon 5⟨silver⟩ are included in the table. The torsion per sector is 1. The flexagon figure is shown in Fig. 10.7. Dynamic properties are a combination of those of first order fundamental triangle even edge flexagons (Section 4.2.3) and degenerate octagon even edge flexagons (Section 8.2.5). In Table 10.3 cycle types shown are for fundamental silver even edge flexagons regarded as partially stellated degenerate octagon even edge flexagons.

A fundamental silver even edge flexagon has much longer hinges than its precursor degenerate octagon even edge flexagon, and is much easier to handle. In turn, the precursor for the degenerate octagon even edge flexagon is a first order fundamental octagon even edge flexagon. The close relationships between these flexagons means that the incomplete cycle descriptions used for degenerate even edge flexagons are also appropriate for fundamental silver even edge flexagons. In Table 10.3, the original meaning of 2/8-cycle is that a subsidiary 8-cycle could be traversed in the precursor first order fundamental octagon even edge flexagon, but in a degenerate octagon even edge flexagon it is an incomplete cycle, and only two subsidiary main positions can be visited (Section 8.2.5). In a fundamental silver even edge flexagon, a transferred meaning is that two of the main positions in the 3-cycle that can be traversed have the same appearance.

The net for the two sector fundamental silver even edge flexagon 2⟨silver⟩ is shown in Fig. 10.8a. This was derived by using the net for a two sector degenerate octagon even edge flexagon (Fig. 8.25) as a precursor, partially stellating the regular octagons to silver triangles (Fig. 2.20), and renumbering. It could also be derived by fan folding the fundamental silver net ⟨silver⟩ (Fig. 3.24) onto the flat regular even edge ring of four silver triangles (Fig. 2.21a) and applying a standard face numbering sequence (Section 4.1.1), or by using the net for the first order fundamental triangle even edge flexagon 2⟨3, 3⟩ (Fig. 4.13) as a precursor and replacing the equilateral triangles by silver triangles.

As assembled, the flexagon is in subsidiary main position 1(2) which is, in appearance, a flat regular even edge ring of four silver triangles (Fig. 2.21a). It can be flexed to subsidiary main positions 2(3) and 1(3), by using the twofold pinch flex, as shown in the intermediate position map. The intermediate position map is

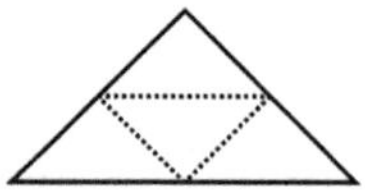

Fig. 10.7 The flexagon figure for fundamental silver even edge flexagons *S*⟨silver⟩

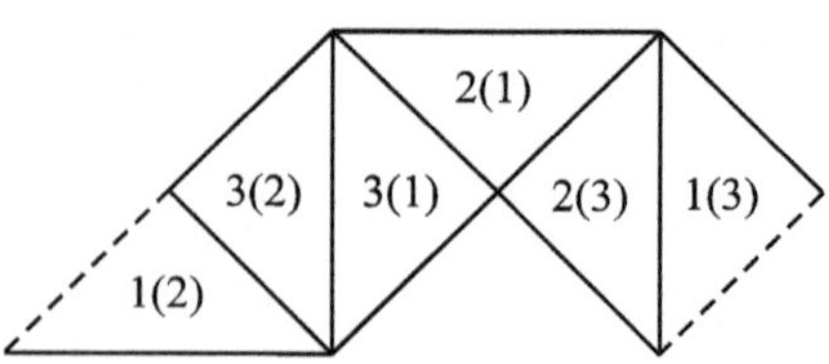

Fig. 10.8 Net for fundamental silver even edge flexagons. (**a**) 2⟨silver⟩. One copy needed. (**b**) 4⟨silver⟩. Two copies needed. Fold together pairs of leaves numbered 2

the same as that for the first order fundamental triangle even edge flexagons $S\langle 3, 3\rangle$ (Fig. 4.7). The two subsidiary main positions are slant regular even edge ring of four silver triangles (Fig. 10.9a). The high curvature (180°) means that they cannot be turned inside out. Hence, the flexagon cannot be flexed directly between subsidiary main positions 2(3) and 1(3). Intermediate positions are silver triangle edge pairs. Intermediate position 3 cannot be reached from subsidiary main position 1(2) without disconnecting a hinge, refolding the flexagon, and reconnecting the hinge. All the subsidiary main positions can be visited so, as assembled, it is not a deficient flexagon. However, if it is assembled by folding together leaves numbered 1 or 2, with leaves numbered 3 visible on the outside of the ring, then it becomes a deficient flexagon because subsidiary main position 1(2) cannot be visited (cf. Section 8.2.5).

The net for the four sector fundamental silver even edge flexagon 4⟨silver⟩ is shown in Fig. 10.8b. This was derived by using the net for the two sector fundamental silver even edge flexagon 2⟨silver⟩ (Fig. 10.8a) as a precursor and increasing the number of sectors from two to four. The net could also be derived by deleting a face from the slit square flexagon described by Mitchell (2002). Other derivations are possible.

As assembled, the flexagon is in subsidiary main position 3(1). The 3-cycle shown in the intermediate position map can be traversed by using the fourfold pinch flex. The intermediate position map is the same as that for the first order fundamental triangle even edge flexagons $S\langle 3, 3\rangle$ (Fig. 4.7). Subsidiary main positions 2(3) and 3(1) are flat regular even edge rings of eight silver triangles (Fig. 2.21b). Subsidiary main position 1(2) is a skew regular even edge ring of eight silver triangles (Fig. 10.9b). Intermediate positions are silver triangle edge quadruples.

Pocket flexes are possible in all three subsidiary main position. In the flat subsidiary main positions, 2(3) and 3(1), one pocket flex leads to a minor main position which is a combination of a slant regular even edge ring of six silver triangles and an edge pair of silver triangles (Fig. 10.10a). A second pocket flex can then result in three different combinations. In skew subsidiary main position 1(2) two pocket flexes can lead to a combination of a flat regular even edge ring of four silver

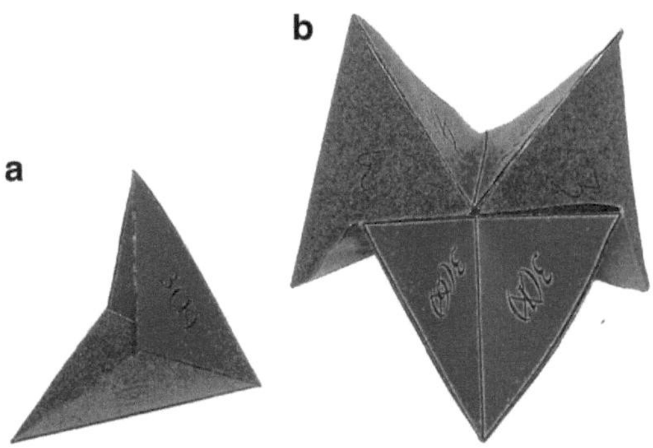

Fig. 10.9 Flexagons as the following. (**a**) A slant regular even edge ring of four 45°–45°–90° isosceles (silver) triangles. (**b**) A skew regular even edge ring of eight 45°–45°–90° isosceles (silver) triangles

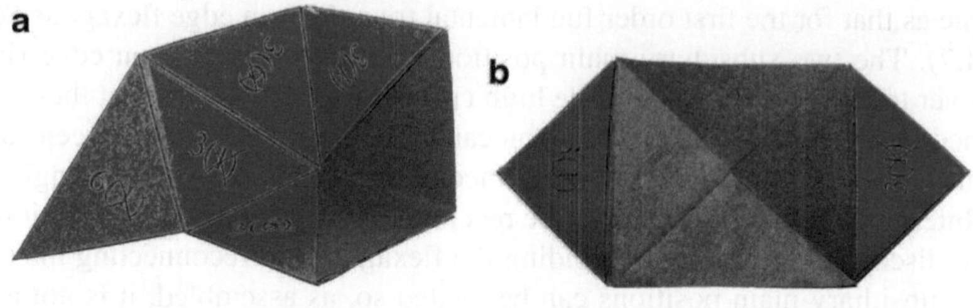

Fig. 10.10 A flexagon as the following combinations. (**a**) A slant regular even edge ring of six 45°–45°–90° isosceles (silver) triangles and an edge pair of 45°–45°–90° isosceles (silver) triangles. (**b**) A flat regular even edge ring of four 45°–45°–90° isosceles (silver) triangles and two edge pairs of 45°–45°–90° isosceles (silver) triangles

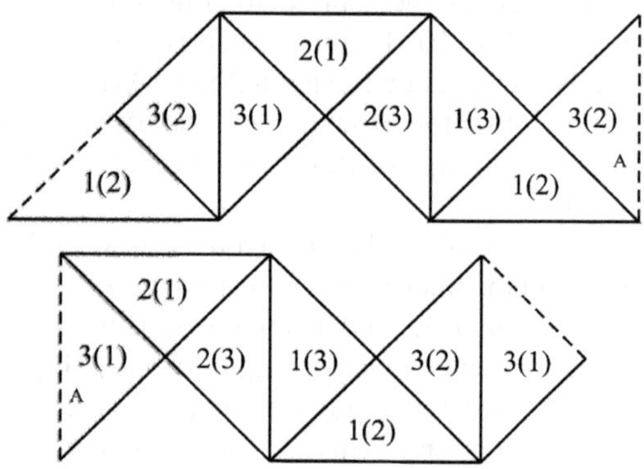

Fig. 10.11 Net for the fundamental silver even edge flexagon 5⟨silver⟩. One copy needed. Join the two parts of the net at A-A. Fold together pairs of leaves numbered 2

triangles and two edge pairs of silver triangles (Fig. 10.10b). Other combinations and flexes are possible.

The net for the five sector fundamental silver even edge flexagon 5⟨silver⟩ is shown in Fig. 10.11. This was derived by using the net for the two sector fundamental silver even edge flexagon 2⟨silver⟩ (Fig. 10.8a) as a precursor and increasing the number of sectors from two to five. Other derivations are possible. As assembled, the flexagon is in subsidiary main position 3(1). The 3-cycle shown in the intermediate position map can be traversed by using the fivefold pinch flex. The intermediate position map is the same as that for the first order fundamental triangle even edge flexagons *S*⟨3, 3⟩ (Fig. 4.7). All three subsidiary main positions are skew regular even edge rings of ten silver triangles. Intermediate positions are silver triangle edge quintuples. Other flexes are possible, including pocket flexes.

10.2.4.2 The Irregular Ring Ten Silver Triangle Even Edge Flexagon

The fundamental silver net ⟨silver⟩ (Fig. 3.24) can be fan folded onto the flat irregular even edge ring of ten silver triangles shown in Fig. 10.5 in one distinct way. This solution is shown in Fig. 10.12 as the flexagon diagram for the fundamental irregular ring ten silver triangle even edge flexagon. A key pat is identified. This is a single leaf which is hinged on its two short edges. Its net is shown in Fig. 10.13. It is not possible to number the faces so that face numbers do not become mixed up, so a standard face numbering sequence is used (Section 4.1.1). The torsion is 5. A transformation between flexagons is possible with the five sector fundamental silver even edge flexagon 5⟨silver⟩ (previous section).

As assembled, the flexagon is in main position 1(2) at the key pat in the flexagon diagram. The main position is, in appearance, a flat irregular even edge ring of ten silver triangles (Fig. 10.5). The 5-cycle shown in the Tuckerman diagram (Fig. 10.14) can be traversed by using a flap flex. It is a local flex because some pats are left unchanged. To flex to main position 4(5) starting from main position 1(2), first fold together the two pairs of leaves numbered 2 and 14. Next, unfold the flap numbered 1(13) and fold together the pair of leaves numbered 1. Then fold

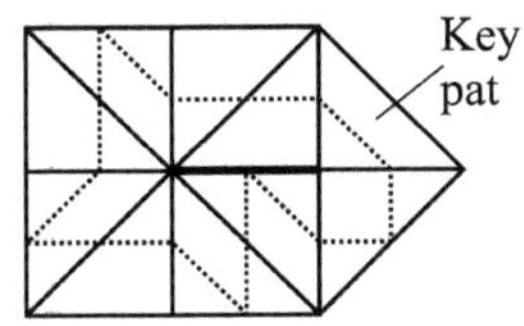

Fig. 10.12 Flexagon diagram for the fundamental irregular ring ten silver triangle even edge flexagon

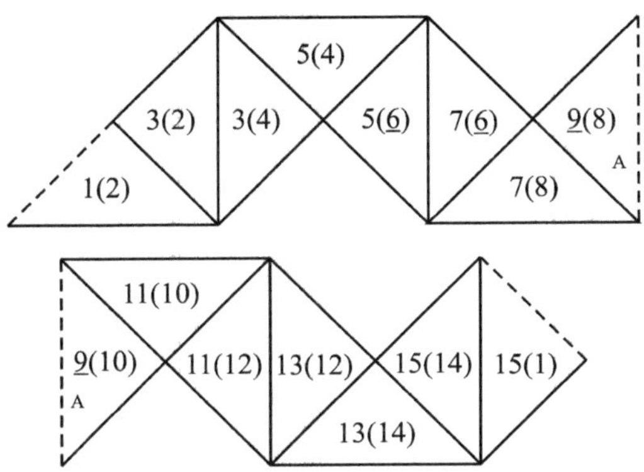

Fig. 10.13 Net for the fundamental irregular ring ten silver triangle even edge flexagon. One copy needed. Join the two parts of the net at A-A. Fold together pairs of leaves numbered 5, 8,11, 13 and 15

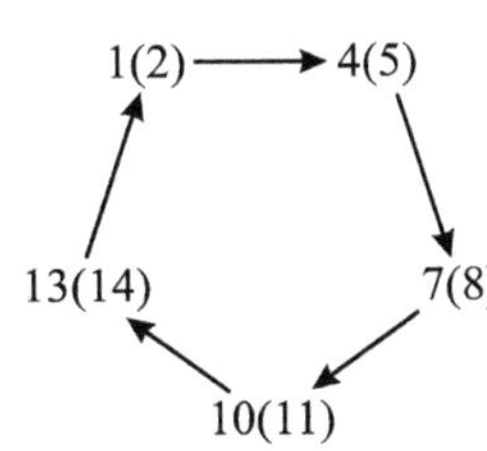

Fig. 10.14 Tuckerman diagram for the 5-cycle of the fundamental irregular ring ten silver triangle even edge flexagon

together the pair of leaves numbered 3. Finally, unfold the triangular flap to reach main position 4(5). The five leaves that are hinged on the two short edges, and hence the main positions, appear in cyclic order as the 5-cycle is traversed.

To transform to the five sector fundamental silver even edge flexagon 5⟨silver⟩, starting from main position 1(2), fold together the two pairs of leaves numbered 2 and 14. Then unfold the flap numbered 1(13). Next, pinch together pairs of pats so that leaves numbered 3, 6, 9 and 12 are folded together. This is an intermediate position of 5⟨silver⟩ and is a silver triangle quintuple. Various other flexes, including pocket flexes, are possible.

10.2.4.3 A Partial Overlap Silver Even Edge Flexagon

The net for a partial overlap silver even edge flexagon is shown in Fig. 10.15. This was derived by using the net for the first order fundamental square even edge flexagon 2⟨4, 4⟩ (Fig. 4.16a) as a precursor and truncating the squares to 45°–45°–90° isosceles (silver) triangles, so It is also a truncated flexagon. It could also be called a truncated first order fundamental square even edge flexagon. The dynamic properties are similar to those of the precursor flexagon. Some of its properties are given in Table 10.4 (cf. Table 4.4). The torsion per sector is 2.

As assembled, the flexagon is in principal main position 1(2). Principal main positions have an irregular hexagonal outline (Fig. 10.16a). Parts of leaves properly belonging to another face are visible, and the twisted band structure of the flexagon shows clearly. The principal 4-cycle shown in the intermediate position map and the Tuckerman diagram can be traversed by using the twofold pinch flex. The intermediate position map and Tuckerman diagram are the same as those for the first order fundamental square even edge flexagons $S\langle 4, 4\rangle$ (Figs. 4.8 and 4.17). Intermediate positions have a rectangular outline (Fig. 10.16b). They can be opened into box positions which have a jagged appearance (Fig. 10.16c).

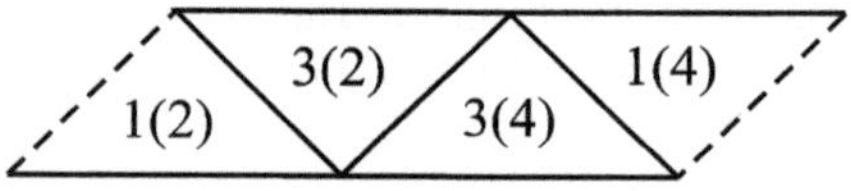

Fig. 10.15 Net for a partial overlap silver even edge flexagon. Two copies needed

Table 10.4 Properties of a partial overlap silver even edge flexagon. The principal cycle is in bold

Typical main position	Cycle type	Number of cycles	Main position type	Ring symbol	Sector symbol	Curvature
1(2)	**4-cycle**	**1**	**Flat**	**4(90°)**	**⟨4, 4, 3, 1⟩**	**0°**
1(3)	None	–	Box	4(0°)	⟨4, 4, 2, 2⟩	360°

10.2.4.4 A Fundamental Cart Wheel Even Edge Flexagon

The net for a fundamental cart wheel even edge flexagon is shown in Fig. 10.17. This was derived by using the net for the first order fundamental triangle even edge flexagon 2⟨3, 3⟩ (Fig. 4.13) as a precursor, replacing the equilateral triangles by 67.5°–67.5°–45° isosceles triangles, and increasing the number of sectors from two to four. It could also be called a fundamental isosceles triangle even edge flexagon.

As assembled, the flexagon is in main position 1(2) which is, in appearance, a cart wheel even edge ring of eight 67.5°–67.5°–45° isosceles triangles (Fig. 10.3c). The 3-cycle shown in the intermediate position map can be traversed by using the fourfold pinch flex. The intermediate position map is the same as that for the first order fundamental triangle even edge flexagons *S*⟨3, 3⟩ (Fig. 4.7). Main positions 2(3) and 3(1) are skew regular even edge rings of eight 67.5°–67.5°–45° isosceles triangles. Intermediate positions are 67.5°–67.5°–45° isosceles triangle edge quadruples.

Starting from main position 1(2), a single pocket flex leads to a combination of a slant regular even edge ring of six isosceles triangles and an isosceles triangle edge pair. Folding the flexagon in two and then unfolding at the centre with a twisting action leads to a combination of a slant regular even edge ring of four isosceles triangles and two isosceles triangle edge pairs, with twofold rotational symmetry. Other flexes are possible.

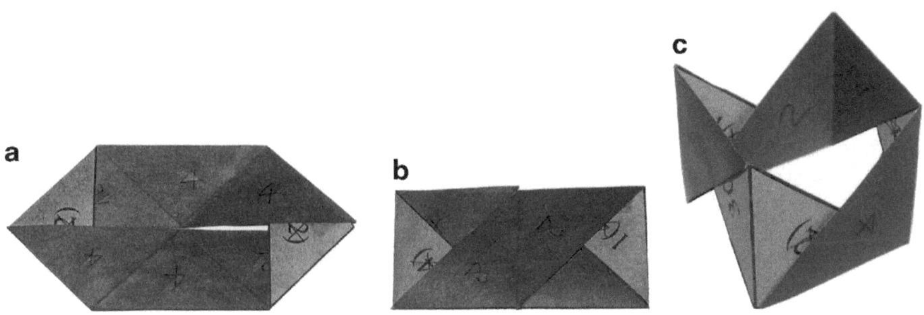

Fig. 10.16 A partial overlap silver even edge flexagon. (**a**) Principal main position. (**b**) Intermediate position. (**c**) Box position

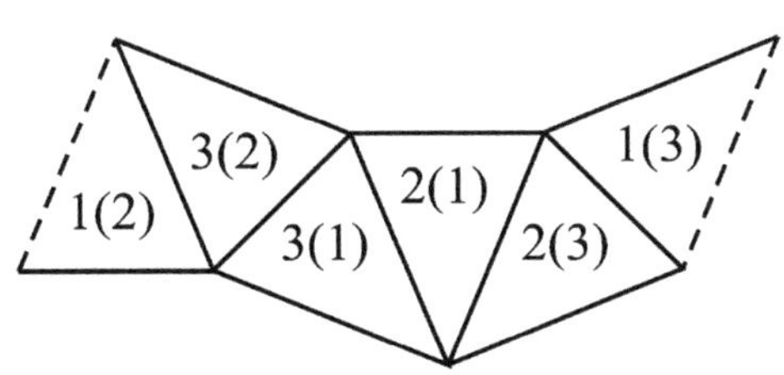

Fig. 10.17 Net for a fundamental cart wheel even edge flexagon. Two copies needed

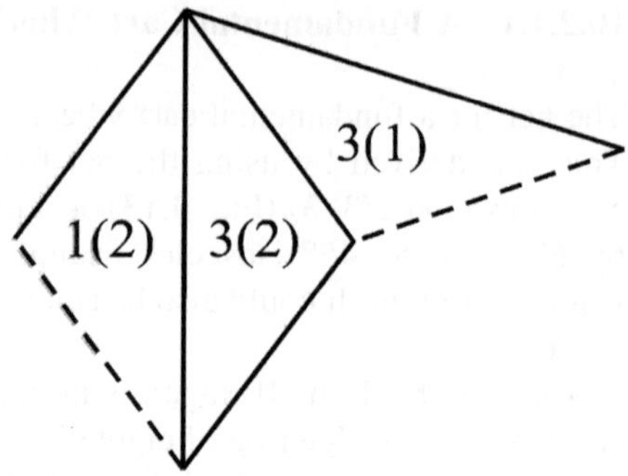

Fig. 10.18 Net for a fundamental star ring even edge flexagon. Five copies needed

10.2.4.5 A Fundamental Star Ring Even Edge Flexagon

The net for a fundamental star ring even edge flexagon is shown in Fig. 10.18. This was derived by using the net for the first order fundamental triangle even edge flexagon 2⟨3, 3⟩ (Fig. 4.13) as a precursor, replacing the equilateral triangles by 36°–36°–108° isosceles triangles and increasing the number of sectors from two to five. It could also be called a fundamental isosceles triangle even edge flexagon.

As assembled, the flexagon is in main position 1(2) which is, in appearance, a star edge ring of ten 36°–36°–108° isosceles triangles (Fig. 10.4a). The 3-cycle shown in the intermediate position map can be traversed by using the fivefold pinch flex. The intermediate position map is the same as that for the first order fundamental triangle even edge flexagons *S*⟨3, 3⟩ (Fig. 4.7). Main position 2(3) is a star edge ring of ten 36°–36°–108° isosceles triangles, and main position 3(1) a skew even edge ring of ten 36°–36°–108° isosceles triangles. Intermediate positions are 36°–36°–108° isosceles triangle edge quintuples. Other flexes are possible, including pocket flexes. For videos of some flexes see Sherman (2008).

10.2.5 Isosceles Triangle Odd Edge Flexagons

A fundamental isosceles triangle odd edge flexagon can be derived by using a second order fundamental odd edge flexagon (Section 4.3.1) as a precursor and replacing the equilateral triangles by isosceles triangles. The torsion per sector is 1. Main position of fundamental isosceles triangle odd edge flexagons are, in appearance, cart wheel odd edge rings of isosceles triangles so they could also be called fundamental cart wheel odd edge flexagons. They were first described by Sherman (2007). Flexing options are similar to those for second order fundamental odd edge flexagons. Two examples with special names, the pentaflexagon and the heptaflexagon, are given below.

The net for the pentaflexagon is shown in Fig. 10.19. 'Penta' refers to the pentagonal outline of a principal main position (Sherman 2007). The net was derived

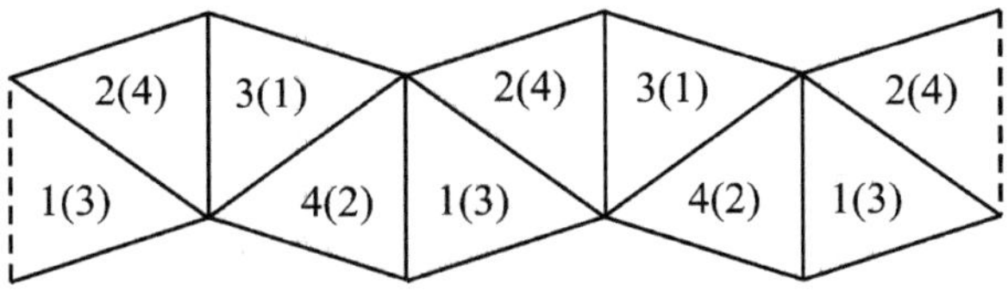

Fig. 10.19 Net for the pentaflexagon. One copy needed. Fold each leaf numbered 3 onto a leaf numbered 4

Fig. 10.20 A flexagon as a combination of a slant regular odd edge ring of three 54°–54°–72° isosceles triangles and an edge pair of 54°–54°–72° isosceles triangles

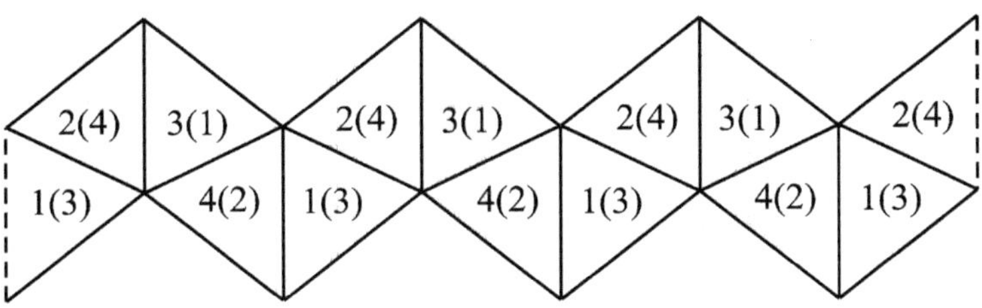

Fig. 10.21 Net for the heptaflexagon. One copy needed. Fold each leaf numbered 3 onto a leaf numbered 4

by using the net for the second order fundamental triangle odd edge flexagon $5\langle 3, 3\rangle_2$ (Fig. 4.40) as a precursor and replacing the equilateral triangles by 54°–54°–72° isosceles triangles.

As assembled, the flexagon is in principal main position 1(2), which is, in appearance, a cart wheel odd edge ring of five 54°–54°–72° isosceles triangles (Fig. 10.3a). The flexagon can be flexed to a minor main position by using a pocket flex. The minor main position is a combination of a slant regular odd edge ring of three 54°–54°–72° isosceles triangles and an edge pair of 54°–54°–72° isosceles triangles (Fig. 10.20). This can be done in ten different ways, whereas there are five different ways for the second order fundamental triangle odd edge flexagon $5\langle 3, 3\rangle_2$ (Section 4.3.2). This difference is because a principal main position is flat (curvature 0°) whereas a principal main position of $5\langle 3, 3\rangle_2$ is slant (curvature 60°).

The net for the heptaflexagon is shown in Fig. 10.21 'Hepta' refers to the heptagonal outline of a principal main position. The net was derived by using the net for the second order fundamental triangle odd edge flexagon $7\langle 3, 3\rangle_2$ (Fig. 4.42) as a precursor and replacing the equilateral triangles by 64.3°–64.3°–51.4° isosceles triangles.

As assembled, the flexagon is in principal main position 1(2), which is a cart wheel odd edge ring of seven 64.3°–64.3°–51.4° isosceles triangles (Fig. 10.3b). The flexagon can be flexed to a minor main position by using a pocket flex. The minor main position is a combination of a slant regular odd edge ring of five 64.3°–64.3°–51.4° isosceles triangles and an edge pair of seven 64.3°–64.3°–51.4° isosceles triangles (Fig. 10.22a). As with $7\langle 3, 3\rangle_2$ this can be done in 14 different ways (Section 4.3.2). A second pocket flex can then be used to reach two further types of subsidiary main positions. One of these is a combination of a slant regular odd edge ring of three 64.3°–64.3°–51.4° isosceles triangles and two edge pairs of 64.3°–64.3°–51.4° isosceles triangles (Fig. 10.22b). The other is a combination of a slant regular odd edge ring of three 64.3°–64.3°–51.4° isosceles triangles and an edge triple of 64.3°–64.3°–51.4° isosceles triangles (Fig. 10.22c). Other flexes are possible, including an asymmetric threefold pinch flex.

10.2.6 Scalene Triangle Even Edge Flexagons

A fundamental scalene triangle even edge flexagon can be derived by using a first order fundamental triangle even edge flexagon (Section 4.2.3) as a precursor and replacing the equilateral triangles by scalene triangles. The torsion per sector is 1 and they are basic even edge flexagons. There are no degenerate versions (Section 8.2.1). Bronze flexagons are scalene triangle flexagons made from 30°–60°–90° scalene (bronze) triangles, and are often regarded as a distinct variety (Mitchell 2002).

If S-fold pinch flexes are used, where *S* is the number of sectors, then the intermediate position map is the same for all fundamental scalene triangle even edge flexagons, and is the same as the intermediate position map for the first order fundamental triangle even edge flexagons $S\langle 3, 3\rangle$ (Fig. 4.7). A 3-cycle can be traversed, so there are three main positions that can be visited. Each of the three main posi-

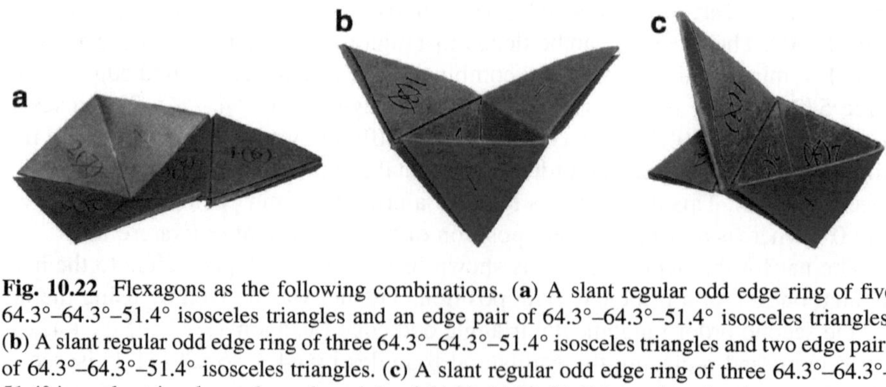

Fig. 10.22 Flexagons as the following combinations. (**a**) A slant regular odd edge ring of five 64.3°–64.3°–51.4° isosceles triangles and an edge pair of 64.3°–64.3°–51.4° isosceles triangles. (**b**) A slant regular odd edge ring of three 64.3°–64.3°–51.4° isosceles triangles and two edge pairs of 64.3°–64.3°–51.4° isosceles triangles. (**c**) A slant regular odd edge ring of three 64.3°–64.3°–51.4° isosceles triangles and an edge triple of 64.3°–64.3°–51.4° isosceles triangles

tions has a different appearance. This is because, due to rotation of leaves (Section 4.2.5.3), the sum of the hinge angles at the centre of the flexagon has three different values. Other flexes are sometimes possible. General features of scalene triangle even edge flexagons are illustrated by the *bronze even edge flexagons* described below.

10.2.6.1 Fundamental Bronze Even Edge Flexagons

A 30°–60°–90° (bronze) triangle is a scalene triangle, so fundamental bronze even edge flexagons could also be called fundamental scalene triangle even edge flexagons. Fundamental bronze even edge flexagons are made from the fundamental bronze edge net ⟨bronze⟩ (Fig. 3.25), and a standard face numbering sequence (Section 4.1.1) is used. A convenient notation is *S*⟨bronze⟩ where *S* is the number of sectors. There are three flat regular even edge rings of bronze triangles (Fig. 2.22). These are main position appearances of the two sector fundamental bronze even edge flexagon 2⟨bronze⟩, the three sector fundamental bronze even edge flexagon 3⟨bronze⟩, and the six sector fundamental bronze even edge flexagon 6⟨bronze⟩. Some of the properties of these flexagons are given in Table 10.5 (cf. Tables 4.3, 4.8 and 8.5). The torsion per sector is 1. Some of the properties of the four sector fundamental bronze even edge flexagon 4⟨bronze⟩ are included in the table. The flexagon figure is shown in Fig. 10.23. Dynamic properties are a combination of those of first order fundamental triangle even edge flexagons (Section 4.2.3) and a degenerate dodecagon even edge flexagon (Section 8.2.6). In Table 10.5 cycle types shown are for fundamental bronze even edge flexagons regarded as partially stellated degenerate dodecagon even edge flexagons.

A fundamental bronze even edge flexagon has much longer hinges than its precursor degenerate dodecagon even edge flexagon, and is much easier to handle. In turn, the precursor for the degenerate dodecagon even edge flexagon is a first order fundamental dodecagon even edge flexagon. The close relationships between these flexagons means that the incomplete cycle descriptions used for degenerate even edge flexagons are also appropriate for fundamental bronze even edge flexagon. In Table 10.5 the original meaning of 1/4-cycle is that a subsidiary 4-cycle could be traversed in the precursor first order fundamental dodecagon even edge flexagon, but in a degenerate dodecagon even edge flexagon it is an incomplete cycle, and only one subsidiary main positions can be visited (Section 8.2.6). In a fundamental bronze even edge flexagon a transferred meaning is that there is only one main position in the3-cycle that can be traversed that has a particular appearance.

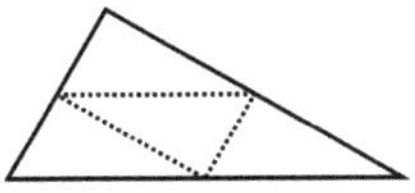

Fig. 10.23 Flexagon figure for fundamental bronze even edge flexagons *S*⟨bronze⟩

Table 10.5 Properties of fundamental bronze even edge flexagons regarded as partially stellated degenerate dodecagon even edge flexagons. Principal incomplete cycles are in bold

Flexagon symbol	Main position	Cycle type	Number of cycles	Main position type	Ring symbol	Curvature
2⟨bronze⟩	1(2)	1/4-cycle	1	Flat	4(90°)	0°
	2(3)	1(3)	1	Slant	4(60°)	120°
	3(1)	**1(8)**	**1**	**Slant**	**4(30°)**	**240°**
3⟨bronze⟩	1(2)	1/4	1	Skew	6(90°)	−180°
	2(3)	1/3	1	Flat	6(60°)	0°
	3(1)	**1/8**	**1**	**Slant**	**6(30°)**	**180°**
4⟨bronze⟩	1(2)	1/4	1	Skew	8(90°)	−360°
	2(3)	1/3	1	Skew	8(60°)	−120°
	3(1)	**1/8**	**1**	**Slant**	**8(30°)**	**120°**
6⟨bronze⟩	1(2)	1/4	1	Skew	12(90°)	−720°
	2(3)	1/3	1	Skew	12(60°)	−360°
	3(1)	1/8	1	Flat	12(30°)	0°

10.2.6.2 The Two Sector Fundamental Bronze Even Edge Flexagon

The net for the two sector fundamental bronze even edge flexagon 2⟨bronze⟩ is shown in Fig. 10.24a. This was derived by using the net for a three sector degenerate dodecagon even edge flexagon (Fig. 8.28) as a precursor, partially stellating the regular dodecagons to bronze triangles, decreasing the number of sectors from three to two, and renumbering. It could also be derived by fan folding the fundamental bronze net ⟨bronze⟩ (Fig. 3.25) onto the flat even edge ring of four bronze triangles (Fig. 2.22a) and applying a standard face numbering sequence (Section 4.1.1), or by using the net for the first order fundamental triangle even edge flexagon 2⟨3, 3⟩ (Fig. 4.13) as a precursor and replacing the equilateral triangles by bronze triangles.

As assembled, the flexagon is in subsidiary main position 1(2) which is, in appearance, a flat regular even edge ring of four bronze triangles (Fig. 2.22a). It can be flexed to subsidiary main position 2(3) and principal main position 1(3), by using the twofold pinch flex, as shown in the intermediate position map. The intermediate position map is the same as that for the first order fundamental triangle even edge flexagons *S*⟨3, 3⟩ (Fig. 4.7). Subsidiary main position 2(3) and principal main position 1(3) are slant regular even edge rings of four bronze triangles types A and B respectively (Fig. 10.25). The high curvatures (120° and 240° respectively) mean that they cannot be turned inside out. Hence, the flexagon cannot be flexed directly between subsidiary main position 2(3) and principal main position 1(3). Intermediate positions are bronze triangle edge pairs. Intermediate position 3 cannot be reached from subsidiary main position 1(2) without disconnecting a hinge, refolding the flexagon, and reconnecting the hinge. All the subsidiary main positions can be visited so, as assembled, it is not a deficient flexagon. However, if it is assembled by folding together leaves numbered 1 or 2, with leaves numbered 3 visible on the outside of the ring, then it becomes a deficient flexagon because subsidiary main position 1(2) cannot be visited (cf. Section 8.2.5 and 10.2.4.1).

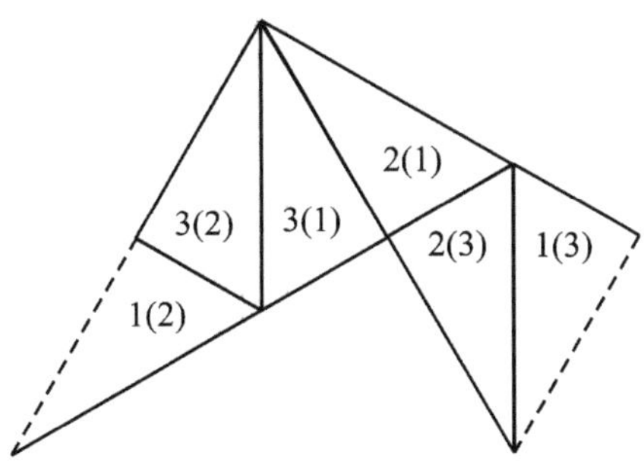

Fig. 10.24 Net for fundamental bronze even edge flexagons. (**a**) 2⟨bronze⟩. One copy needed. (**b**) 4⟨bronze⟩. Two copies needed. Fold together pairs of leaves numbered 1

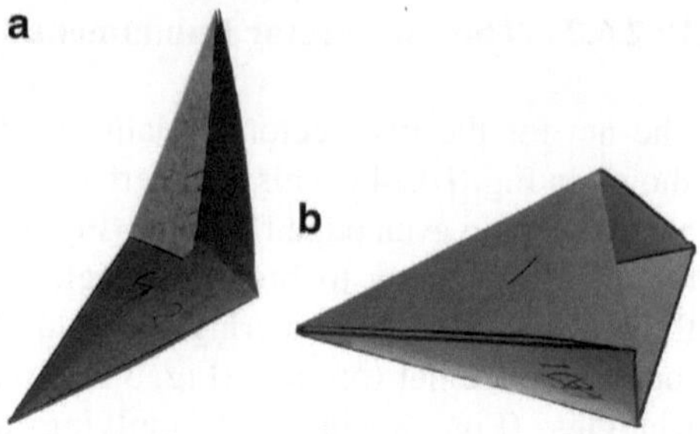

Fig. 10.25 A flexagon as slant regular even edge rings of four 30°–60°–90° (bronze) triangles. (**a**) Type A. (**b**) Type B

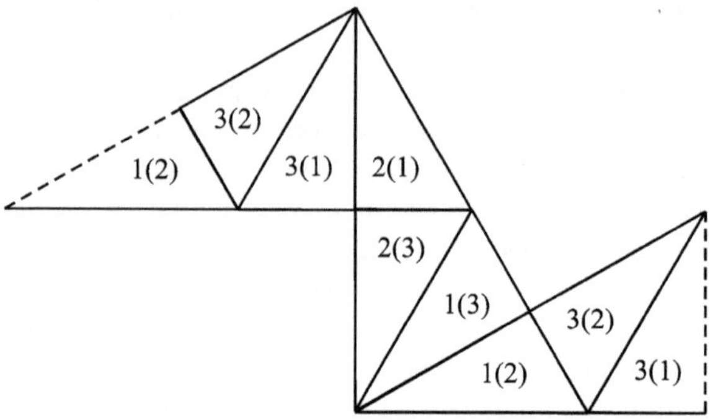

Fig. 10.26 Net for the three sector fundamental bronze even edge flexagon 3⟨bronze⟩. One copy needed. Fold together pairs of leaves numbered 1

10.2.6.3 The Three Sector Fundamental Bronze Even Edge Flexagon

The net for the three sector fundamental bronze even edge flexagon 3⟨bronze⟩ is shown in Fig. 10.26. This was derived by using the net for the two sector fundamental bronze even edge flexagon 2⟨bronze⟩ (Fig. 10.24a) as a precursor and increasing the number of sectors from two to three. Other derivations are possible.

As assembled, the flexagon is in subsidiary main position 2(3) which is, in appearance, a flat regular even edge ring of six bronze triangles (Fig. 2.22b). It can be flexed to subsidiary main position 1(2), which is a skew regular even edge ring of six bronze triangles (Fig. 10.27a), by using the threefold pinch flex. It can also be flexed to principal main position 3(1), which is a slant regular even edge ring of six bronze triangles (Fig. 10.27b). The high curvature (180°) means that this cannot be turned inside out, so the 3-cycle shown in the intermediate position map cannot be traversed completely. The intermediate position map is the same as that for the first order fundamental triangle even edge flexagons *S*⟨3, 3⟩ (Fig. 4.7).

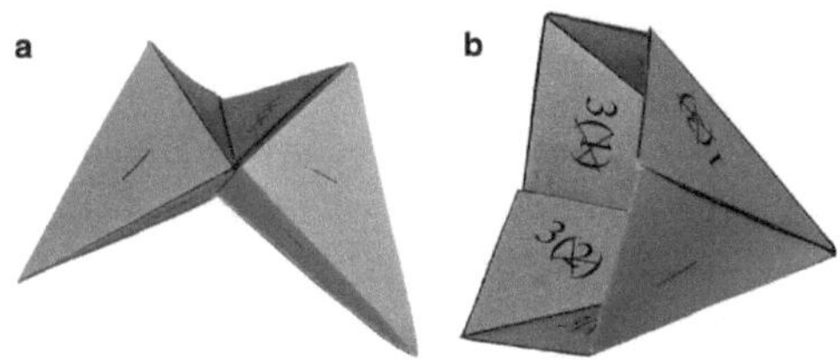

Fig. 10.27 A flexagon as regular even edge rings of six 30°–60°–90° (bronze) triangles. (**a**) Skew. (**b**) Slant

All three main positions can be visited so it is not a deficient flexagon. The intermediate positions are bronze triangle even edge triples.

Pocket flexes are possible in all three main positions, and lead to three different combinations of a slant regular even edge ring of four bronze triangles and a bronze triangle edge pair. A second pocket flex leads back to a different main position.

10.2.6.4 The Four Sector Fundamental Bronze Even Edge Flexagon

The net for the four sector fundamental bronze even edge flexagon 4⟨bronze⟩ is shown in Fig. 10.24b. This was derived by using the net for the two sector fundamental bronze even edge flexagon 2⟨bronze⟩ (Fig. 10.24a) as a precursor and increasing the number of sectors from two to four. Other derivations are possible.

As assembled, the flexagon is in subsidiary main position 2(3) which is, in appearance, a skew regular even edge ring of eight bronze triangles, curvature −120°. It can be flexed to subsidiary main position 1(2), which is a different skew regular even edge ring of eight bronze triangles, curvature −360°, by using the fourfold pinch flex. It can also be flexed to principal main position 3(1), which is a slant regular even edge ring of eight bronze triangles. The high curvature (120°) means that this cannot be turned inside out so the 3-cycle shown in the intermediate position map cannot be traversed completely. The intermediate position map is the same as that for the first order fundamental triangle even edge flexagons $S\langle 3, 3\rangle$ (Fig. 4.7). All three main positions can be visited so it is not a deficient flexagon. The intermediate positions are bronze triangle even edge quadruples. Other flexes are possible, including pocket flexes and flexes with twofold rotational symmetry.

10.2.6.5 The Six Sector Fundamental Bronze Even Edge Flexagon

The net for the six sector fundamental bronze even edge flexagon 6⟨bronze⟩ was derived by using the net for the two sector fundamental bronze even edge flexagon 2⟨bronze⟩ (Fig. 10.24a) as a precursor and increasing the number of sectors from two to 6. Other derivations are possible. The flexagon is a solitary version of the hexa-dodeca-flexagon described by Schwartz (2008). The dynamic properties are complicated because six has factors 2 and 3. Two different dual marked versions of the net are shown in Figs. 10.28 and 10.29. These nets illustrate some of the flexing options. Face numbers are the same in both nets. Face letters differ, and have been chosen to facilitate flexing to various positions. For videos of some flexes see Sherman (2008).

A dual marked net for both sixfold and threefold pinch flexing of the six sector fundamental bronze even edge flexagon 6⟨bronze⟩ is shown in Fig. 10.28. As assembled, the flexagon is in principal main position 3(1) which is, in appearance, a flat regular even edge ring of 12 bronze triangles (Fig. 2.22c). The 3-cycle shown in the intermediate position map can be traversed by using the sixfold pinch flex.

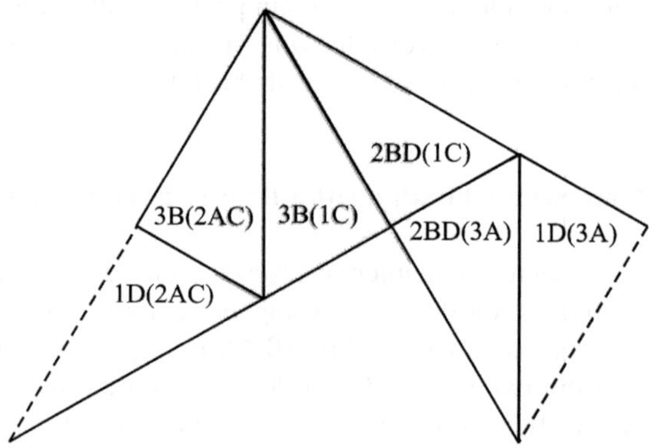

Fig. 10.28 A dual marked net for both sixfold and threefold pinch flexing of the six sector fundamental bronze even edge flexagon 6⟨bronze⟩. Three copies needed. Fold together pairs of leaves numbered 2

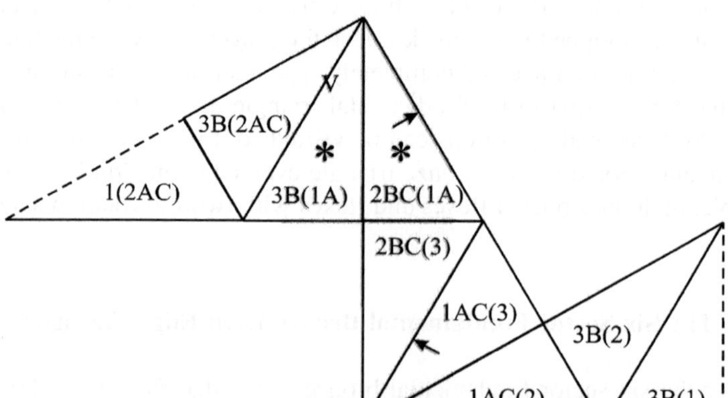

Fig. 10.29 A dual marked net for twofold flexing of the six sector fundamental bronze even edge flexagon 6⟨bronze⟩. Two copies needed. Fold together pairs of leaves numbered 2

This is difficult because of the need to maintain approximate sixfold rotational symmetry during flexing, and the flexagon is easily muddled. The intermediate position map is the same as that for the first order fundamental triangle even edge flexagons $S\langle 3, 3\rangle$ (Fig. 4.7). Subsidiary main positions 1(2) and 2(3) are two different types of skew regular even edge rings of 12 bronze triangles, curvatures −720° and −360° respectively. The intermediate positions are bronze triangle even edge sextuples. Various pocket flexes are possible. These are fairly easy starting from principal main position 3(1) but, because of the high curvatures they are difficult starting from subsidiary main positions 1(2) and 2(3).

There are two types of threefold pinch flex in which threefold rotational symmetry is maintained during flexing. In the triangle-threefold pinch flex groups of four pats are pinched together so that the flaps in intermediate positions have a triangular outline (Fig. 10.30a) and the two minor 3-cycles shown in the Tuckerman diagram (Fig. 10.31a) can be traversed. Face numbers become mixed up, and face letters have been chosen so that all leaves on a face have either the same number or the same letter. Principal main position 1(3) is common to both cycles, and is a common main position. The other main positions are minor main positions. In the kite-threefold pinch flex groups of four pats are pinched together so that the flaps in intermediate positions have a kite shaped outline (Fig. 10.30b), and the two minor 3-cycles shown in the Tuckerman diagram (Fig. 10.31b) can be traversed.

In a Tuckerman traverse, a flexagon is rotated between pinch flexes only when this is necessary for flexing to be continued. This ensures that all the main positions shown on a Tuckerman diagram are visited. The arrows on a Tuckerman diagram

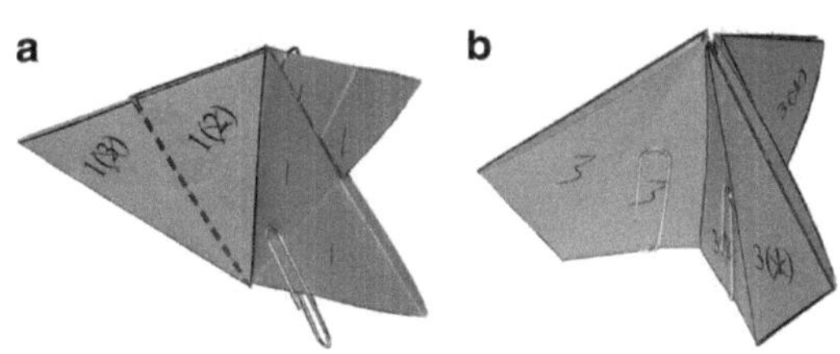

Fig. 10.30 Intermediate positions for threefold pinch flexing of the six sector fundamental bronze even edge flexagon 6⟨bronze⟩. (**a**) Triangular flaps. (**b**) Kite shaped flaps

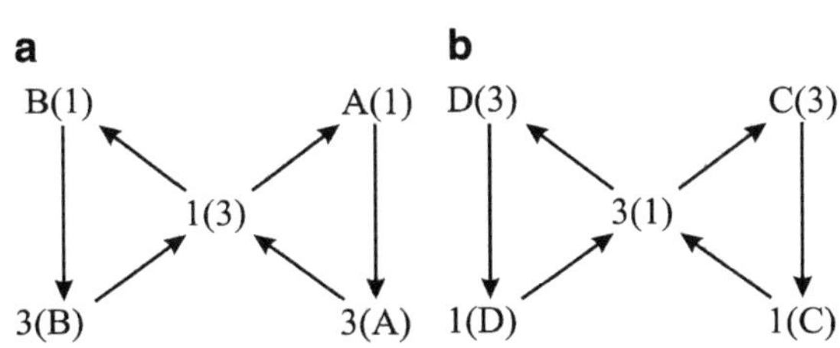

Fig. 10.31 Tuckerman diagram for threefold pinch flexing of the six sector fundamental bronze even edge flexagon 6⟨bronze⟩. (**a**) Triangular intermediate position flaps. (**b**) Kite shaped intermediate position flaps

show the direction of the traverse. The cycles are oriented so that if the flexagon is not rotated at a common main position, then the corresponding direction arrows are in a straight line. If the flexagon is turned over, the direction of traverse is reversed. Some properties for flexing using threefold pinch flexes are given in Table 10.6 (cf. Table 10.5).

Minor main positions have four different appearances. Minor main positions A(1) and B(1) are flat irregular even edge rings of 12 bronze triangles with a hexagonal outline (Fig. 10.32). The fundamental bronze net ⟨bronze⟩ (Fig. 3.25) can be fan folded onto this ring in one distinct way. The minor main positions can therefore be regarded as the two main positions of an irregular ring 12 bronze triangle even edge flexagon, and hence as the result of transformation between flexagons.

Minor main positions 3(A) and 3(B) are flat and have a propeller outline (Fig. 10.33a). Starting at these minor main positions, three pocket flexes can be used to flex the flexagon into a flat regular ring of six bronze triangles (Fig. 2.22b). Minor main positions 1(C) and 1(D) are skew and have a propeller outline. (Fig. 10.33b). Pocket flexes are not possible. Minor main positions C(3) and D(3)

Table 10.6 Properties of the six sector fundamental bronze even edge flexagon 6⟨bronze⟩ using threefold pinch flexes. The principal main position is in bold

Typical main position	Intermediate position type	Main position type
1(3)	**Triangular**	**Flat**
A(1)	Triangular	Hexagonal
3(A)	Triangular	Propeller
3(1)	**Kite shaped**	**Flat**
C(3)	Kite shaped	Slant
1(C)	Kite shaped	Skew

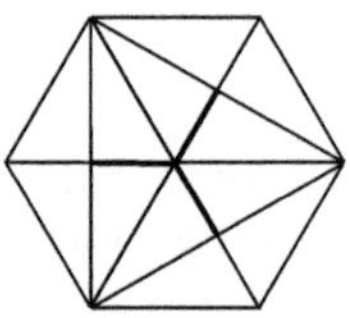

Fig. 10.32 A flat irregular even edge ring of 12 30°–60°–90° (bronze) triangles

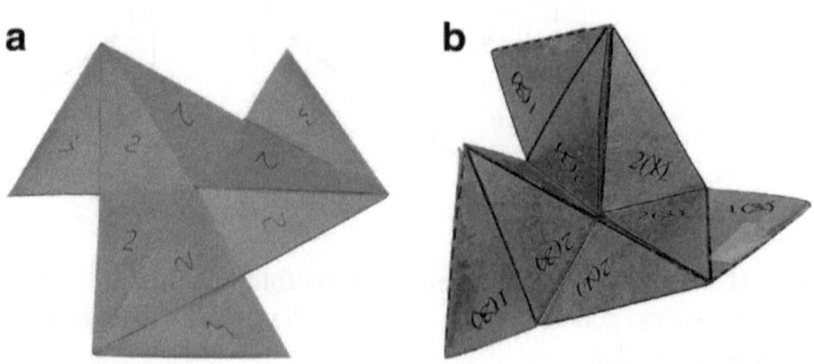

Fig. 10.33 Minor main positions of the six sector fundamental bronze even edge flexagon 6⟨bronze⟩. (**a**) Propeller. (**b**) Skew

are slant irregular even edge rings of 12 bronze triangles (Fig. 10.34). The curvature is high (180°), but the open centres mean that they can be turned inside out, so the minor 3-cycles shown in Fig. 10.31b can be traversed completely. Care is needed to avoid muddling.

A dual marked net for flexing the six sector fundamental bronze even edge flexagon 6⟨bronze⟩ with twofold rotational symmetry is shown in Fig. 10.29. The flexing sequence for one of the minor cycles that can be traversed by flexing with twofold rotational symmetry, a minor 4-cycle, is given in Table 10.7, and is also shown in the Tuckerman diagram (Fig. 10.35). Face letters have been chosen so that, in this particular minor 4-cycle, all the leaves visible on a face either have the same number or the same letter.

As assembled, the flexagon is in principal main position 1(3). The minor 4-cycle shown in the table and in the Tuckerman diagram can be traversed as follows. The sequence is difficult. Using a twofold pinch flex, fold the flexagon in two, making a mountain fold at the arrowheads, to reach first intermediate position 1 (Fig. 10.36a). Then unfold about the opposite hinges to reach minor main position A(1) (Fig. 10.37a). Next, fold in two, along the longer hinges, so that pairs of leaves numbered 1 are folded together to reach a second intermediate position (Fig. 10.36b). Then unfold about the opposite hinges to reach second minor main position B(A), which is a skew irregular even edge ring of 12 bronze triangles (Fig. 10.37b). Twist

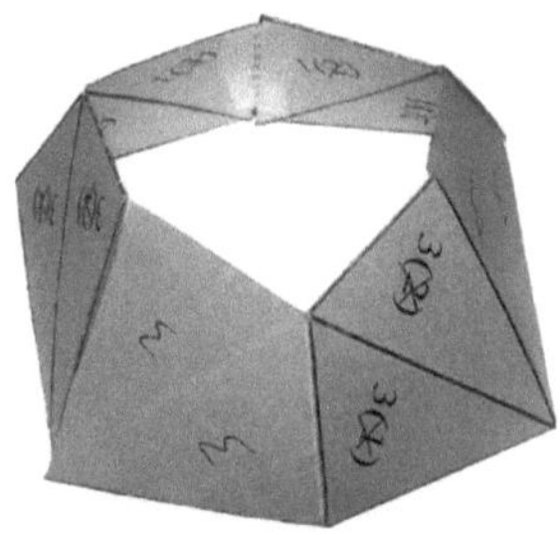

Fig. 10.34 A flexagon as a slant irregular even edge ring of 12 30°–60°–90° (bronze) triangles

Table 10.7 Flexing sequence using flexing with twofold rotational symmetry for a minor 4-cycle of the six sector fundamental bronze even edge flexagon 6⟨bronze⟩. The principal main position is in bold

Typical position	Intermediate position type	Main position type
1(3)		**Flat**
1	First	
A(1)		First (flat)
A	Second	
B(A)		Second (skew)
Undefined	Third	
C(1)		Third (flat)
3	Fourth	

Fig. 10.35 Tuckerman diagram for a twofold flexing 4-cycle of the six sector fundamental bronze even edge flexagon 6⟨bronze⟩

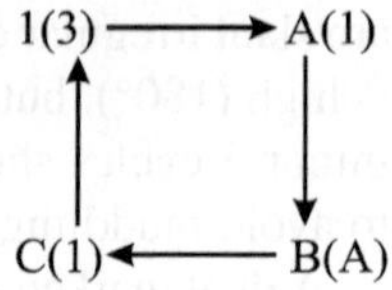

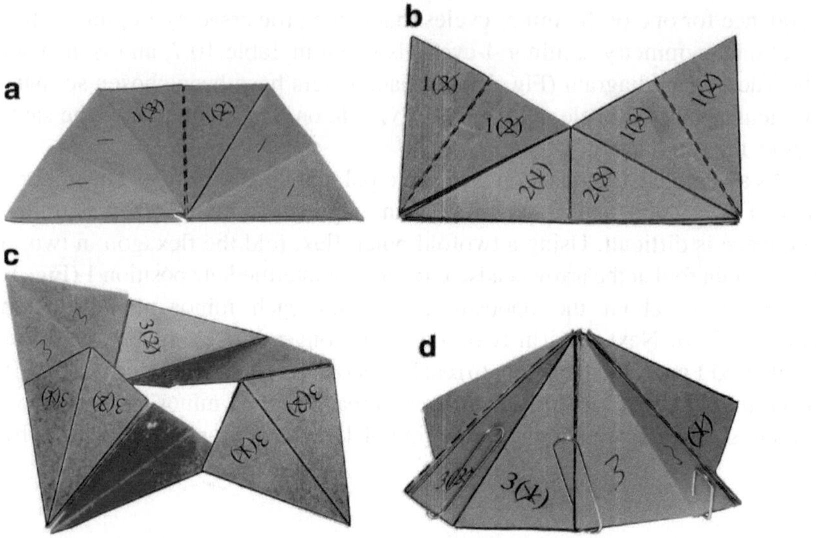

Fig. 10.36 Intermediate positions for twofold flexing of the six sector fundamental bronze even edge flexagon 6⟨bronze⟩. (**a**) First. (**b**) Second. (**c**) Third.(**d**) Fourth

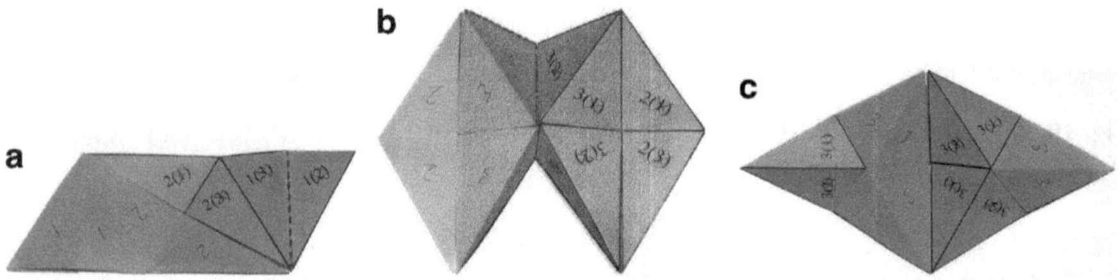

Fig. 10.37 Minor main positions for twofold flexing of the six sector fundamental bronze even edge flexagon 6⟨bronze⟩. (**a**) First. (**b**) Second. (**c**) Third

flex to third minor main position C(1) (Fig. 10.37c) by folding together the two pairs of leaves marked with asterisks, with the asterisks on the outside, rotating outwards and flattening the flexagon. The third intermediate position is not clearly defined (Fig. 10.36c). Next, with face 3 uppermost, fold the pairs of pats at the ends of the flexagon downwards, making valley folds at the hinges marked V, to reach fourth intermediate position 3 (Fig. 10.36d). Finally, open the flexagon into principal main position 1(3), and a traverse of the minor 4-cycle is complete.

10.2.6.6 The Irregular Ring Eight Bronze Triangle Even Edge Flexagon

The fundamental bronze net ⟨bronze⟩ (Fig. 3.25) can be fan folded onto the flat irregular even edge ring of eight bronze triangles type A (Fig. 10.6a) in one distinct way. This solution is shown in Fig. 10.38 as the flexagon diagram for the fundamental irregular ring eight bronze triangle even edge flexagon. A key pat is identified. This is a single leaf which is hinged on its two shorter edges. Its net is shown in Fig. 10.39. It is not possible to number the faces so that face numbers do not become mixed up, so a standard face numbering sequence is used (Section 4.1.1). The torsion is 5. A transformation between flexagons is possible with the four sector fundamental bronze even edge flexagon 4⟨bronze⟩ (Section 10.2.6.4).

As assembled, the flexagon is in main position 1(2) at the key pat in the flexagon diagram (Fig. 10.38). The main position is, in appearance, a flat irregular even edge ring of eight bronze triangles type A (Fig. 10.6a). The 4-cycle shown in the Tuckerman diagram (Fig. 10.40) can be traversed by using a flap flex. It is a local flex because some pats are left unchanged. To flex to main position 4(5) starting from main position 1(2), first fold together the two pairs of leaves numbered 1 and 4. Next, unfold the flap numbered 6(3) and fold together the pair of leaves numbered 6. Then fold together the pair of leaves numbered 8. Finally, unfold the triangular flap to reach

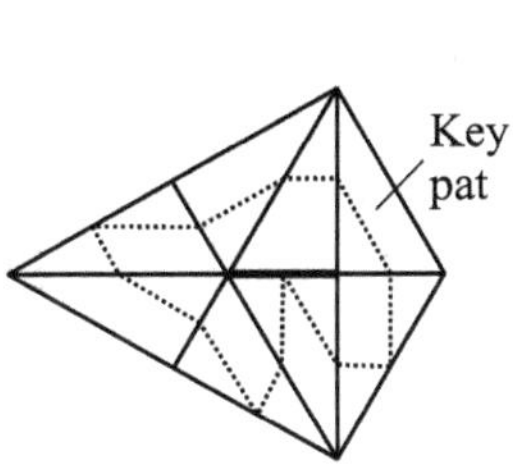

Fig. 10.38 Flexagon diagram for the fundamental irregular ring eight bronze triangle even edge flexagon

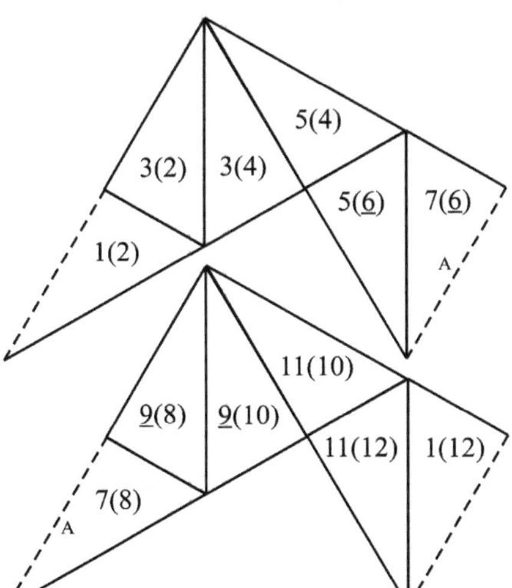

Fig. 10.39 Net for the fundamental irregular ring eight bronze triangle even edge flexagon. One copy needed. Join the two parts of the net at A-A Fold together pairs of leaves numbered 3, 5, 7 and 10

Fig. 10.40 Tuckerman diagram for the 4-cycle of the fundamental irregular ring eight bronze triangles even edge flexagon

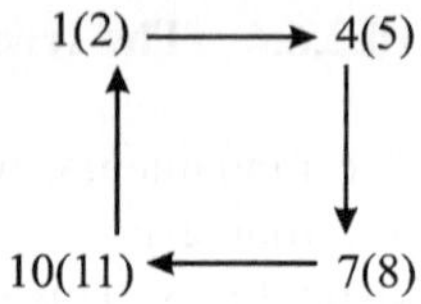

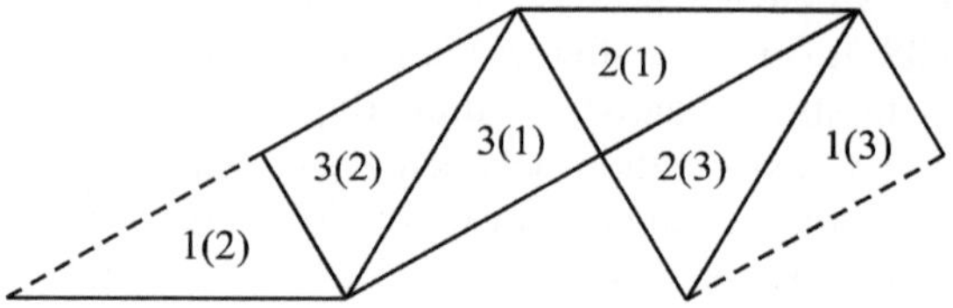

Fig. 10.41 Net for a partial overlap bronze even edge flexagon. Two copies needed. Fold together pairs of leaves numbered 2

main position 4(5). The four leaves that are hinged on the two shorter edges, and hence the main positions, appear in cyclic order as the 4-cycle is traversed.

To transform to the four sector fundamental bronze even edge flexagon 4⟨bronze⟩, starting from main position 1(2), fold together the two pairs of leaves numbered 1 and 4. Then unfold the flap numbered 2(5). Next, pinch together pairs of pats so that leaves numbered 6, 9 and 12 are folded together. This is an intermediate position of 4⟨bronze⟩ which is, in appearance, a bronze triangle quadruple. Various other flexes, including pocket flexes, are possible.

10.2.7 A Partial Overlap Bronze Even Edge Flexagon

The net for a partial overlap bronze even edge flexagon is shown in Fig. 10.41. This was derived by using the two sector fundamental silver even edge flexagon 2⟨silver⟩ (Section 10.2.4.1) as a precursor and stretching it in subsidiary main position 3(1) so that the silver triangles became bronze triangles. Before stretching subsidiary main position 3(1) is, in appearance, a flat regular even edge ring of eight 45°–45°–90° (silver) triangles (Fig. 2.21b). After stretching it is, in appearance, a flat irregular even edge rings of eight 30°–60°–90° (bronze) triangles type B (Fig. 10.6b), which is a stretch polygon ring. The leaves are identical, but do not always overlap exactly, so it is a partial overlap flexagon and a second outcome stretch flexagon (Section 10.1.2). It could also be called a scalene triangle even edge flexagon. The net could also be derived by deleting a face from the slit diamond flexagon described by Mitchell (2002). The net cannot be derived by using the net for a dodecagon flexagon as a precursor and partially stellating the regular dodecagons because the long edges would not match correctly. The dynamic properties are similar to those of the precursor flexagon.

As assembled, the flexagon is in subsidiary main position 3(1). The 3-cycle shown in the intermediate position map can be traversed by using the fourfold pinch flex. The intermediate position map is the same as that for the first order fundamental triangle even edge flexagons *S*⟨3, 3⟩ (Fig. 4.7). Subsidiary main positions 2(3) and 3(1) are flat irregular even edge rings of eight bronze triangles type B (Fig. 10.6b), and leaves overlap exactly. Subsidiary main position 1(2) is a skew irregular even edge ring of eight bronze triangles, and some leaves do not overlap exactly (Fig. 10.42a). Intermediate positions may be either a bronze triangle edge quadruple (Fig. 10.42b), or have the appearance shown in Fig. 10.43. In Fig. 10.43a, leaves have been bent to clarify the structure, and in Fig. 10.43b the intermediate position has been laid flat.

10.2.8 A Scalene Triangle Even Edge Flexagon

The net for a scalene triangle even edge flexagon is shown in Fig. 10.44. This was derived by using the three sector fundamental bronze even edge flexagon 3⟨bronze⟩ (Section 10.2.6.3) as a precursor and stretching it by 20% in subsidiary main position 2(3). Before stretching subsidiary main position 2(3) is, in appearance, a flat regular even edge ring of six 30°–60°–90 (bronze) triangles (Fig. 2.22b). After stretching, it is a flat irregular even edge ring of six scalene triangles, which is a stretch polygon ring (Fig. 10.45). The leaves are three different types of scalene

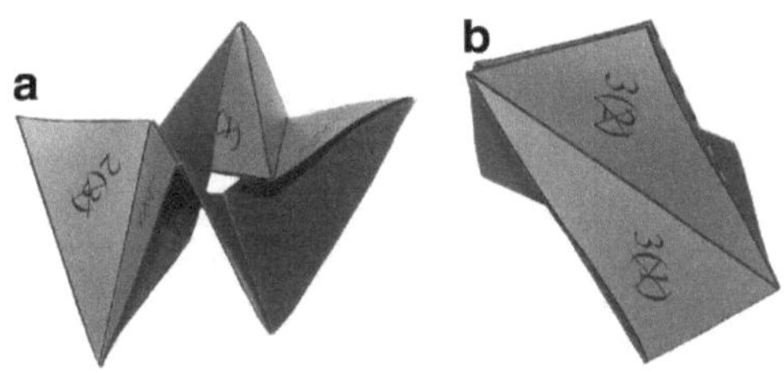

Fig. 10.42 A flexagon as the following configurations. (**a**) A skew irregular even edge ring of eight 30°–60°–90° (bronze) triangles. (**b**) A 30°–60°–90° (bronze) triangle edge quadruple

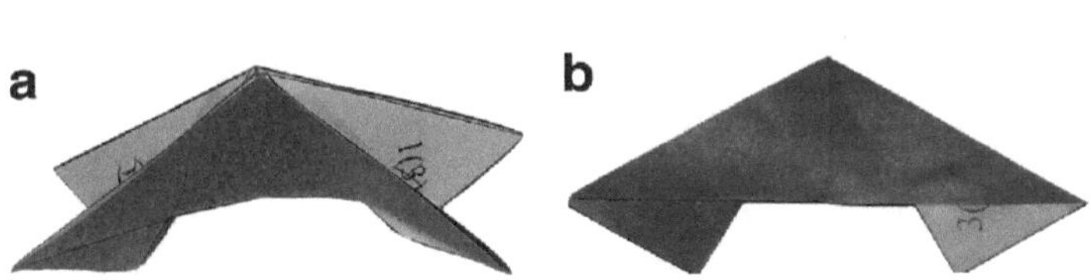

Fig. 10.43 Appearances of intermediate positions of a partial overlap bronze even edge flexagon. (**a**) Opened. (**b**) Flattened

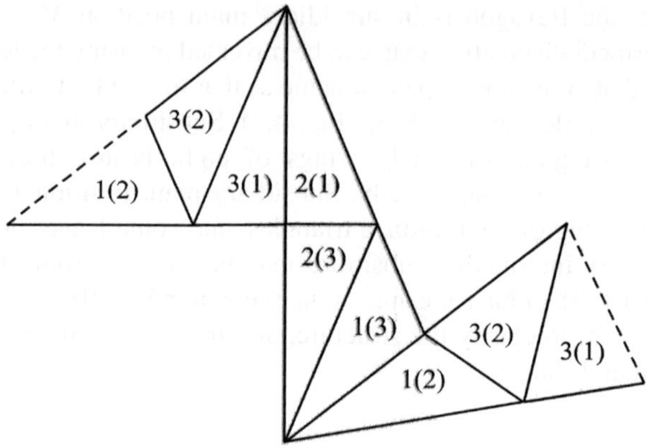

Fig. 10.44 Net for a scalene triangle even edge flexagon. One copy needed. Fold together pairs of leaves numbered 1

Fig. 10.45 A flat irregular even edge ring of six scalene triangles

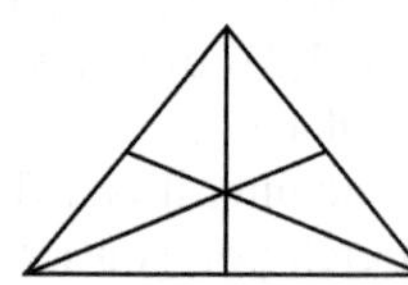

triangles. These do not always overlap exactly during flexing, so it is a third outcome stretch flexagon Section (10.1.2). It could also be called a partial overlap flexagon. The dynamic properties are similar to those of the precursor flexagon.

As assembled, the flexagon is in subsidiary main position 2(3) which is, in appearance, a flat irregular even edge ring of six scalene triangles (Fig. 10.45). It can be flexed to subsidiary main position 1(2) by using the threefold pinch flex. This is skew, the triangles do not overlap exactly and the position is not clearly defined (Fig. 10.46a). The flexagon can also be flexed to principal main position 3(1). This is slant, the leaves do not overlap exactly. The position is again not clearly defined, and cannot be photographed convincingly (Fig. 10.46b). Principal main position 3(1) cannot be turned inside out, so the 3-cycle shown in the intermediate position map cannot be traversed completely. The intermediate position map is the same as that for the first order fundamental triangle even edge flexagons $S\langle 3, 3\rangle$ (Fig. 4.7). All three main positions can be visited, so it is not a deficient flexagon. Leaves do not overlap exactly in intermediate positions and these are apparently not clearly defined. However, all three intermediate positions can be flattened into clearly defined shapes (Fig. 10.46c, d, e). Non overlapping leaves are not visible externally in intermediate position 2 (Fig. 10.46d).

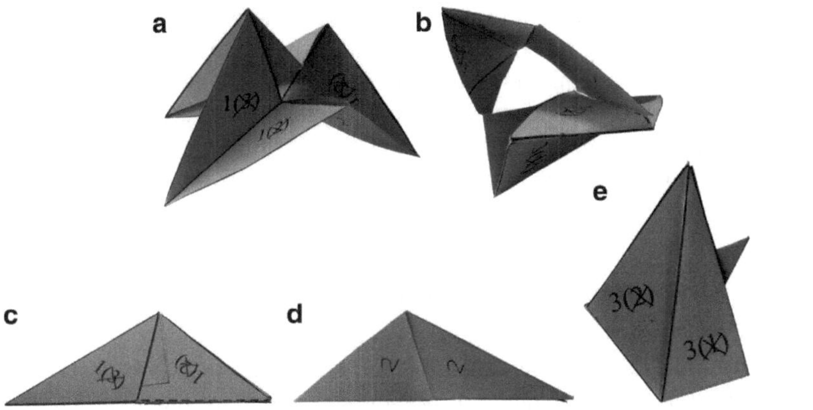

Fig. 10.46 A scalene triangle even edge flexagon. (**a**) Subsidiary main position 1(2). (**b**) Principle main position 3(1). (**c**) Intermediate position 1.(**d**) Intermediate position 2. (**e**) Intermediate position 3

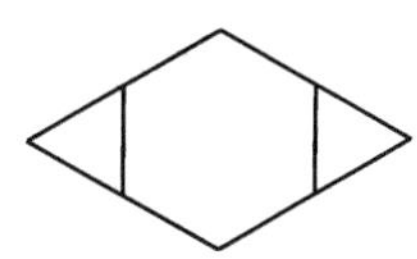

Fig. 10.47 A 60°–120° rhombus as a partially stellated regular hexagon (Les Pook, Flexagons inside out, 2003, © Cambridge University Press 2003, reprinted with permission)

10.3 Irregular Quadrilateral Even Edge Flexagons

10.3.1 Irregular Quadrilaterals

Only a limited range of irregular quadrilaterals leads to irregular quadrilateral even edge flexagons whose paper models are reasonably easy to handle. These include rectangle even edge flexagons, rhombus even edge flexagons and trapezium even edge flexagons.

A rectangle can be regarded as a square that has been transformed by altering relative edge lengths, without altering vertex angles. Rectangles are the only irregular quadrilaterals that are equiangular polygons in which all the vertex angles are equal (Section 10.1.1).

Any rhombus can be regarded as a square that has been transformed by altering vertex angles, without altering relative edge lengths. A 60°–120° rhombus can be regarded as a partially stellated regular hexagon (Fig. 10.47). This stellation reduces the number of edges from six to four.

Any 60°–120° trapezium can be regarded either as an equilateral triangle with one vertex truncated (Fig. 10.48a) or as a square in which all four vertex angles, and at least one edge length, have been altered. Partially stellating a regular hexagon leads to a 60°–120° trapezium with edge lengths in the ratio 1:2:3 (Fig. 10.48b). Bisecting a regular hexagon leads to 60°–120° trapezia with three equal edge lengths, which are half the longest edge length (Fig. 10.48c).

10.3.2 Irregular Quadrilateral Even Edge Rings

Some possible flat irregular quadrilateral even edge rings are described below. These illustrate the wide range of possibilities. Many of them can be derived by stretching flat even edge rings of regular convex polygons. These are stretch polygon rings. Each of the stretch polygon even edge rings described below consists of identical (congruent) irregular quadrilaterals, so they are first outcome stretch polygon rings (Section 10.1.2). A flat regular even edge ring of four 2:1 rectangles is shown in Fig. 10.49. This was derived by stretching a flat regular even edge ring of four squares (Fig. 1.1b) so that the squares became 2:1 rectangles.

Stellating the regular hexagons of the flat regular even edge ring of six regular hexagons (Fig. 4.31b) to 60°–120° rhombi leads to a flat regular even edge ring of six 60°–120° rhombi (Fig. 10.50a). Similarly, stellating the flat irregular even edge ring of eight regular hexagons (Fig. 9.28a), leads to a flat irregular even edge ring of eight 60°–120° rhombi (Fig. 10.50b). This ring can also be derived by replacing the squares of the flat compound edge ring of eight squares (Fig. 1.11) by 60°–120° rhombi. Replacing the squares of the flat compound edge ring of eight squares by 75°–105° rhombi results in a flat irregular even edge ring of eight 75°–105° rhombi (Fig. 10.50c). Stretching the flat compound edge ring of eight squares (Fig. 1.11)

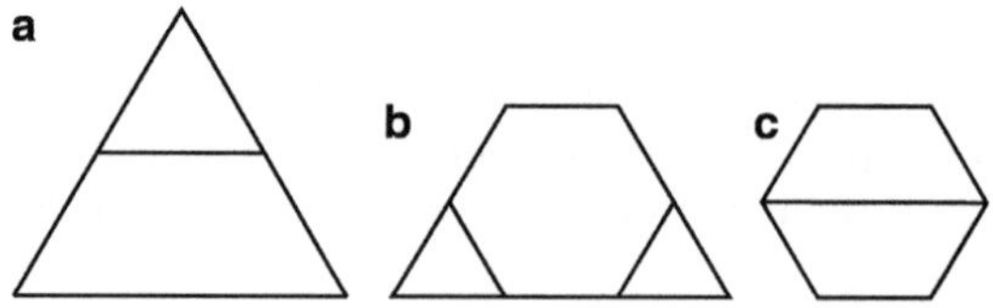

Fig. 10.48 Three 60°–120° trapezia. (**a**) As a truncated equilateral triangle. (**b**) As a partially stellated regular hexagon. (**c**) As a bisected regular hexagon

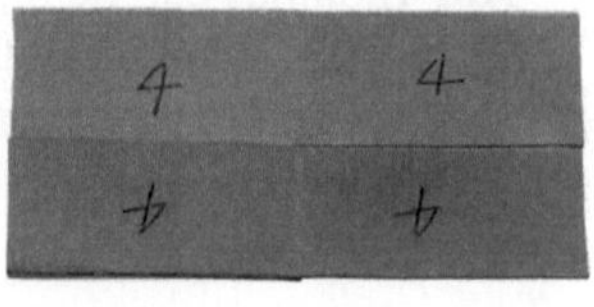

Fig. 10.49 A flexagon as a flat regular even edge ring of four 2:1 rectangles

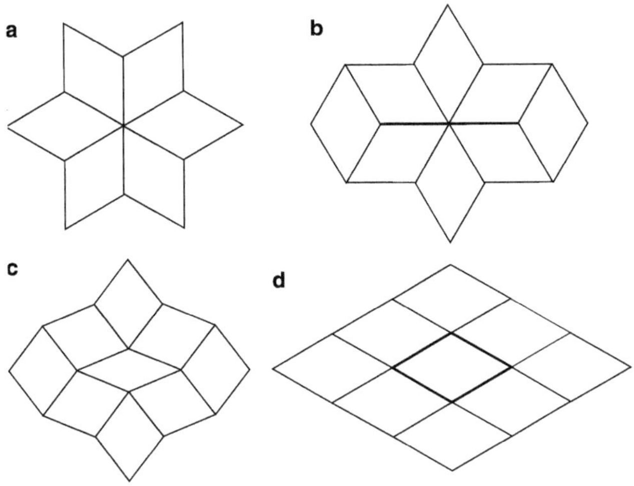

Fig. 10.50 Flat even edge rings. (**a**) A regular ring of six 60°–120° rhombi. (**b**) An irregular ring of eight 60°–120° rhombi. (**c**) An irregular ring of eight 75°–105° rhombi. (**d**) A stretch irregular ring of eight 60°–120° rhombi

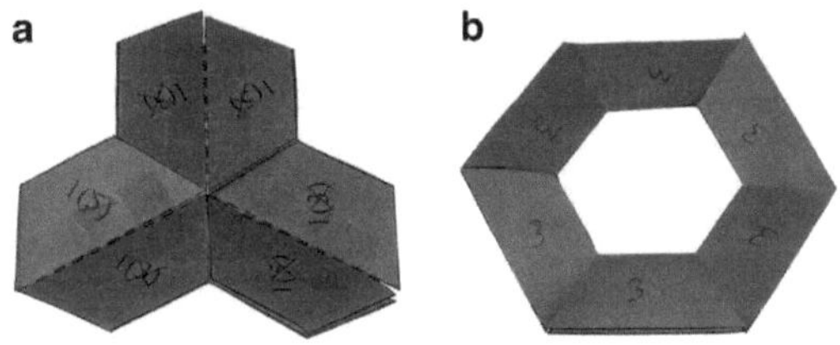

Fig. 10.51 A flexagon as flat regular even edge rings of six 60°–120° trapezia. (**a**) Type A. (**b**) Type B

so that the squares become 60°–120° rhombi results in a flat irregular even edge ring of eight 60°–120° rhombi (Fig. 10.50d). The two possible flat regular even edge rings of six 60°–120° trapezia are shown in Fig. 10.51.

10.3.3 A Rectangle Even Edge Flexagon

The net for a rectangle even edge flexagon is shown in Fig. 10.52. This was derived by using the first order fundamental square even edge flexagon 2⟨4, 4⟩ (Section 4.2.4) as a precursor and stretching a principal main position so that the squares became 2:1 rectangles. The leaves are identical and overlap exactly, so it is a first outcome stretch flexagon (Section 10.1.2). The flexagon figure is shown in Fig. 10.53. Some of its properties are given in Table 10.8 (cf. Table 4.4). Except for the appearance of positions, the dynamic properties are the same as those of the first order fundamental square even edge flexagon 2⟨4, 4⟩ (Sections 4.2.4 and 4.2.5).

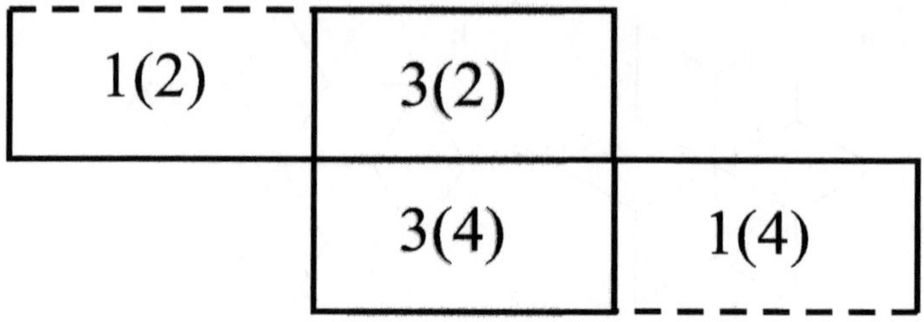

Fig. 10.52 Net for a rectangle even edge flexagon. Two copies needed

Fig. 10.53 Flexagon figure for a rectangle even edge flexagon

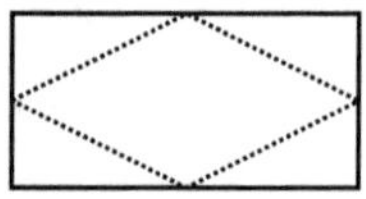

Table 10.8 Properties of a rectangle even edge flexagon. The principal cycle is in bold

Typical main position	Cycle type	Number of cycles	Main position type	Ring symbol	Sector symbol	Curvature
1(2)	**4-cycle**	**1**	**Flat**	**4(90°)**	**⟨4, 4, 3, 1⟩**	**0°**
1(3)	None	–	Box	4(0°)	⟨4, 4, 2, 2⟩	360°

Fig. 10.54 A flexagon as box edge rings of four 2:1 rectangles. (**a**) Shallow box. (**b**) Tall box

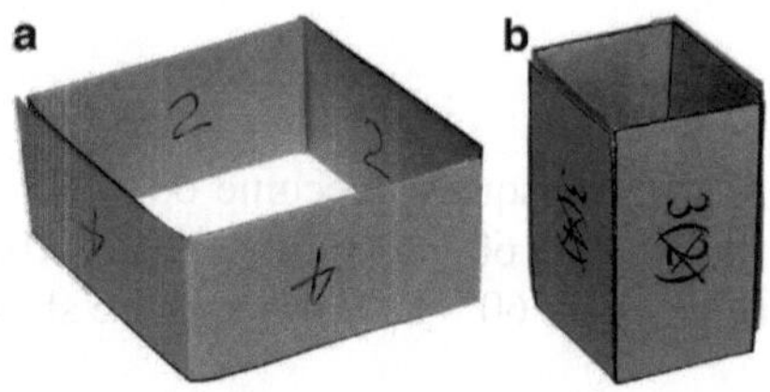

As assembled, the flexagon is in principal main position 1(2). All the principal main positions are, in appearance, flat regular even edge rings of four 2:1 rectangles (Fig. 10.49). The principal 4-cycle shown in the intermediate position map can be traversed by using the twofold pinch flex. The intermediate position map is the same as that for the first order fundamental square even edge flexagons $S\langle 4, 4\rangle$ (Fig. 4.8). Intermediate positions are 2:1 rectangle edge pairs. From intermediate positions, the flexagon can be opened into positions. Box positions 2(4) and 4(2) are shallow box edge rings (Fig. 10.54a), and box positions 1(3) and 3(1) are tall box edge rings (Fig. 10.54b).

10.3.4 Rhombus Even Edge Flexagons

Any square even edge flexagon can be transformed into a corresponding rhombus even edge flexagon by transforming the squares into rhombi, ensuring that adjacent leaves in the net overlap exactly when folded together. Some hexagon even edge flexagons can be stellated to produce corresponding 60°–120° rhombus even edge flexagons, but the range of hexagon even edge flexagons that can be stellated to60°–120° rhombus even edge flexagons is restricted by the reduction in the number of edges. In practice, and this is always possible, it is easier to use square even edge flexagons as precursors. Some examples are given below. Descriptions of dynamic properties are based on the dynamic properties of precursor square even edge flexagons.

10.3.4.1 A 60°–120° Rhombus Even Edge Flexagon

The net for a 60°–120° rhombus even edge flexagon is shown in Fig. 10.55. This was derived by using the net for the two sector first order fundamental square even edge flexagon 2⟨4, 4⟩ (Figs. 1.2 and 4.16a) as a precursor, replacing the squares by 60°–120° rhombi and increasing the number of sectors from two to three. The dynamic properties are similar to those of the precursor flexagon. Some of its properties are given in Table 10.9 (cf. Table 4.4). The torsion per sector is 2. The flexagon figure is shown in Fig. 10.56 (cf. Fig. 4.2).

As assembled, the flexagon is in principal main position 1(2). It can be flexed around the principal 4-cycle shown in the intermediate position map by using the threefold pinch flex. The intermediate position map is the same as that for the first order fundamental square even edge flexagons *S*⟨4, 4⟩ (Fig. 4.8). There are two different types of principal main position, which appear alternately. Principal main positions 1(2) and 3(4) are, in appearance, flat regular even edge rings of six

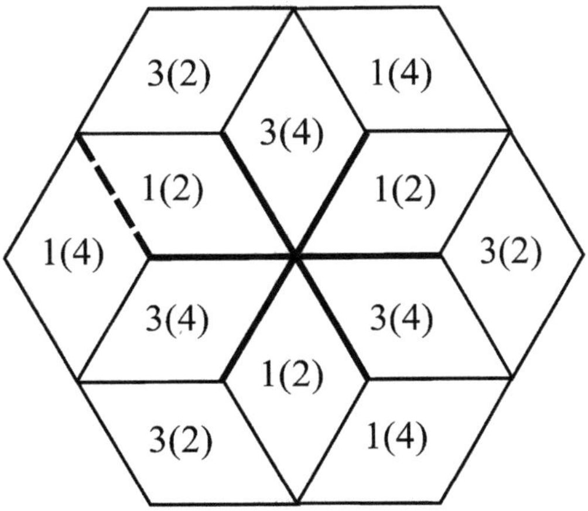

Fig. 10.55 Net for a 60°–120° rhombus even edge flexagon. One copy needed

Table 10.9 Properties of a 60°–120° rhombus even edge flexagon. The principal cycle is in bold

Typical main position	Cycle type	Number of cycles	Main position type	Ring symbol	Curvature
1(2)	**4-cycle**		**Flat**	**6(60°)**	**0°**
2(3)			**Skew**	**6(120°)**	**–360°**
1(3)	None	–	Box	6(0°)	360°

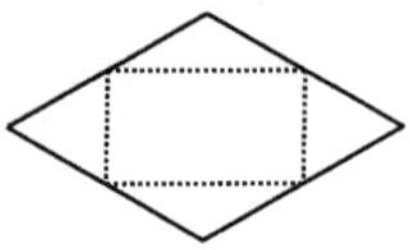

Fig. 10.56 Flexagon figure for a 60°–120° rhombus even edge flexagon

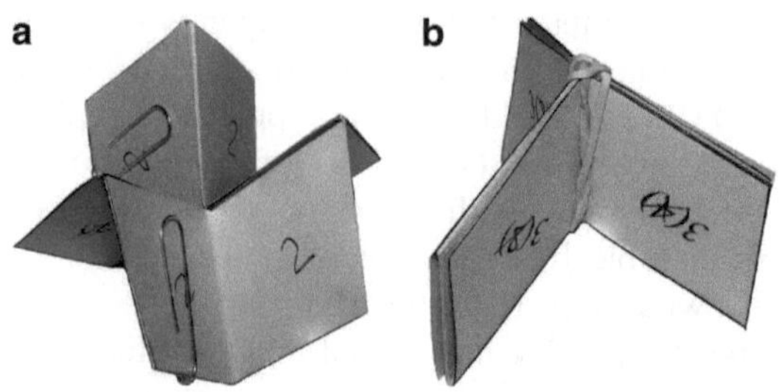

Fig. 10.57 A flexagon as the following. (**a**) Skew regular even edge ring of six 60°–120° rhombi. (**b**) 60°–120° rhombus edge triple

60°–120° rhombi (Fig. 10.50a). Principal main positions 2(3) and 4(1) are skew regular even edge rings of six 60°–120° rhombi (Fig. 10.57a). Intermediate positions are 60°–120° rhombus edge triples (Fig. 10.57b). Intermediate positions can be opened into box positions which are regular box even edge rings of six 60°–120° rhombi.

10.3.4.2 Irregular Ring Rhombus Even Edge Flexagons

Nets for two irregular ring rhombus even edge flexagons are shown in Figs. 10.58 and 10.59. These were derived by using the net for the fundamental compound square flexagon 4⟨4, 4, 3⟩ (Fig. 6.3) as a precursor and replacing the squares by 60°–120° rhombi and 75°–105° rhombi respectively. The dynamic properties of both flexagons are similar to those of the precursor flexagon. Some of their properties are given in Table 10.10 (cf. Table 6.3). There are two sectors and the torsion per sector is 2.

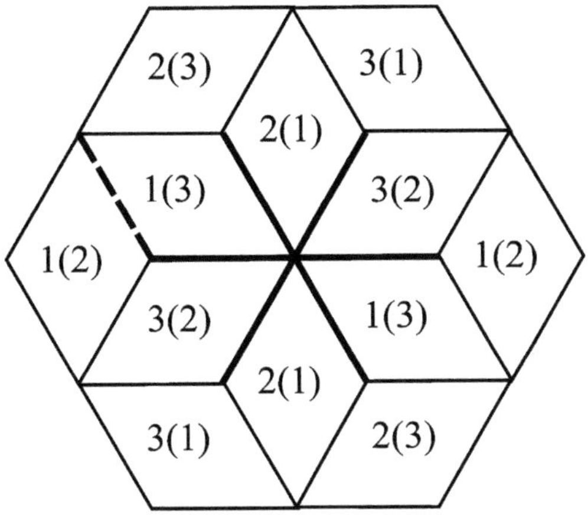

Fig. 10.58 Net for an irregular ring 60°–120° rhombus even edge flexagon. One copy needed (Les Pook, Flexagons inside out, 2003, © Cambridge University Press 2003, reprinted with permission)

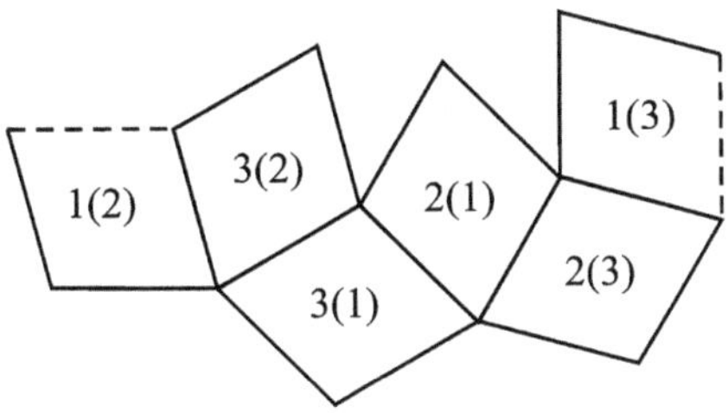

Fig. 10.59 Net for an irregular ring 75°–105° rhombus even edge flexagon. Two copies needed

Table 10.10 Properties of irregular ring rhombus even edge flexagons. Principal cycles are in bold

Rhombus type	Typical main position	Cycle type	Number of cycles	Main position type
60°–120°	**1(2)**	**3-cycle**	**1**	**Flat**
75°–105°	**1(2)**	**3-cycle**	**1**	**Flat**

As assembled, the flexagons are in principal main position 2(1). The principal 3-cycle shown in the intermediate position map can be traversed by using a twist flex. The intermediate position map is the same as that for the first order fundamental triangle even edge flexagons *S*⟨3, 3⟩ (Fig. 4.7). For both flexagons, the twist flex used is the same as that which can be used for the precursor flexagon (Section 6.3.2). Starting from principal main position 2(1) to flex to principal main position 3(2) hold a diametrically opposite pair of leaves numbered 3(2), turn them over and pull gently apart. Principal main positions of an irregular ring 60°–120° rhombus even edge flexagon are, in appearance, flat irregular even edge rings of eight 60°–120° rhombi (Fig. 10.50b). Principal main positions of an irregular ring 75°–105° rhombus even edge flexagon are flat irregular even edge rings of eight 75°–105° rhombi (Fig. 10.50c).

10.3.4.3 A Partial Overlap 60°–120° Rhombus Even Edge Flexagon

The net for a partial overlap 60°–120° rhombus even edge flexagon is shown in Fig. 10.60. This was derived by using the fundamental compound square flexagon 4⟨4, 4, 3⟩ (Section 6.3.2) as a precursor and stretching a main position so that the squares became 60°–120° rhombi. The leaves are identical, but do not always overlap exactly, so it is a second outcome stretch flexagon (Section 10.1.2). The dynamic properties are similar to those of the precursor flexagon. There are two sectors and the torsion per sector is 2.

As assembled the flexagon is in main position 1(2). This is, in appearance, a stretch polygon ring which is a flat irregular even edge ring of eight 60°–120° rhombi (Fig. 10.50d). The 3-cycle shown in the intermediate position map can be traversed by using twist flexes. The intermediate position map is the same as that for the first order fundamental triangle even edge flexagons $S\langle 3, 3\rangle$ (Fig. 4.7). Starting from main position 1(2), to flex to main position 2(3), hold a diametrically opposite pair of leaves numbered 2(3), turn outwards and pull gently. Next, to flex to main position 3(1), hold the pair of leaves at the ends of the flexagon numbered 3(1) and turn outwards. To complete the 3-cycle, hold the pair of leaves at the ends of the flexagon numbered 1(2) and turn outwards. Main positions 2(3) and 3(1) are flat and have an irregular outline (Fig. 10.61). In these main positions, some leaves do not overlap exactly so it is a partial overlap flexagon. The flexagon can be flexed into a number of other configurations in which face numbers are mixed up.

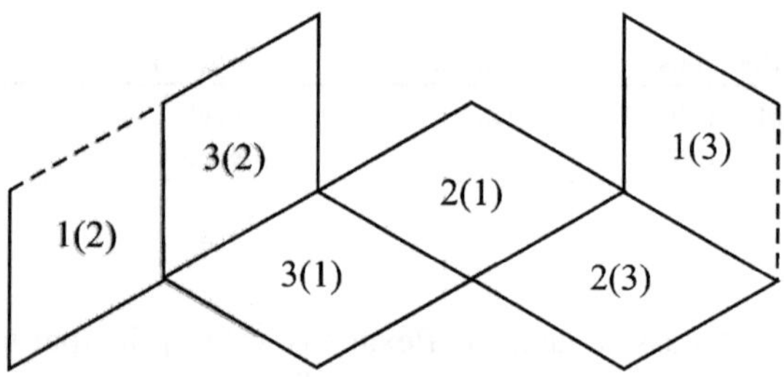

Fig. 10.60 Net for a partial overlap 60°–120° rhombus even edge flexagon. Two copies needed

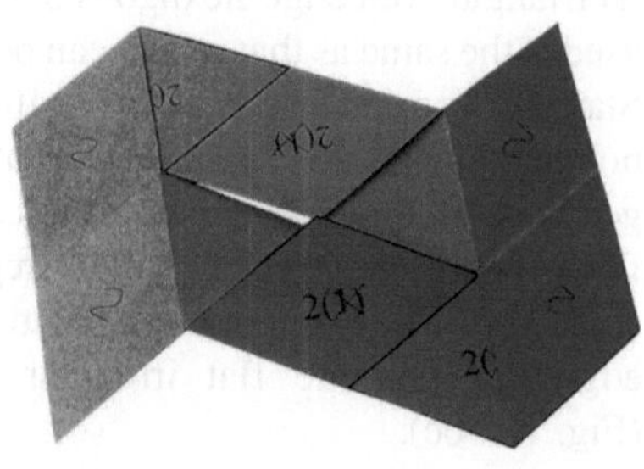

Fig. 10.61 Flat irregular outline main position of a partial overlap 60°–120° rhombus even edge flexagon

10.3.5 A Trapezium Even Edge Flexagon

Trapezium even edge flexagons were first described by Conrad and Hartline (1962). The net for a 60°–120° trapezium even edge flexagon is shown in Fig. 10.62. This was derived by using the net for the degenerate hexagon even edge flexagon type K (Fig. 8.17) as a precursor, replacing the hexagons with 60°–120° trapezia of the type shown in Fig. 10.48c and increasing the number of sectors from two to three. Some of its properties are given in Table 10.11 (cf. Table 8.3). There are three sectors and the torsion per sector is 2. The flexagon figure is shown in Fig. 10.63. The close relationship between the flexagon and its precursor means that the incomplete cycle descriptions used for the degenerate hexagon even edge flexagon are also appropriate for the rhombus even edge flexagon. In Table 10.11 the original meaning of 2/6 cycle (Section 8.2.4, cf. Table 8.3) is that a principal 6-cycle could be traversed in the precursor first order fundamental hexagon even edge flexagon, but in a degenerate hexagon even edge flexagon it is an incomplete cycle, and only two principal main positions can be visited. In the rhombus even edge flexagon a transferred meaning is that there are two principal main positions, both of which have the same appearance.

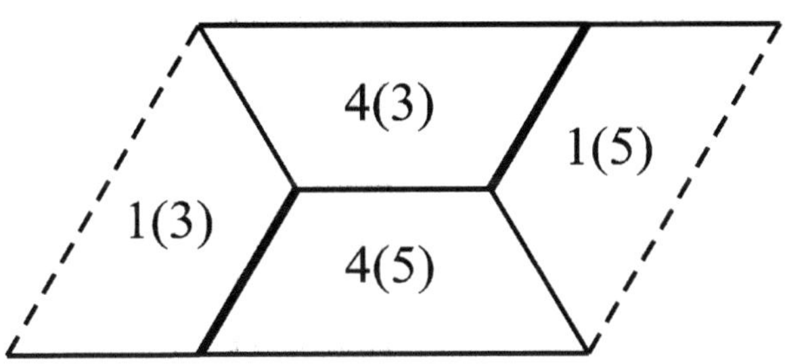

Fig. 10.62 Net for a 60°–120° trapezium even edge flexagon. Three copies needed. Fold together pairs of leaves numbered 4 and 5

Table 10.11 Properties of a 60°–120° trapezium even edge flexagon.. The incomplete principal cycle is in bold

Typical main position	Cycle type	Number of cycles	Main position type	Ring symbol	Curvature
3(4)	**2/6 cycle**	**1**	**Skew**	**6(120°)**	**–360°**
1(3)	3-cycle	1	Flat type A	6(60°)	0°
3(5)			Flat type B	6(60°)	0°
1(4)	None	–	Box	6(0°)	360°

Fig. 10.63 Flexagon figure for a 60°–120° trapezium even edge flexagon

As assembled, the flexagon is in subsidiary main position 1(3). It can be flexed around the subsidiary 3-cycle shown in the intermediate position map by using the threefold pinch flex. The intermediate position map is the same as that for the two sector degenerate hexagon even edge flexagon type K (Fig. 8.18). Subsidiary main positions 1(3) and 5(1) are, in appearance, flat regular even edge rings of six 60°–120° trapezia type A (Fig. 10.51a). In subsidiary main position 3(5) the trapezia are differently oriented (Type B, Fig. 10.51b). It is also possible to flex to principal main positions 3(4) and 4(5). These are skew regular even edge ring of six 60°–120° trapezia (Fig. 10.64). Intermediate positions are 60°–120° trapezium edge pairs. Intermediate positions 1 and 4 can be opened into a box position, which is a regular box edge ring of six 60°–120° trapezia.

10.4 Irregular Pentagon Even Edge Flexagons

10.4.1 Irregular Pentagons

Only a limited range of irregular pentagons leads to irregular pentagon even edge flexagons whose paper models are reasonably easy to handle. Two of these are described below.

An equiangular polygon is an irregular polygon in which all the vertex angles are equal but the edge lengths differ (Section 10.1.1). An equiangular irregular pentagon is an equiangular polygon that can have 3, 4 or 5 different edge lengths. The example shown in Fig. 10.65 has three different edge lengths.

Fig. 10.64 A flexagon as a skew regular even edge ring of six 60°–120° trapezia

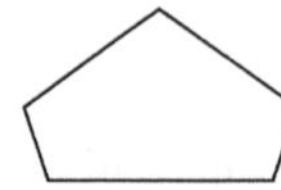

Fig. 10.65 An equiangular irregular pentagon with three different edge lengths

There are numerous ways in which a regular dodecagon can be partially stellated to produce an irregular pentagon. One of these is shown in Fig. 10.66. This particular example has widely varying edge lengths. The version shown in Fig. 10.67, in which relative edge lengths have been altered without altering the vertex angles, is more practical for the construction of flexagons. Altering the relative edge lengths makes flexagons easier to handle but does not otherwise significantly affect their dynamic properties.

10.4.2 Irregular Pentagon Even Edge Rings

Some possible flat irregular pentagon even edge rings are described below. These illustrate the wide range of possibilities.

There are six possible regular even edge rings of four of the equiangular irregular pentagons shown in Fig. 10.65. Three of these are skew (Fig. 10.68) and three are slant (Fig. 10.69).

Replacing the regular dodecagons in the flat irregular even edge ring of eight dodecagons (Fig. 9.40) with irregular pentagons with altered relative edge lengths (Fig. 10.67) leads to a flat irregular even edge ring of eight irregular pentagons (Fig. 10.70). Two skew irregular even edge rings of the same irregular pentagon are shown in Fig. 10.71.

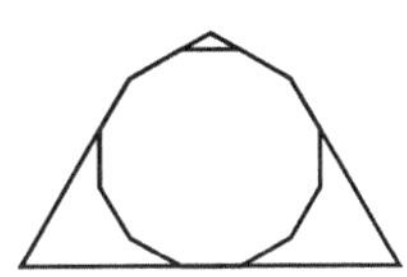

Fig. 10.66 An irregular pentagon as a partially stellated regular dodecagon

Fig. 10.67 An irregular pentagon with altered relative edge lengths

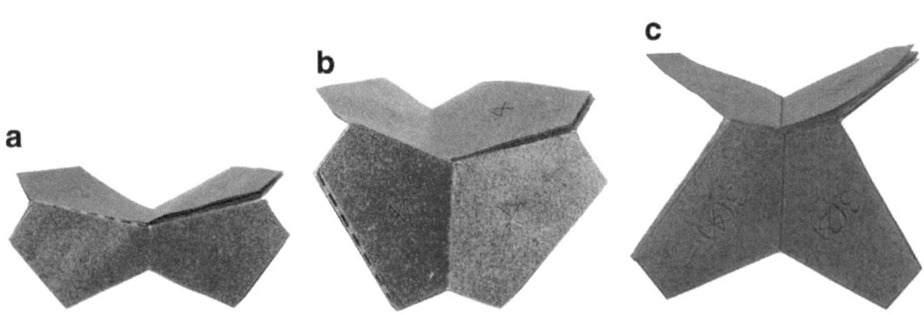

Fig. 10.68 A flexagon as skew regular even edge rings of four equiangular irregular pentagons. (**a**) Type A. (**b**) Type B. (**c**) Type C

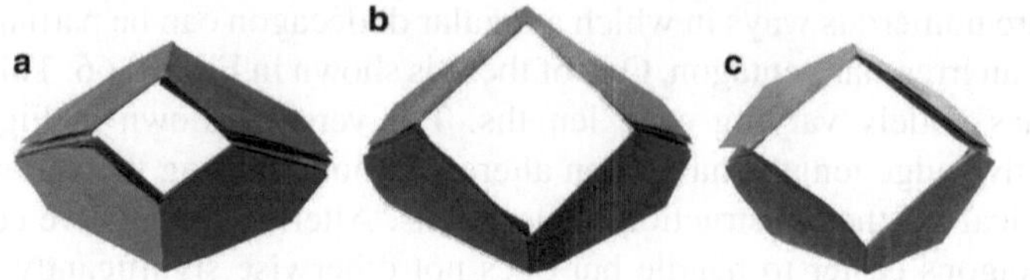

Fig. 10.69 A flexagon as slant regular even edge rings of four equiangular irregular pentagons. (**a**) Type A. (**b**) Type B. (**c**) Type C

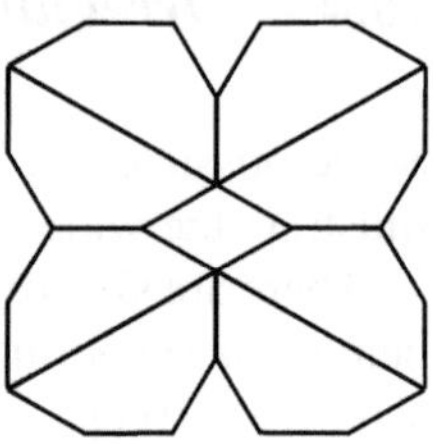

Fig. 10.70 A flat irregular even edge ring of eight irregular pentagons

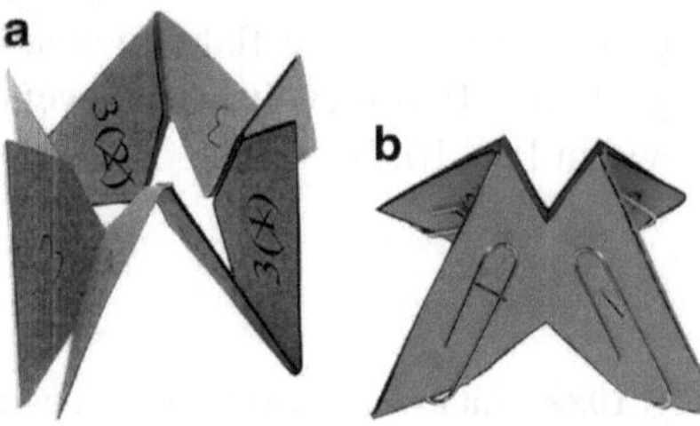

Fig. 10.71 A flexagon as skew irregular even edge rings. (**a**) Eight irregular pentagons. (**b**) Four irregular pentagons

10.4.3 An Equiangular Irregular Pentagon Even Edge Flexagon

The net for an equiangular irregular pentagon even edge flexagon is shown in Fig. 10.72. This was derived by using the net for the first order fundamental pentagon even edge flexagon 2⟨5, 5⟩ (Fig. 4.23a) as a precursor and replacing the regular pentagons by equiangular irregular pentagons with three different edge lengths (Fig. 10.65). Some of its properties are given in Table 10.12 (cf. Table 4.5). There are two sectors and the torsion per sector is 3. The flexagon figure is shown in Fig. 10.73. Except for the appearance of positions, the dynamic properties are the same as those of the precursor flexagon (Section 4.2.6).

As assembled, the flexagon is intermediate position 1. Intermediate positions are equiangular irregular pentagon edge pairs. The principal 5-cycle shown in the intermediate position map can be traversed by using the twofold pinch flex. The intermediate position map is the same as that for the first order fundamental pentagon even edge flexagons *S*⟨5, 5⟩ and *S*⟨5, 5/2⟩ (Fig. 4.9). Principal main

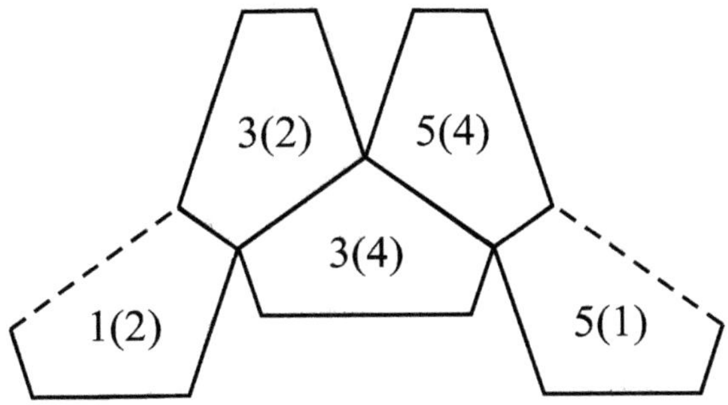

Fig. 10.72 Net for an equiangular irregular pentagon even edge flexagon. Two copies needed. Fold until leaves numbered 1 are visible

Fig. 10.73 Flexagon figure for an equiangular irregular pentagon even edge flexagon

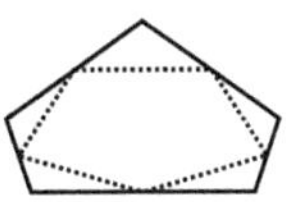

Table 10.12 Properties of an equiangular irregular pentagon even edge flexagon. The principal cycle is in bold

Typical main position	Cycle type	Number of cycles	Main position type	Ring symbol	Sector symbol	Curvature
1(2)	**5-cycle**	**1**	**Skew type A**	**4(108°)**	**⟨5, 5, 4, 1⟩**	**–72°**
2(3)			**Skew type C**	**4(108°)**	**⟨5, 5, 4, 1⟩**	**–72°**
3(4)			**Skew type B**	**4(108°)**	**⟨5, 5, 4, 1⟩**	**–72°**
1(3)	5-cycle	1	Slant type A	4(36°)	⟨5, 5, 3, 2⟩	216°
2(4)			Slant type B	4(36°)	⟨5, 5, 3, 2⟩	216°
5(2)			Slant type C	4(36°)	⟨5, 5, 3, 2⟩	216°

positions are skew regular even edge rings of four irregular pentagons, and there are three different types. Principal main positions 1(2) and 5(1) are type A (Fig. 10.68a), principal main position 3(4) is type B (Fig. 10.68b), and principal main positions 2(3) and 4(5) are type C (Fig. 10.68c). The high curvature of the subsidiary main positions (216°) means that subsidiary 5-cycle shown in the intermediate position map cannot be traversed directly. All the subsidiary main positions can be visited by flexing via principal main positions. Subsidiary main positions are slant regular even edge rings of four irregular pentagons, and there are three different types. Subsidiary main positions 1(3) and 4(1) are type A (Fig. 10.69a), subsidiary main positions 2(4) and 3(5) are type B (Fig. 10.69b), and subsidiary main position 5(2) is type C (Fig. 10.69c).

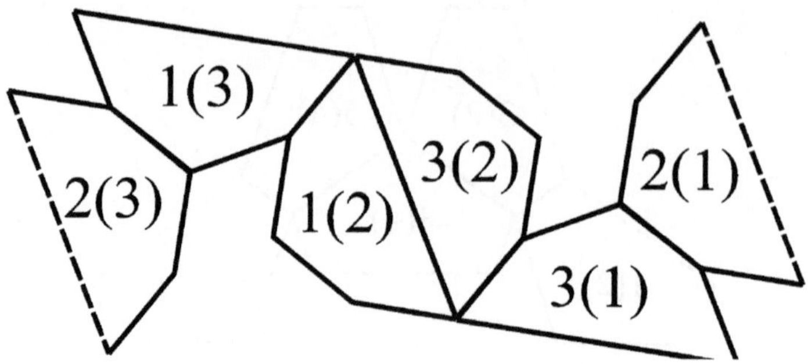

Fig. 10.74 Net for an irregular ring eight irregular pentagon even edge flexagon. Two copies needed

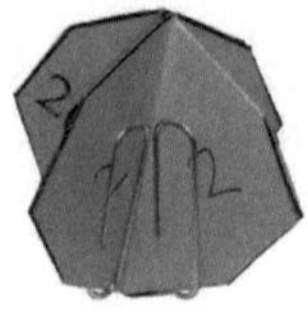

Fig. 10.75 A flexagon as an irregular pentagon edge quadruple

10.4.4 An Irregular Ring Eight Irregular Pentagon Even Edge Flexagon

The net for an irregular ring eight irregular pentagon even edge flexagon is shown in Fig. 10.74. This was derived by using the net for a degenerate irregular ring eight dodecagon even edge flexagon (Fig. 9.43) as a precursor, and replacing the regular dodecagons with irregular pentagons with altered relative edge lengths (Fig. 10.67). There are two sectors and the torsion per sector is 2. The dynamic properties are similar to those of the precursor flexagon (Section 9.6.2).

As assembled, the flexagon is in main position 2(1). It can be flexed round the 3-cycle shown in the Tuckerman diagram by using a fourfold pinch flex. The Tuckerman diagram is the same as that for the minitwist flexagon (Fig. 9.39). Main positions 2(1) and 3(2) are, in appearance, flat irregular even edge rings of eight irregular pentagons (Fig. 10.70). Main position 1(3) is a skew irregular even edge ring of eight irregular pentagons (Fig. 10.71a). Intermediate positions 1 and 3 are skew regular even edge rings of four irregular pentagons (Fig. 10.71b). Intermediate position 2 is an irregular pentagon edge quadruple (Fig. 10.75).

References

Ball D (2002) Equiangular polygons. Mathematical Gazette 86:396–407

Conrad AS, Hartline DK (1962) *Flexagons*. RIAS Technical Report 62-11. Baltimore, MD: RIAS

Mitchell D (2002) Silverflexagons 2. The slit-square silverflexagons and related forms. Water Trade. Distributed by the British Origami Society

Pook LP (2003) *Flexagons Inside Out*. Cambridge University Press, Cambridge

Schwartz A (2008) Flexagon discovery: the shape shifting 12-gon. http://www.eighthsquare.com/12-gon.html. Accessed 22 April 2008

Sherman S (2007) Triangle flexagon bestiary. http://loki3.com/flex/triangles.html. Accessed 3 June 2007

Sherman S (2008) Dodecaflexagon videos. Flexagon Lovers Group posting 493. http://tech.groups.yahoo.com/group/Flexagon_Lovers/. Accessed 25 October 2008

Chapter 11
Complex Flexagons

11.1 Introduction

The trihexaflexagon (Section 4.2.3), in which one cycle can be traversed, was discovered by Stone in 1939. He immediately realised that more complicated flexagons were possible, and discovered the hexahexaflexagon, in which four cycles can be traversed (Conrad and Hartline 1962; Pook 2003). The hexahexaflexagon is an example of a complex flexagon.

A complex flexagon consists of two or more solitary flexagons (Section 4.2.1). Its dynamic properties include features of the dynamic properties of the precursor flexagons. The torsion of a complex flexagon is the algebraic sum of the torsions of the constituent flexagons. Complex flexagons can also incorporate parts of solitary flexagons. The characteristic flex for a complex flexagon is the same as that for the precursor flexagons. Most of the more interesting flexagons for which nets have been published are complex flexagons, and include some spectacular examples. For this reason it would have been better to have introduced the concept of a complex flexagon earlier in the book. However, material on solitary flexagons in Chapters 4–10 is needed as a preliminary to the discussion of complex flexagons. Solitary flexagons are broadly equivalent to single polyhedra, whereas complex flexagons are broadly equivalent to compound polyhedra, such as the well known stella octangula, which is a compound of two regular tetrahedra (Fig. 1.14, Cromwell 1997). There are several ways in which two solitary flexagons, with the same type of leaf, can be joined together to form a complex flexagon.

To link together two solitary even edge flexagons to form a linked even edge flexagon both are flexed into main positions that have the same appearance, and in which alternate pats are single leaves. The single leaves are removed from both Flexagons, and the remaining pats are assembled into a single, linked, flexagon. An equivalent procedure, which he calls splitting, is described by Belenky (2009). The relationship between even skeletal flexagons and even edge flexagons (Section 5.2.1) means that, theoretically, linked even skeletal flexagons are possible. However, in practice they are so unstable they are not described.

Linking can be extended indefinitely to create very complicated even edge flexagons. In general, the dynamic properties of a linked even edge flexagon are a

L. Pook, *Serious Fun with Flexagons*, Solid Mechanics and Its Applications 164,
DOI 10.1007/978-90-481-2503-6_11,

combination of those of the constituent flexagons. However, additional minor cycles, not present in the precursor flexagons, sometimes appear. The dynamic behaviour of some linked even edge flexagons is very complicated, for example the hexa-dodeca-flexagon discovered by Schwartz (2008). The nets for some linked even edge flexagons have an attractive symmetry.

The characteristic feature of a linked even edge flexagon, as opposed to all solitary even edge flexagons, is that, in some main positions, pats include reverse folded piles of leaves. For some flexes these reverse folded piles of leaves behave as if they were single leaves. Not all even edge flexagons that can be constructed by linking have this feature, so they are not linked even edge flexagons. An example is the first order fundamental square even edge flexagon 2⟨4, 4⟩ (Section 4.2.4), which can be constructed by linking two degenerate square even edge flexagons. Some linked even edge flexagons are interleaved so interleaf flexes may be possible, but these are difficult. Whether these interleaf flexes are regarded as legitimate is a matter of taste. A square even edge flexagon that could be interleaf flexed was first described by Mitchell (1999).

Some interleaved even edge flexagons are related to interleaved point flexagons (Sections 5.4.1, 5.6.1 and 7.3.1). This is in the sense that one sector of the net for the even edge flexagon is the dual of the net for the point flexagon, and the face numbering sequence is the same. They should not be regarded as direct equivalents because the even edge flexagons are complex flexagons whereas the related interleaved point flexagons are solitary flexagons. This difference arises because an edge hinge has one less degree of freedom than a point hinge (Section 1.2).

Point flexagons (Sections 5.3–5.6 and 7.3) can be linked to form linked point flexagons. The method used is analogous to that used for even edge flexagons. Again, linking can be extended indefinitely to create very complicated point flexagons. Some of the nets given by Sherman (2007a) are for linked point flexagons, although he does not use this term. Point flexagons can also be linked by *end links*, without removing leaves, but this does not lead to point flexagons that cannot be constructed in other ways.

In a conjoined point flexagon, leaves at the ends of intermediate positions of simple band point flexagons are conjoined. In paper models this can be done by gluing leaves together to form conjoined leaves. In general, point flexagons are simple bands, that is the band is a single loop of hinged polygons. Conjoined point flexagons are never simple bands. There are always two or more separate loops, such as a figure-of-eight band, and a conjoined leaf is sometimes a loose flap attached by a pair of point hinges.

Linking that involves removal of leaves requires the presence, in a main position, of alternate pats that are single leaves. This is impossible in main positions of odd edge flexagons (Section 4.3.1), so linking is also impossible. However, it is possible to replace individual leaves in an odd edge flexagon with pairs of leaves, and hence create a bundled odd edge flexagon.

In a slipagon (Engel 1993), two or more even edge flexagons are interlinked during assembly, and they cannot be separated by flexing. The name appears to have been coined because the constituent flexagons sometimes have to be slipped relative to each other in order to reach some positions.

A coupled point flexagon consists of two or more point flexagons, with the same type of leaf, interleaved with each other during assembly. A coupled point flexagon can be flexed as if it were a single point flexagon. Unlike slipagons, the individual point flexagons do not have to be slipped relative to each other during flexing. Some examples of complex flexagons are given in this chapter in order to illustrate the huge range of possibilities.

11.2 Linked Even Edge Flexagons

11.2.1 Methods of Linking

To link together two even edge flexagons to form a linked even edge flexagon, both are flexed into main positions that have the same appearance, and in which alternate pats are single leaves. The single leaves are removed from both flexagons, and the remaining pats are assembled into a single, linked, even edge flexagon. This is a main position link. There are four types of main position link: flat main position links, box position links, skew main position links, and slant main position links. There is no essential difference between the four types of link, but it is convenient to consider them separately. A linked flexagon can contain more than one type of link (Pook 2003).

Torsion (Section 4.2.1) is an important factor in selecting even edge flexagons for linking. Some even edge flexagons are twisted bands, so exist as enantiomorphic pairs, which are not usually regarded as distinct (Section 1.2). However, when linking even edge flexagons, whether the torsion is positive or negative, that is the hand, does have to be considered. In general, the two even edge flexagons being linked must be of the same hand, but there are exceptions. The torsion per sector of a linked even edge flexagon is the algebraic sum of the torsions per sector of the precursor even edge flexagons.

The trihexaflexagon (Section 4.2.3) is a solitary flexagon that has only one type of main position, so only one type of link is possible. This is a flat main position link. To construct a flat main position link between two trihexaflexagons, make two paper models (Fig. 4.14). They must both be of the same hand. Flex each trihexaflexagon to the principal main position which is to be linked. Cut the flexagons along the hinges between pats, and remove the single leaves. Keep the remaining pats in their original relationship to one another and mark the cut edges of leaves for future reference. Assemble what will be a principal main position of a tetrahexaflexagon by placing the remaining pats from each of the precursor trihexaflexagons alternately. Join adjacent marked edges. Disconnect a hinge and unfold the flexagon. Renumbering is needed to avoid duplication and, if needed, to ensure that all the numbers on a face are the same. The resulting net for the tetrahexaflexagon

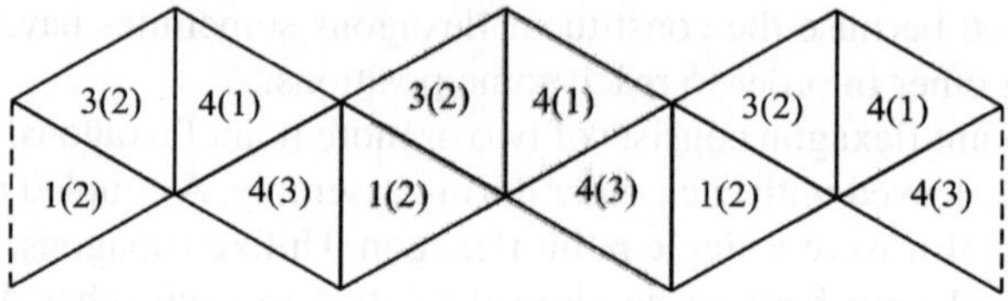

Fig. 11.1 Net for the tetrahexaflexagon. One copy needed. Fold together pairs of leaves numbered 2 and 4

Fig. 11.2 Construction of box position links (Les Pook, Flexagons inside out, 2003, © Cambridge University Press 2003, reprinted with permission)

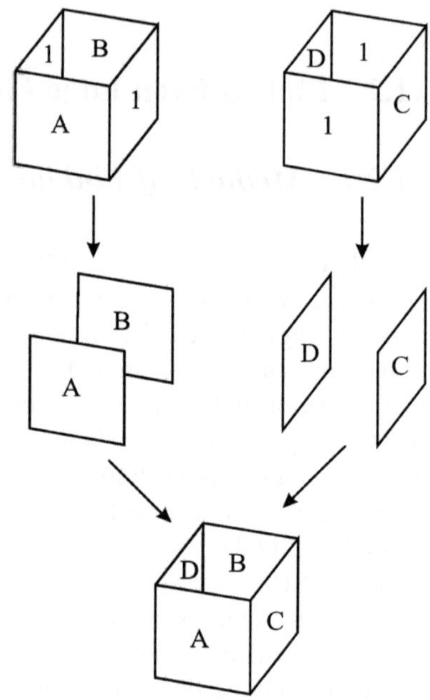

is shown in Fig. 11.1. In linking the two trihexaflexagons, six leaves are removed and the six faces in the two original trihexaflexagons are reduced to four in the tetrahexaflexagon.

The same procedure is used, with due attention to detail, when constructing flat main position links between other types of flexagon. It is sometimes possible to visualise the net for a desired linked flexagon without constructing a paper model. Purely theoretical methods of designing nets for flexagons with flat main position links are available (Moseley 2008). Skew main position links are constructed in the same way as flat main position links, with due attention to detail.

Two sector square even edge flexagons have two types of main position (Sections 4.2.4 and 4.2.5.3) so two types of link are possible. These are flat main position links and box position links. The method of constructing a box position link between two square even edge flexagons with two sectors, using paper models, is similar to that used for flat main position links between flexagons, and is shown in Fig. 11.2. Flex each of the flexagons to a box position in which alternate pats are single leaves. These pats are marked 1 in Fig. 11.2, top. Cut the flexagons along the hinges between pats, and remove the single leaves. Keep the remaining pats in their original relationship to one another as shown by the letters (Fig. 11.2, top and centre). Mark the cut edges of leaves for future reference. Assemble what will be a box position of the linked square even edge flexagon by placing the remaining pats from each of the precursor flexagons as opposite pats. This is indicated by the letters in Fig. 11.2, bottom. Join adjacent marked edges. It is usually easy to see what needs to be done in order to achieve this. Slant main position links are constructed in the same way as box position links, with due attention to detail.

Examples of even edge flexagons with all four types of link are given below in order to illustrate various aspects of their dynamic behaviour. In most of the examples two solitary even edge flexagons are linked.

11.2.2 Linked Hexaflexagons

The simplest hexaflexagon is the trihexaflexagon (Section 4.2.3). It is a solitary flexagon. All other hexaflexagons are linked even edge flexagons formed by linking two or more trihexaflexagons using flat main position links. Hexaflexagons have three sectors, and are flexed by using the threefold pinch flex. Principal main positions are, in appearance, flat regular even edge rings of six equilateral triangles (Fig. 1.1a). Intermediate positions are equilateral triangle edge triples (Fig. 2.1).

A trihexaflexagon has three faces; one principal cycle that can be traversed, as shown in the intermediate position map (Fig. 4.7). Each trihexaflexagon linked to a precursor hexaflexagon increases the number of faces by one and also increases the number of principal cycles by one. The number of distinct types of hexaflexagon, neglecting enantiomorphs, increases rapidly with the number of faces, as shown in Table. 11.1 (Conrad and Hartline 1962). Kosters (1999) later obtained the

Table 11.1 Numbers of distinct types of hexaflexagon

Number of faces	Number of types
3	1
4	1
5	1
6	3
7	4
8	12
9	27
10	82
11	228
12	733
13	2,282
14	7,528
15	24,834
16	83,898
17	285,357
18	983,244

same results independently. The number of trihexaflexagons linked, the number of cycles, and the torsion per sector are all two less than the number of faces. The number of flat main position links is three less than the number of faces.

The tetrahexaflexagon is the only possible hexaflexagon with four faces. It is also the simplest example of a flat main position link between two solitary flexagons. 'Tetra' refers to its four faces (Gardner 1965, 2008). Its net is shown in Fig. 11.1. As assembled, the tetrahexaflexagon is in principal main position 3(1). The two principal 3-cycles, marked A and B in the Tuckerman diagram (Fig. 11.3, cf. Fig. 10.31) can be traversed by using the threefold pinch flex. Within each principal 3-cycle, the dynamic properties are the same as those of a trihexaflexagon, except that there are reverse folded piles of leaves that behave as if there were single leaves. Principal main position 3(1) is a common main position that appears in both principal cycles. For videos of hexaflexagons see Moseley (2008) and Sherman (2008).

Pocket flexes are possible in all the principal main positions, including the common principal main position. One pocket flex leads to a combination of a slant regular even edge ring of four equilateral triangles and an equilateral triangle edge pair. A second pocket flex leads to another principal main position. Using pocket flexes treats the tetrahexaflexagon as the second order fundamental even edge flexagon $6\langle 3, 3\rangle_2$ (Sections 4.3.1 and 4.3.2) for which the second order fundamental triangle edge net $\langle 3\rangle_2$ (Fig. 3.13) is used.

Table. 11.1 shows that there are three types of hexaflexagon with six faces. One of these is the well known hexahexaflexagon. This was the second type of flexagon to be discovered (Conrad and Hartline 1962; Pook 2003). The first 'hexa' refers to its six faces (Gardner 1965, 2008). Its net is shown in Fig. 11.4. This is a straight strip of equilateral triangles. As assembled, the flexagon is in common principal main position 1(3). The Tuckerman diagram (Fig. 11.5) can be traversed by using the threefold pinch flex. The net is the dual of the net for the augmented irregular cycle interleaved triangle point flexagon type A (Fig. 7.28) with the number of sectors increased from one to three. The dynamic properties are similar in that the intermediate position map for the augmented irregular cycle interleaved triangle

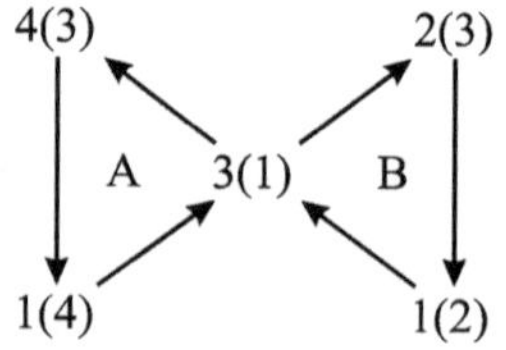

Fig. 11.3 Tuckerman diagram for the tetrahexaflexagon

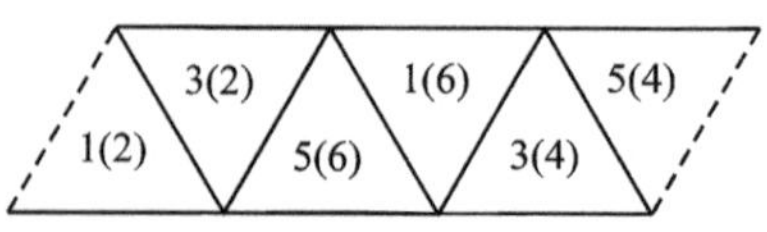

Fig. 11.4 Net for the hexahexaflexagon. Three copies needed. Fold pairs of leaves together in the order 6, 4, 2 and 5

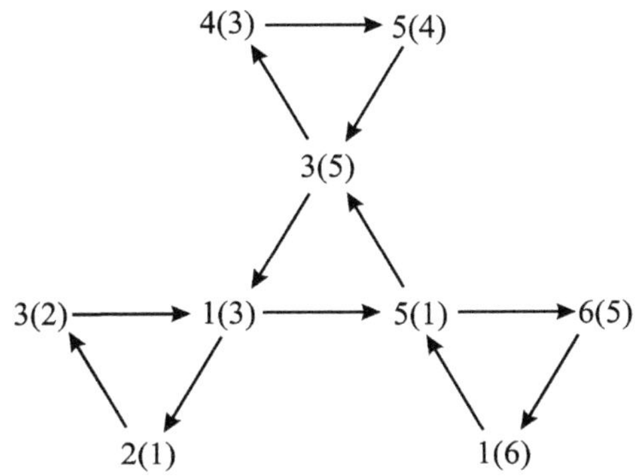

Fig. 11.5 Tuckerman diagram for the hexahexaflexagon

Fig. 11.6 Net for a crescent hexaflexagon with six faces. One copy needed. Fold pairs of leaves together in the order 6, 5, 4 and 3

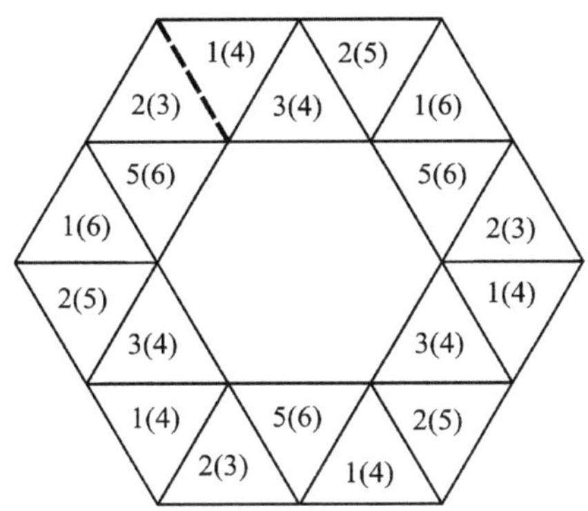

point flexagon type A (Fig. 7.27) can also be used to describe dynamic properties of the hexahexaflexagon. Other flexes are possible, including pocket flexes and the V-flex, but interleaf flexes are not possible. For an animation of the hexahexaflexagon being flexed see Highland Games (Highland 2008).

The V-flex (McLean 1979; Belenky 2009) is a complicated asymmetric flex which exploits the large number of degrees of freedom of all but the simplest types of hexaflexagon. If one uses the V-flex, face numbers become mixed up. A lot of practice is needed to operate the V-flex smoothly and without errors. The hexahexaflexagon can be flexed to display 3420 different main position faces by using both the V-flex and the threefold pinch flex (McLean 1979, 2008). Individual leaves have to be identified in order to characterise these main position faces. For videos of the V-flex see Moseley (2008) and Sherman (2008). Sherman also includes videos of other possible flexes.

The net for a crescent hexaflexagon is shown in Fig. 11.6. This is another of the possible hexaflexagons with six faces. The net has an attractive symmetry. 'Crescent' is a reference to the shape of the Tuckerman diagram (Fig. 11.7). As assembled, the flexagon is in common principal main position 1(2). The Tuckerman

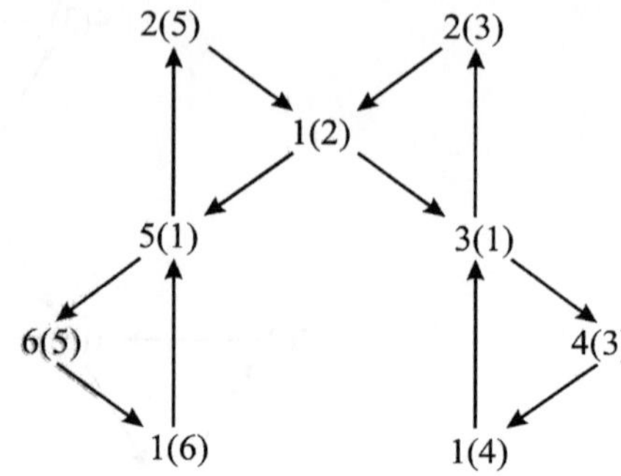

Fig. 11.7 Tuckerman diagram for a crescent hexaflexagon with six faces

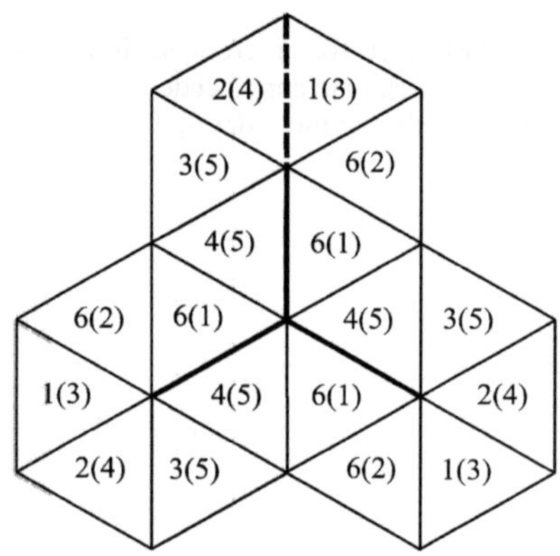

Fig. 11.8 Net for a street hexaflexagon with six faces. One copy needed

diagram can be traversed by using the threefold pinch flex. Face I appears in five of the principal main positions, three of which are common principal main positions. The leaves of the face have different arrangements in each of the five principal main positions. Pook (2003) gives the net for a seven faced crescent hexaflexagon in which one of the faces appears in all six possible arrangements. Other flexes are possible, including pocket flexes.

The net for a street hexaflexagon is shown in Fig. 11.8. This is the third possible hexaflexagon with six faces. As assembled, the flexagon is in common principal main position 2(1). The Tuckerman diagram (Fig. 11.9) can be traversed by using the threefold pinch flex. The faces are numbered so that faces 1 to 5 appear in numerical order, on the upper face, as the flexagon is flexed without turning it between flexes. This is the characteristic feature of a street hexaflexagon. Gardner (1965, 2008) gives a slightly different definition of a street flexagon. Other flexes are possible, including pocket flexes.

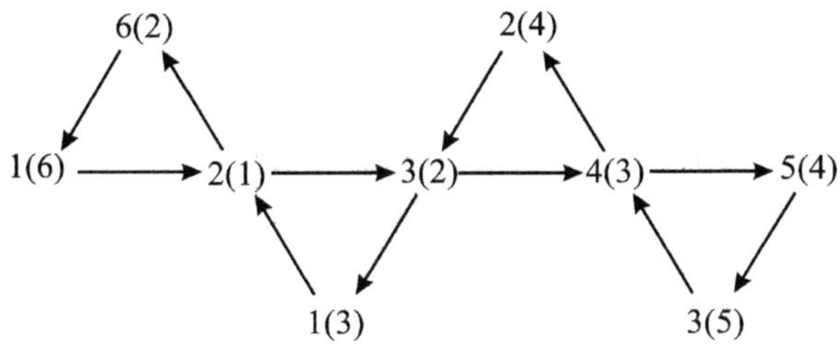

Fig. 11.9 Tuckerman diagram for a street hexaflexagon with six faces

11.2.3 Linked Square Even Edge Flexagons

There are three types of two sector solitary square even edge flexagons. These are the two sector first order fundamental square even edge flexagon 2⟨4, 4⟩ (Section 4.2.4), the two sector irregular square even edge flexagon (Section 7.2.3), and the two sector degenerate square even edge flexagon (Section 8.2.2). Square even edge flexagons have two types of main position (Sections 4.2.4 and 4.2.5.3). Principal main positions are, in appearance, flat regular even edge rings of four squares (Fig. 1.1b), and subsidiary main positions are box edge rings of four squares (Fig. 2.5). Hence, two types of link are possible (Section 11.2.1) flat main position links, and box position links. These complexities make enumeration of possible distinct types of linked square even edge flexagons difficult. Intermediate positions are square edge pairs (Fig. 1.15).

All three types of two sector solitary square even edge flexagons can be linked using flat main position links, hence there are six different ways in which two can be selected for linking. In addition, there are two different ways in which two degenerate square even edge flexagons can be linked, making a total of seven known types with flat main position links (Pook 2003). This total excludes linking two degenerate square even edge flexagons to recover the two sector first order fundamental square even edge flexagon 2⟨4, 4⟩ and the two sector irregular square even edge flexagon. The latter has zero torsion so, unusually, precursor flexagons of opposite hands are used.

Box position links are impossible with the two sector first order fundamental square even edge flexagon 2⟨4, 4⟩. This leaves two types of two sector solitary square even edge flexagons that can be linked using box position links, hence there are three different ways in which two can be selected for linking. Furthermore, there are two different ways in which two degenerate square even edge flexagons can be linked, making a total of four known types with box position links, Hence in total, there are 11 known distinct types of flexagons that can be constructed by linking two two sector solitary square even edge flexagons. Linking two two sector degenerate square even edge flexagons leads to the simplest possible linked flexagons. There are four distinct ways of doing this, two with flat main position links, and two with box position links.

A difference between flat main position links and box position links is that the former occur singly, whereas the latter always occur in pairs. This means that it is possible to traverse between cycles of the precursor flexagons in two different ways. Flat main position links and box position links are mutually exclusive in the sense that it may be possible to link two precursor flexagons by one or by the other, but not by both.

11.2.3.1 Square Even Edge Flexagons with Flat Main Position Links

There are two types of two sector square even edge flexagon with two incomplete cycles and a flat main position link. These are the simplest square even edge flexagons with a flat main position link. The net for one of these is shown in Fig. 11.10. This was derived by linking two two sector degenerate square even edge flexagons (Fig. 8.11) by a flat main position link. Other derivations are possible. The torsion per sector is 2. An incomplete principal 2/4 cycle can be traversed in each of the precursor flexagons (Section 8.2.2). This means that a principal 4-cycle could be traversed in a precursor flexagon, but in the degenerate square even edge flexagon only two principal main positions can be visited. These two incomplete principal 2/4 cycles appear in the linked flexagon, as shown in the Tuckerman diagram (Fig. 11.11).

As assembled, the flexagon is in common principal main position 1(2). From here it can be flexed to principal main positions 2(4) and 3(1) by using the twofold pinch flex. The flexagon does not have to be turned between flexes when traversing from one cycle to the other. For this reason the two incomplete 2/4-cycles are in line in the Tuckerman diagram.

There are four possible arrangements of the leaves on a face of a two sector square even edge flexagon (not eight as stated in Pook [2003]). The simplest two sector square even edge flexagon on which all four arrangements appear has three incomplete cycles and two flat main position links. Its net is shown in Fig. 11.12. This was derived by linking three degenerate square even edge flexagons (Fig. 8.11)

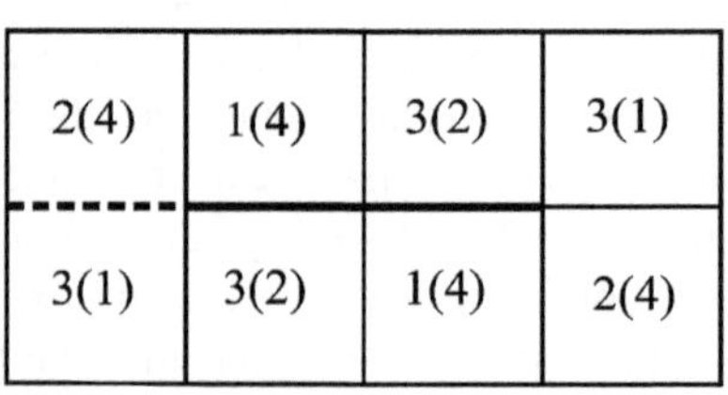

Fig. 11.10 Net for a two sector square even edge flexagon with two incomplete cycles and a flat main position link. One copy needed

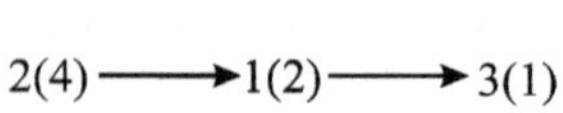

Fig. 11.11 Tuckerman diagram for a two sector square even edge flexagon with two incomplete cycles and a flat main position link

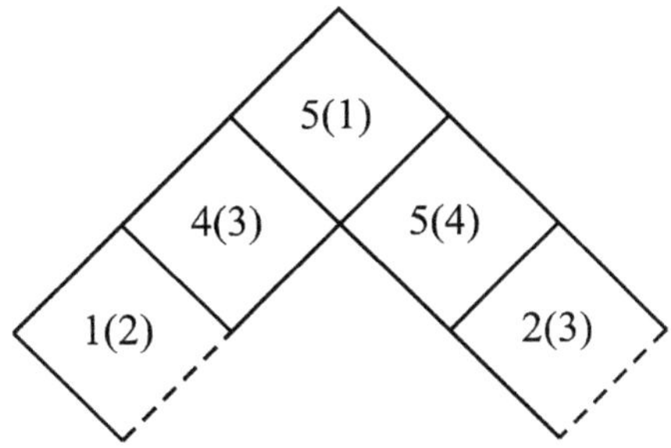

Fig. 11.12 Net for a two sector square even edge flexagon with three incomplete cycles and two flat main position links. Two copies needed. Fold pairs of leaves together in the order 5, 4 and 3

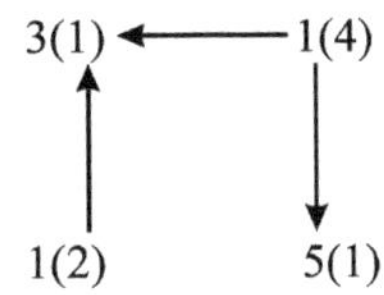

Fig. 11.13 Tuckerman diagram for a two sector square even edge flexagon with three incomplete cycles and two flat main position links (Les Pook, Flexagons inside out, 2003, © Cambridge University Press 2003, reprinted with permission)

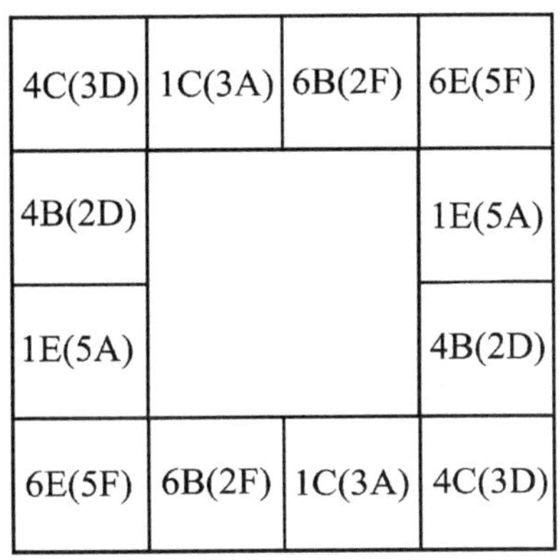

Fig. 11.14 Dual marked net for the Janus flexagon. Two complete cycles and a flat main position link. One copy needed. For the numbered enantiomorph, fold together pairs of leaves numbered 3, 4, 5 and 6. Do each corner of the net in turn. The last corner can be difficult, but it usually snaps into place. For the lettered enantiomorph, fold together pairs of leaves marked C, D, E and F

by two flat main position links The torsion per sector is 3. The three incomplete principal 2/4 cycles appear in the linked flexagon, as shown in the Tuckerman diagram (Fig. 11.13).

As assembled, the flexagon is in principal main position 1(2). From here it can be flexed to the other three principal main positions by using the twofold pinch flex. Face 1 appears in all four principal main positions in the four possible arrangements of its leaves. The flexagon has to be turned through 90° between flexes when traversing from one cycle to the other. For this reason adjacent incomplete 2/4-cycles are at 90° in the Tuckerman diagram.

The Janus flexagon is a well known type of square even edge flexagon that has not previously been given a special name. It has two complete 4-cycles and a flat main position link. A dual marked net is shown in Fig. 11.14. This was derived by linking two two sector irregular square even edge flexagons (Fig. 7.8) of opposite hand by a flat main position link. Each flexagon being linked has to be in a principal main position in which pats that are folded piles of leaves have the same hand. There are two sectors and the torsion is zero. A flexagon with zero torsion exists as an enantiomorphic pair (Section 7.2.1). Using the face numbers during assembly of the Janus flexagon leads to one enantiomorph, whilst using the face letters leads to the other. The net is a flat ring of squares; it is an uncut flexagon that can be assembled without cutting and rejoining the net. Leaves have to be bent to do this.

As assembled using face numbers, the flexagon is in common principal main position 2(1). Face letters are mixed up. The two principal 4-cycles marked A and B in the Tuckerman diagram for the numbered enantiomorph (Fig. 11.15a) can be traversed by using the twofold pinch flex. Within each principal 4-cycle the dynamic properties are the same as those of the precursor flexagons, except that there are reverse folded piles of leaves that behave as if they were single leaves. As assembled using face numbers, the flexagon is in common principal main position B(A). Face numbers are mixed up. The Tuckerman diagram for the lettered enantiomorph (Fig. 11.15b) can be traversed.

It is not possible to traverse from one enantiomorph to the other by using the twofold pinch flex. However, it is possible to traverse from one enantiomorph to the other, in a transformation between flexagons, by using the band flex. In other words the flexagon is disassembled back to the net and re-assembled as the other enantiomorph.

11.2.3.2 Square Even Edge Flexagons with Box Position Links

The linked square even edge flexagons described in this section have the same precursors as those in the previous section. The difference is that they are linked by box links rather than flat main position links. There are two two sector degenerate square even edge flexagons with two incomplete cycles and a pair of box position links. These are the simplest square even edge flexagons with a pair of box position links.

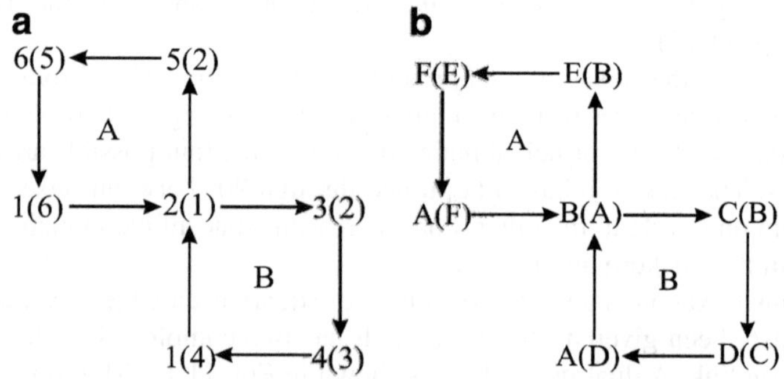

Fig. 11.15 Tuckerman diagrams for the Janus flexagon. (**a**) Numbered enantiomorph. (**b**) Lettered enantiomorph

The net for the two sector square even edge flexagon with two incomplete cycles and a pair of box position links type A is shown in Fig. 11.16. This was derived by linking two degenerate square even edge flexagons (Fig. 8.11) of the same hand, different ways up, by a pair of box position links. The net is the dual of the net for the irregular cycle interleaved square point flexagon (Fig. 7.17) with the number of sectors increased from one to two. Hinge letters have been added to the net. The torsion per sector is 2.

As assembled, the flexagon is in principal main position 2(1) which is, in appearance, a flat regular even edge ring of four squares (Fig. 1.1b). The two incomplete principal 2/4 cycles of the precursor flexagons, which are linked by a pair of box position links, become a single 6-cycle in the linked flexagon, as shown in the simplified map (Fig. 11.17). A simplified map is a Tuckerman diagram with intermediate position codes added. The simplified map can be traversed by using twofold pinch flexes and box flexes. The box flex is sometimes called the *tube flex*, and a flexagon capable of being box flexed a tubulating flexagon.

Starting from principal main position 2(1), use a twofold pinch flex to reach principal main position 3(2) via intermediate position 2. This is an incomplete principal 2/4-cycle of one of the precursor flexagons. Then use a box flex to reach principal main position 4(3). To do this, fold the flexagon in two to reach intermediate position 3A, which is a square edge pair (Fig. 1.15). The code 3A means that hinges marked A are at the centre of the flexagon (Fig. 11.18, left). Then open the flexagon into a box position.(Fig. 11.18, top). This is subsidiary main position 3(1), which is

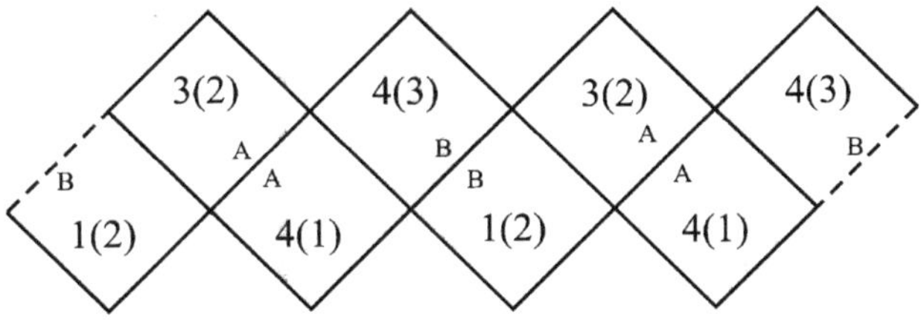

Fig. 11.16 Net for the two sector square even edge flexagon with two incomplete cycles and a pair of box position links type A. One copy needed. Fold pairs of leaves together in the order 4 and 3

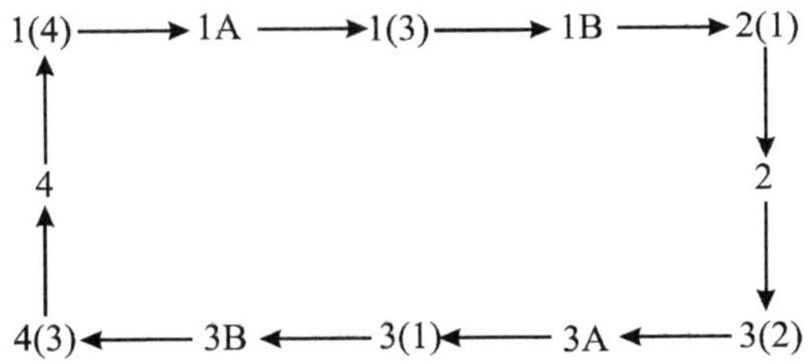

Fig. 11.17 Simplified map for the two sector square even edge flexagon with two incomplete cycles and a pair of box position links type A, numbered faces

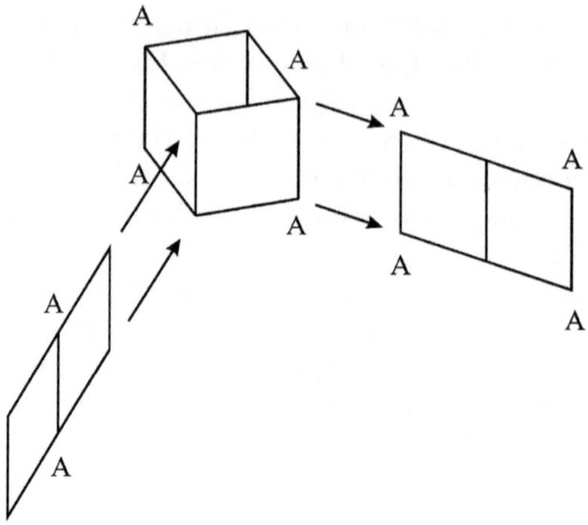

Fig. 11.18 The box flex

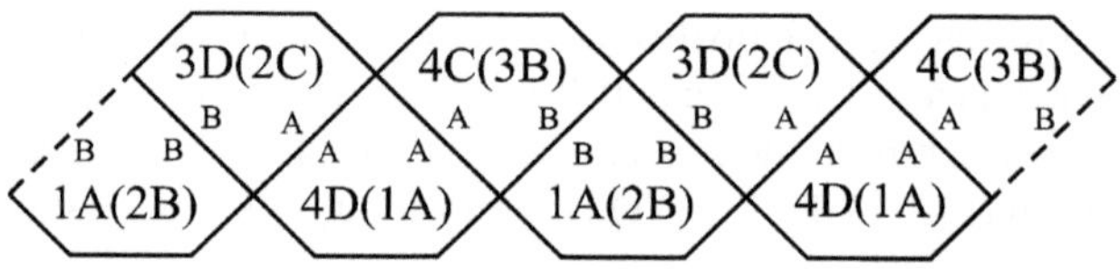

Fig. 11.19 Dual marked net for the two sector square (irregular pentagon) even edge flexagon with two incomplete cycles and a pair of box position links type A. One copy needed. Fold pairs of leaves together in the order 4 and 3

a box edge ring of four squares (Fig. 2.5). Then close the box position so that hinges A are at the ends of the flexagon (Fig. 11.18, right). This is intermediate position 3B with hinges marked B at the centre. Next, open the flexagon into principal main position 4(3) and so on round the 6-cycle.

The flexagon is analogous to the irregular cycle interleaved square point flexagon (Section 7.3.3) in that two distinct face numbering sequences are possible. A dual marked net, in which numbers are used for one distinct face numbering sequence, and letters for another, is shown in Fig. 11.19. Face numbers are the same as those shown in Fig. 11.16 so the simplified map for flexing using face numbers is as shown in Fig. 11.17. A single interleaf flex is used to transform from one face numbering sequence to the other. The squares have been truncated to irregular pentagons to make interleaf flexing easier, but the leaves overlap reasonably well. The flexagon could also be called an irregular pentagon even edge flexagon or a partial overlap flexagon.

As assembled, the flexagon is in principal main position 2(1) and face letters on face 2 are mixed up. To transform from the numbered 6-cycle to the lettered 6-cycle

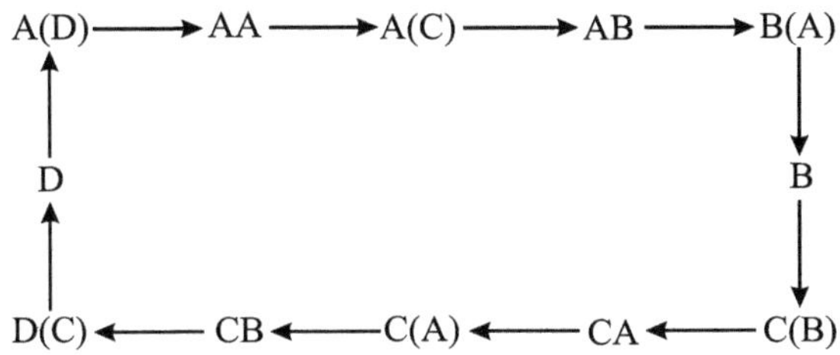

Fig. 11.20 Simplified map for the two sector square (irregular pentagon) even edge flexagon with two incomplete cycles and a pair of box position links type A, lettered faces

Fig. 11.21 Net for the two sector square even edge flexagon with two incomplete cycles and a pair of box position links type B. One copy needed. Fold pairs of leaves together in the order 4 and 3

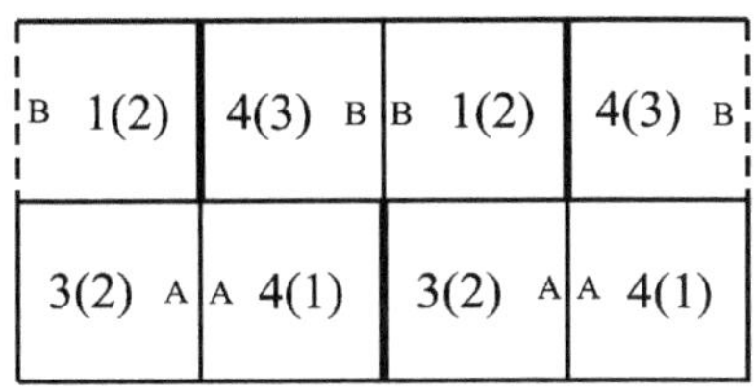

shown in Fig. 11.20 use a single interleaf flex at principal main position 2(1). To do this, interchange the top two leaves on each of those pats that consist of folded plies of three leaves, to reach principal main position B(A). The leaves have to be bent. In intermediate position codes, the first or only letter is a face letter and the second letter is a hinge letter.

The net for the two sector square even edge flexagon with two incomplete cycles and a pair of box position links type B is shown in Fig. 11.21. This was derived by linking two degenerate square even edge flexagons (Fig. 8.11) of the same hand with both precursors the same way up. Hinge letters have been added to the net. The torsion per sector is 2. As assembled, the flexagon is in principal main position 2(1). The two incomplete principal 2/4 cycles of the precursor flexagons become a single 6-cycle in the linked flexagon, as shown in the simplified map. The simplified map is the same as that for type A, numbered faces (Fig. 11.17). The simplified map can be traversed by using twofold pinch flexes and box flexes as for type A (above). For this traverse, apart from differences in pat structure, the dynamic properties are the same as type A. A significant difference is that interleaf flexes are not possible.

The net for a square even edge flexagon with two complete cycles and a pair of box position links is shown in Fig. 11.22. This was derived by linking two irregular square even edge flexagons (Fig. 7.8). Hinge letters have been added to the net. The torsion is zero. As assembled, the flexagon is in principal main position 2(1). The two principal 4-cycles shown by single headed arrows in the simplified map (Fig. 11.23) can be traversed by using the twofold pinch flex. The Tuckerman diagram for one of the principal 4-cycles is the same as that for the principal 4-cycle of the first order fundamental square even edge flexagons $S\langle 4, 4\rangle$ (Fig. 4.17).

The two principal 4-cycles are linked by a pair of box position links, and the box flexes used to flex between them are shown by double headed arrows in the simplified map.

Fig. 11.22 Net for a square even edge flexagon with two complete cycles and a pair of box position links. One copy needed. Fold pairs of leaves together in the order 6, 5, 4 and 3 (Les Pook, Flexagons inside out, 2003, © Cambridge University Press 2003, reprinted with permission)

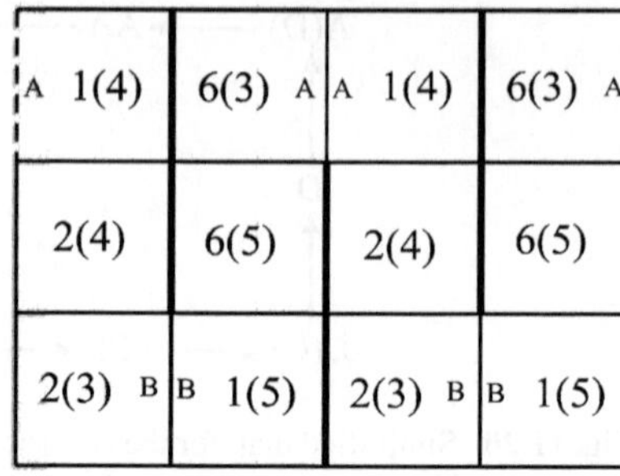

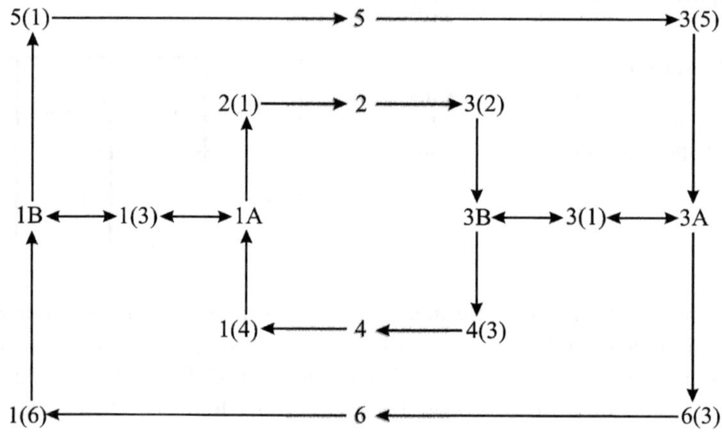

Fig. 11.23 Simplified map for a square even edge flexagon with two complete cycles and a pair of box position links

The pair of box position links means that there are two ways of traversing between the two principal 4-cycles by using box flexes. As a result there are several different routes for traversing from, say, principal main position 5(1) to principal main position 6(3).

11.2.4 Linked Pentagon Even Edge Flexagons

There are nine types of two sector solitary pentagon even edge flexagons. These are the two sector first order fundamental pentagon even edge flexagons 2⟨5, 5⟩ and 2⟨5, 5/2⟩ (Section 4.2.6), the two sector irregular cycle pentagon even edge flexagons, types A and B (Section 7.2.4), and the two sector degenerate pentagon even edge flexagons types A–E (Section 8.2.3). Two sector pentagon even edge flexagons have two types of main position appearance (Section 4.2.6), so two types of link are possible. These complexities make enumeration of possible distinct types of linked pentagon even edge flexagons very difficult.

Skew main positions are, in appearance, skew regular even edge rings of four regular pentagons (Fig. 2.7). They have two possible configurations (Section 2.2.1).

A snap flex, in which leaves are bent, may be used to transform from one configuration to the other. A skew main position link is similar to a flat main position link, and is constructed in a similar way (Section 11.2.1). Slant main position are slant regular even edge rings of four regular pentagons (Fig. 4.24). A slant main position link is similar to a box position link and is constructed in a similar way (Section 11.2.1).

11.2.4.1 A Pentagon Even Edge Flexagon with a Skew Main Position Link

The net for a pentagon even edge flexagon with a skew main position link is shown in Fig. 11.24. This was derived by linking two degenerate pentagon even edge flexagons type D (Fig. 8.15). Hinge letters have been added to the net. The torsion per sector is 2.

As assembled, the flexagon is in intermediate position 1A, with hinges A at the centre. The simplified map (Fig. 11.25) can be traversed by using twofold pinch flexes and a snap flex. Direction arrows are not shown because it is not possible to enter them in a consistent fashion. The codes 1(3) and 3(1) are the two possible forms of the common principal main position. There is no convenient way of marking the leaves to distinguish between them so the codes are assigned arbitrarily. Use a snap flex to traverse between the two forms. Leaves have to be bent. The common principal main position, and principal main positions 2(1) and 1(4) are, in appearance, skew regular even edge ring of four regular pentagons (Fig. 2.7). The subsidiary main positions are slant regular even edge rings of four regular pentagons (Fig. 4.24). The number outside the brackets indicates the numbers visible on the outside of the flexagon so the codes 2(3), 3(2) and 3(4), 4(3) distinguish between the two forms of each of the two subsidiary main positions. The high curvature (216°) means that these cannot

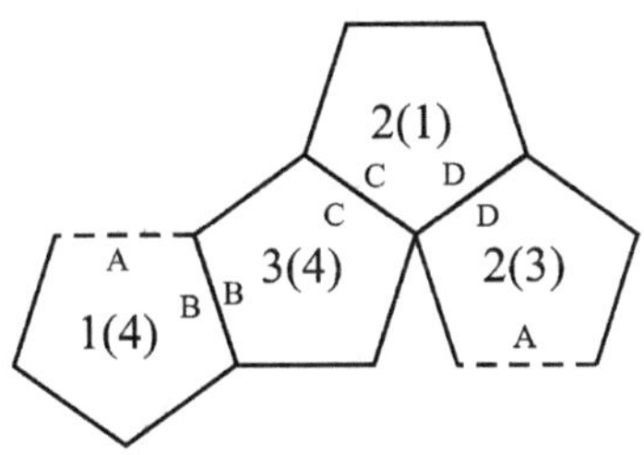

Fig. 11.24 Net for a pentagon even edge flexagon with a skew main position link. Two copies needed. Fold together pairs of leaves in the order 2, 3 and 4. Assemble with hinges A at the centre

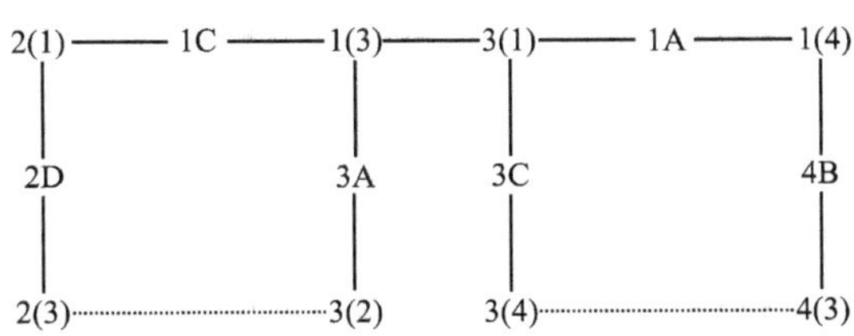

Fig. 11.25 Simplified map for a pentagon even edge flexagon with a skew main position link

be turned inside out so it is not possible to traverse between the two forms, They are connected by dotted lines on the simplified map. The left and right halves of the simplified map correspond to the intermediate position maps for the two precursor flexagons (Fig. 8.16), with a snap flex used to traverse between them.

11.2.4.2 A Pentagon Even Edge Flexagon with a Pair of Slant Main Position Links

The net for a pentagon even edge flexagon with a pair of slant main position links is shown in Fig. 11.26. This was derived by linking two degenerate pentagon even edge flexagons type D (Fig. 8.15). Hinge letters have been added to the net. It could also be derived by using the net for the two sector square even edge flexagon with two incomplete cycles and a pair of box position links type B (Fig. 11.21) as a precursor and replacing the squares by regular pentagons. The torsion per sector is 2.

As assembled, the flexagon is in intermediate position 1A, with hinges A at the centre. The 6-cycle shown in the simplified map can be traversed by using twofold pinch flexes and slant flexes. The simplified map is the same as that for the two sector square even edge flexagon with two incomplete cycles and a pair of box position links type A, numbered faces (Fig. 11.17). A slant flex is similar to a box flex. Principal main positions 2(1), 3(2), 4(3) and 1(4) are skew regular even edge rings of four regular pentagons (Fig. 2.7). Subsidiary main positions 1(3) and 3(1) are slant regular even edge rings of four regular pentagons (Fig. 4.24). The high curvature (216°) means that these cannot be turned inside out so it is not possible to traverse directly between the two forms.

In a slant flex, starting from intermediate position 1A, open it to reach subsidiary main position 1(3) and close the flexagon with hinges B at the centre to reach intermediate position 1B. To traverse the 6-cycle, next open the flexagon into principal main position 1(2), close the flexagon into intermediate position 2, open it into principal main position 3(2) and so round the 6-cycle. Leaves do not have to be bent. Traversing the 6-cycle turns subsidiary main position 1(3) inside out.

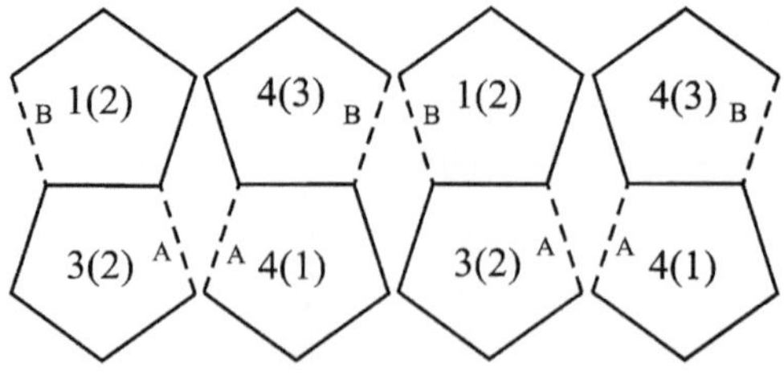

Fig. 11.26 Net for a pentagon even edge flexagon with a pair of slant main position links. One copy needed. Join the net at corresponding hinge letters. Fold together pairs of leaves in the order 2, 3 and 4. Assemble with hinges A at the centre

11.2.5 *Linked Silver Even Edge Flexagons*

11.2.5.1 General Properties

Numbers of distinct types of linked triangle even edge flexagons of the same hand increase if the equilateral triangles are replaced by isosceles triangles or scalene triangles. This is illustrated in Table. 11.2 by data for linked trihexaflexagons 3⟨3, 3⟩ (Table. 11.1), linked four sector fundamental silver even edge flexagons 4⟨silver⟩ (Pook 2008a), and linked six sector fundamental bronze even edge flexagons 6⟨bronze⟩ (Pook 2008b). Silver even edge flexagons are made from 45°–45°–90° (silver) triangles, so are examples of *isosceles* triangle even edge flexagons (Section 10.2.4.1). Bronze even edge flexagons are made from 30°–60°–90° (bronze) triangles so are examples of scalene triangle even edge flexagons (Section 10.2.6.1).

Four sector fundamental silver even edge flexagons of the same hand can be linked by either flat main position links or skew main position links. Two four sector fundamental silver even edge flexagons of opposite hand can be linked, but only a flat main position link is possible, so there is one distinct type. The three possible distinct types of flexagon formed by linking two four sector fundamental silver even edge flexagons are described below.

11.2.5.2 A Silver Even Edge Flexagon with a Flat Main Position Link

A dual marked net for a silver even edge flexagon with a flat main position link is shown in Fig. 11.27. This was derived by linking two fundamental silver even edge flexagons 4⟨silver⟩ (Fig. 10.8b) of the same hand. Face letters have been chosen so that all the leaves in each face either have the same number or the same letter. Mitchell (2002) calls this flexagon the slit-square flexagon, a reference to the shape of its net.

As assembled, the flexagon is in common subsidiary main position 3(1). There are four sectors, and the torsion per sector is 2. The two 3-cycles, marked A and B in the Tuckerman diagram can be traversed by using the fourfold pinch flex. The Tuckerman diagram is the same as that for the tetrahexaflexagon (Fig. 11.3). Within each 3-cycle the dynamic properties are the same as those of the precursor flexagons (Section 10.2.4.1). Common subsidiary main position 3(1) and subsidiary

Table 11.2 Numbers of distinct linked triangle even edge flexagons with precursors of the same hand

Number of flexagons linked	Type of flexagon linked: 3⟨3, 3⟩ (Equilateral triangles)	4⟨silver⟩ (Isosceles triangles)	6⟨bronze⟩ (Scalene triangles)
2	1	2	3
3	1	2	3
4	3	7	19

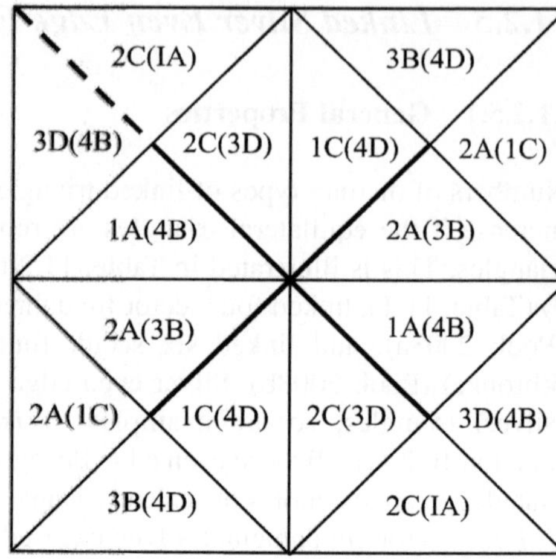

Fig. 11.27 Dual marked net for a silver even edge flexagon with two cycles and a flat main position link. One copy needed. Fold together pairs of leaves numbered 2 and 4

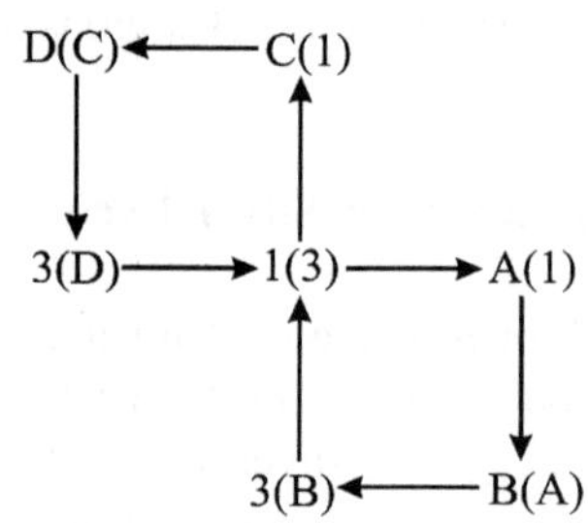

Fig. 11.28 Tuckerman diagram for minor cycles of a silver even edge flexagon with a flat main position link

main positions 2(3) and 1(4) are, in appearance, flat regular even edge rings of eight silver triangles (Fig. 2.21b). Subsidiary main positions 1(2) and 4(3) are skew regular even edge rings of eight silver triangles (Fig. 10.9b). Intermediate positions are silver triangle edge quadruples.

Flexes with twofold rotational symmetry are possible. Two minor 4-cycles can be traversed by using the twofold pinch flex, as shown in the minor cycle Tuckerman diagram (Fig. 11.28, cf. Fig. 11.15). Minor 4-cycles with twofold rotational symmetry do not appear in the precursor flexagons. Minor main positions B(A) and D(C) are flat regular even edge rings of eight silver triangles (Fig. 2.21b), but the pat structure differs from that of common main position 3(1). Minor main positions A(1), 3(B), C(1) and D(3) are combinations of a flat regular even edge ring of four 45°–45°–90° isosceles (silver) triangles and two edge pairs of 45°–45°–90° isosceles (silver) triangles (Fig. 10.10b). Minor intermediate positions have a rectangular outline (Fig. 11.29). Other flexes are possible, including pocket flexes and box flexes.

Fig. 11.29 Appearance of minor intermediate positions of a silver even edge flexagon with a flat main position link, and of an uncut silver even edge flexagon with a flat main position link

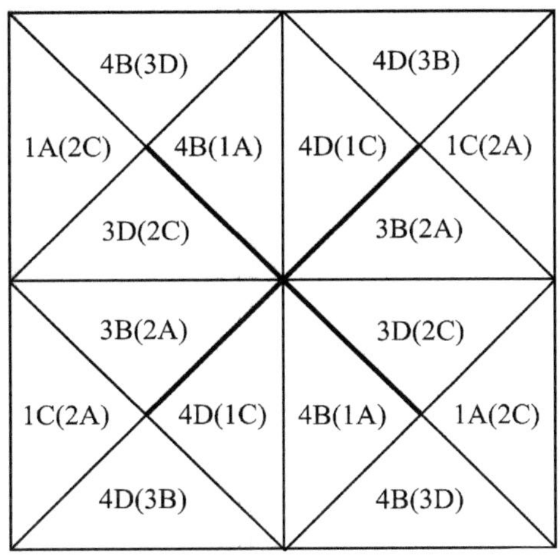

Fig. 11.30 Dual marked net for an uncut silver even edge flexagon with a flat main position link. One copy needed. On one side of the flexagon, turn over the pair of leaves numbered 1(4) twice so that pairs of leaves numbered 4 are folded together. Then do the same on the other side of the flexagon. See Fig. 11.31

11.2.5.3 An Uncut Silver Even Edge Flexagon with a Flat Main Position Link

A dual marked net for an uncut silver even edge flexagon with a flat main position link is shown in Fig. 11.30. This was derived by linking two fundamental silver even edge flexagons 4⟨silver⟩ (Fig. 10.8b) of opposite hands. Face letters have been chosen so that all the leaves in each face either have the same number or the same letter. It can be assembled without cutting and rejoining the net, so it is an *uncut flexagon*. The assembly sequence is shown in Fig. 11.31. As with the Janus flexagon (Section 11.2.3.1) it can also be assembled as the enantiomorph, but a different set of face markings is needed.

As assembled, the flexagon is in common subsidiary main position 3(1). There are four sectors and the torsion is zero. The two incomplete cycles shown in the Tuckerman diagram (Fig. 11.32, cf. Fig. 11.11) can be traversed by using the fourfold pinch flex. The 3-cycles of the precursor flexagons have become incomplete 2/3 cycles. The notation 2/3 means that only two main positions can be visited

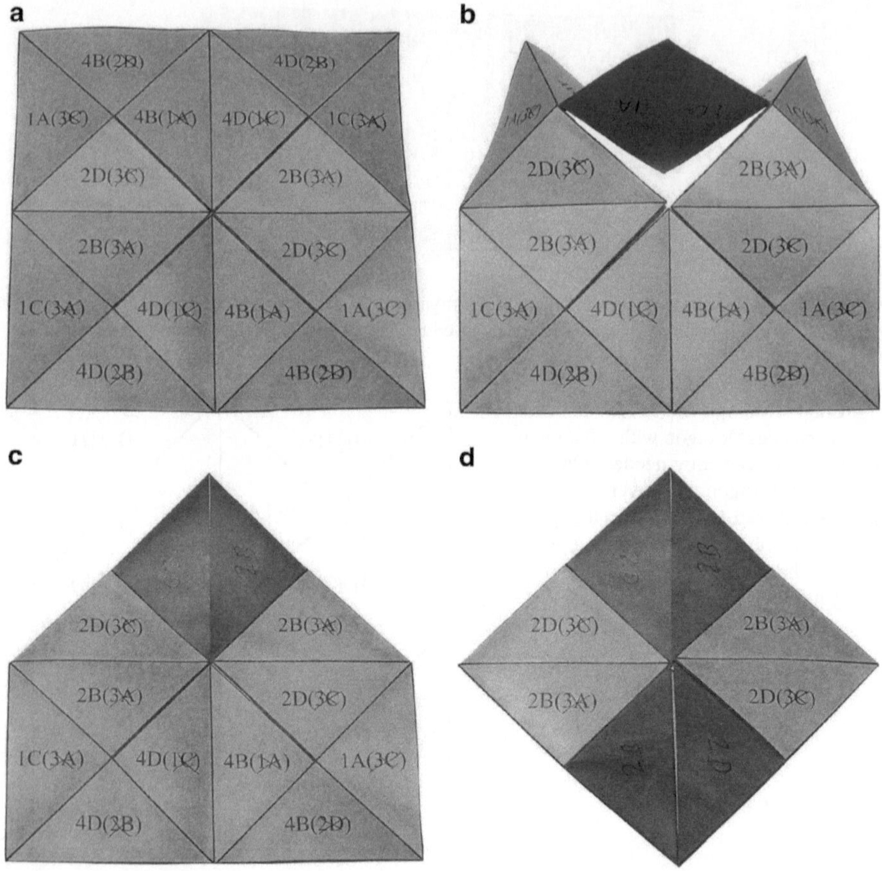

Fig. 11.31 Assembly of an uncut silver even edge flexagon with a flat main position link. (**a**) Net cut and creased. (**b**) One of the pairs of leaves numbered 1(4) turned over. (**c**) The pair of leaves turned over again. This folds together pairs of leaves numbered 3 and 4. (**d**) Sequence repeated with the other pair of leaves numbered 1(4), to reach main position 2(1)

2(3) ⟶ 1(3) ⟶ 4(1)

Fig. 11.32 Tuckerman diagram for fourfold pinch flexing of an uncut silver even edge flexagon with a flat main position link

in what was a 3-cycle in a precursor flexagon. The dynamic properties therefore differ from those of silver even edge flexagon with a flat main position link (previous section). Common subsidiary main position 3(1) and subsidiary main positions 2(3) and 4(1) are, in appearance, flat regular even edge rings of eight silver triangles (Figs. 2.21b and 11.31d). Intermediate positions are silver triangle edge quadruples.

Flexes with twofold rotational symmetry are possible. Two minor 4-cycles can be traversed by using the twofold pinch flex, as shown in the minor cycle Tuckerman diagram. This is the same as that for a silver even edge flexagon with a flat main position link (Fig. 11.28), and positions have the same appearances but different pat structures. The common main position is the same as that for the two 2/3-cycles. Minor 4-cycles with twofold rotational symmetry do not appear in the precursor flexagons. Minor main positions B(A) and D(C) are flat regular even edge rings of eight silver triangles (Fig. 2.21b). Minor main positions A(1), 3(B), C(1) and D(3) are combinations of a flat regular even edge ring of four 45°–45°–90° isosceles (silver) triangles and two edge pairs of 45°–45°–90° isosceles (silver) triangles (Fig. 10.10b). Minor intermediate positions have a rectangular outline (Fig. 11.29). Other flexes are possible, including box flexes, but pocket flexes are not possible.

11.2.5.4 A Silver Even Edge Flexagon with a Skew Main Position Link

A dual marked net for a silver even edge flexagon with a skew main position link is shown in Fig. 11.33. This was derived by linking two fundamental silver even edge flexagons 4⟨silver⟩ (Fig. 10.8b) of the same hand. Face letters have been chosen so that all the leaves in each face either have the same number or the same letter.

As assembled, the flexagon is in subsidiary main position 1(2). There are four sectors and the torsion per sector is 2. The two cycles, marked A and B in the Tuckerman diagram can be traversed by using the fourfold pinch flex. The Tuckerman diagram is the same as that for the tetrahexaflexagon (Fig. 11.3). Within each 3-cycle, the dynamic properties are the same as those of the precursor flexagons (Section 10.2.4.1). A snap flex is needed to traverse between the cycles, but there are sufficient degrees of freedom for this to be done without bending the leaves. Common subsidiary main position 3(1) is a skew regular even edge ring of eight silver triangles (Fig. 10.9b). Subsidiary main positions 1(2), 2(3), 4(3) and 1(4) are flat regular even edge rings of eight silver triangles (Fig. 2.21b). Intermediate positions are silver triangle edge quadruples. Other flexes are possible, including pocket flexes, but there are no minor cycles that can be traversed with just the twofold pinch flex.

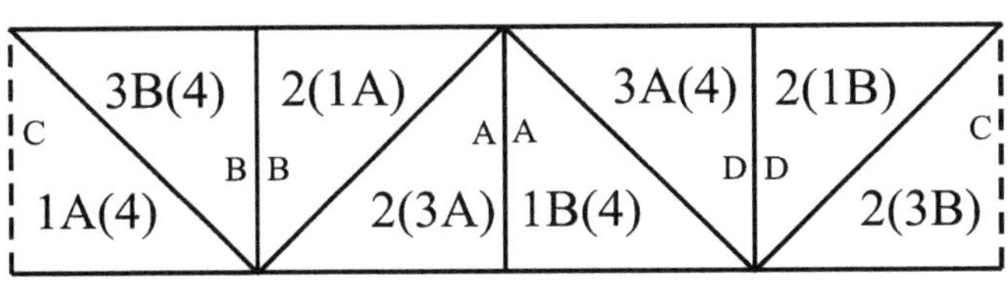

Fig. 11.33 Dual marked net for a linked silver even edge flexagon with a skew main position link. Two copies needed. Fold together pairs of leaves in the order 4 and 3

Two minor 8-cycles can be traversed by using twofold pinch flexes and box flexes. The simplified map for one of these is shown in Fig. 11.34. Subsidiary main positions 1(2) and 3(2) are flat regular even edge rings of eight silver triangles (Fig. 2.21b). Minor main positions A(1), 2(A), 3(B) and B(1) are combinations of a flat regular even edge ring of four 45°–45°–90° isosceles (silver) triangles and two edge pairs of 45°–45°–90° isosceles (silver) triangles (Fig. 10.10b). Minor main positions A(2) and B(2) are antibox edge rings of eight 45°–45°–90° (silver) triangles (Fig. 11.35). They can be flattened into flat regular even edge rings of four 45°–45°–90° (silver) triangles (Fig. 2.21a). Minor intermediate positions have a rectangular outline (Fig. 11.29).

Starting from subsidiary main position 1(2), fold the flexagon in two, so that hinges marked D are at the centre of the flexagon, to reach intermediate position

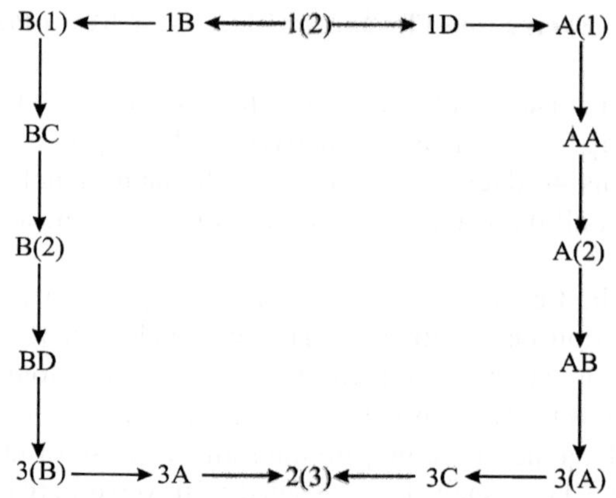

Fig. 11.34 Simplified map for a minor 8-cycle of a linked silver even edge flexagon with a skew main position link

Fig. 11.35 A flexagon as an antibox edge ring of eight 45°–45°–90° (silver) triangles

1D. The 1 means that leaves numbered 1 are visible, and the D that hinges marked D are at the centre. Then complete the twofold pinch flex by unfolding the flexagon to reach minor main position A(1). Next, fold in two with leaves marked A visible and hinges marked A at the centre to reach intermediate position AA. Then open into minor main position A(2) with leaves marked A visible on the outside of the antibox edge ring and leaves numbered 2 visible on the inside. Complete the box flex by closing so that hinges B are at the centre, to reach intermediate position AB. Unfold into minor main position 3(A) and so on round the minor 8-cycle.The direction of the arrows changes at subsidiary main position 3(2). This means that the flexagon has to be turned over to continue on round the cycle. The minor 8-cycle provides an alternative to the fourfold pinch flex for traversing between subsidiary main positions 1(2) and 2(3). The other minor 8-cycle provides a similar alternative for traversing between subsidiary main positions 4(3) and 1(4). Similar minor 8-cycles can be traversed in the four sector fundamental silver even edge flexagon 4⟨silver⟩ (Section 10.2.4.1), and a silver even edge flexagon with a flat main position link (Section 11.2.5.2).

11.2.6 Linked Bronze Even Edge Flexagons

11.2.6.1 General Properties

Two 6 sector fundamental bronze even edge flexagons 6⟨bronze⟩ (Section 10.2.6.5) of the same hand can be linked by either flat main position links, or by one of two different types of skew main position links. Two six sector fundamental bronze even edge flexagons of opposite hand can be linked, but only flat main position links are possible. Hence, there are four possible distinct types of flexagon formed by linking two six sector fundamental bronze even edge flexagons (Section 11.2.5.1). The dynamic properties of all four flexagons are complex, but they are all unstable. and easily muddled. The most easily handled has a skew main position link, and is described below.

In a stretch flexagon, a flat main position of a precursor edge flexagon is stretched uniformly in one direction (Section 10.1.2). Two examples of linked bronze even edge flexagons, which can be derived stretching linked silver even edge flexagons, are given below.

11.2.6.2 A Bronze Even Edge Flexagon with a Skew Main Position Link

A dual marked net for a bronze even edge flexagon with a skew main position link is shown in Fig. 11.36. This was derived by linking two fundamental bronze even edge flexagons 6⟨bronze⟩ (Fig. 10.28), of the same hand. The dynamic properties are identical to those of the hexa-dodeca-flexagon described by Schwartz (2008). She used a straight strip net with additional faces that do not appear during flexing. The dynamic properties of the flexagon are complex. Flexing possibilities are an

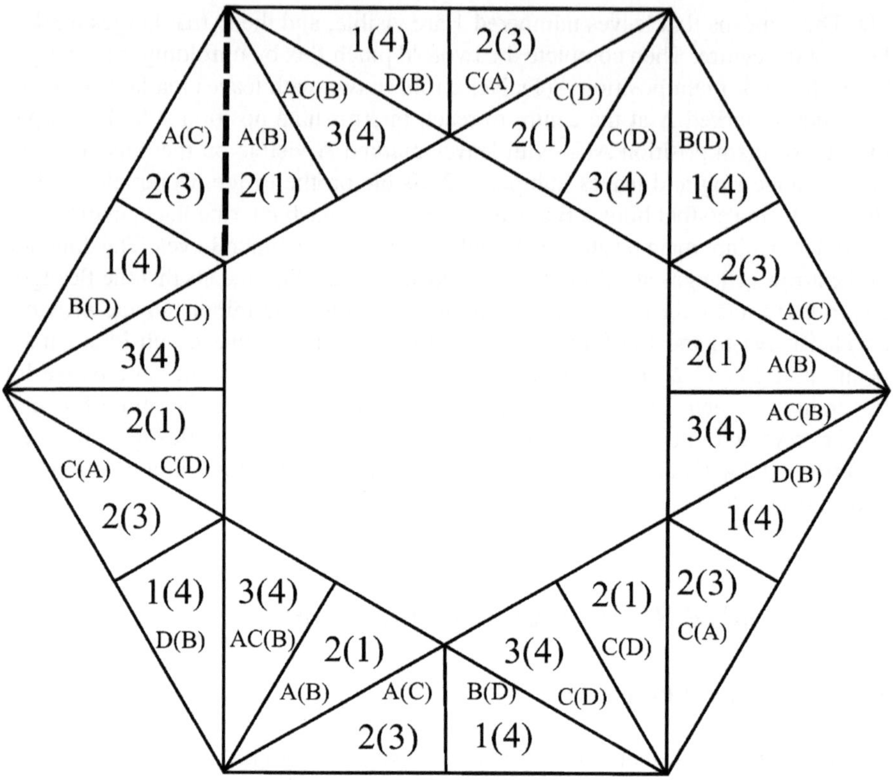

Fig. 11.36 Dual marked net for a bronze even edge flexagon with a skew main position link. One copy needed. Fold together pairs of leaves in the order 4 and 3

extension of those for the precursors. All the flexes described for the fundamental bronze even edge flexagon 6⟨bronze⟩ (Section 10.2.6.5) are theoretically possible, but the instability of the flexagon makes some of them impractical. However, both the sixfold pinch flex and the threefold pinch flex can be used. The flexagon is stable. when flexed using the threefold pinch flex.

As assembled, the flexagon is in principal main position 1(2). There are six sectors, and the torsion per sector is 2. The two cycles, marked A and B in the Tuckerman diagram can be traversed by using the sixfold pinch flex. The Tuckerman diagram is the same as that for the tetrahexaflexagon (Fig. 11.3). A snap flex is needed to traverse between the two cycles. There are sufficient degrees of freedom for this to be done without bending the leaves, but the flexagon is easily muddled. It is advisable to make a hinge easy to disconnect so that, if it becomes muddled, the hinge can be disconnected and the flexagon can be refolded into a desired position. Principal main positions 1(2) and 4(3) are, in appearance, flat regular even edge rings of 12 bronze triangles (Fig. 2.22c). Common subsidiary main position 1(3) is a skew regular even edge ring of 12 bronze triangles, curvature −720°. It is this very high curvature that makes the snap flex, in which mountain folds become valley folds and vice versa (Section 2.2.1), difficult. Subsidiary main positions 2(3) and 1(4) are skew regular even edge rings of 12 bronze triangles of a different type, curvature −360°. Intermediate positions are bronze triangle even edge sextuples.

Flexing using the threefold pinch flex, with intermediate positions with triangular flaps (Fig. 10.30a), works well. Face numbers become mixed up during flexing. In the dual marked net, face letters have been chosen so that all the leaves visible on a face either have the same number or the same letter. Loose flaps appear during flexing. Schwartz (2008) calls these loose flaps rogue triangles. Loose flaps do not appear in the precursor flexagons. In a flap flex the three loose flaps on a face are turned over. Lines with arrow heads at both ends are used to indicate flap flexes on the Tuckerman diagram (Fig. 11.37). The two minor 3-cycles from the two precursor flexagons (Fig. 10.31a) are at the top and bottom of the Tuckerman diagram. Two additional minor 3-cycles appear, and these are shown at the sides of the Tuckerman diagram. Principal main position 1(2) and 4(3) are flat regular even edge rings of 12 bronze triangles (Fig. 2.22c). Minor main positions 3(B), 3(D), A(1) and C(1) are flat irregular even edge rings of 12 bronze triangles with a hexagonal outline (Fig. 10.32). The other minor main positions are flat and have a propeller outline (Fig. 10.33a). The Tuckerman diagram shows that it is possible to traverse between principal main position 1(2) and 4(3), via flat minor main positions, without using the snap flex which is needed if sixfold pinch flexing is used.

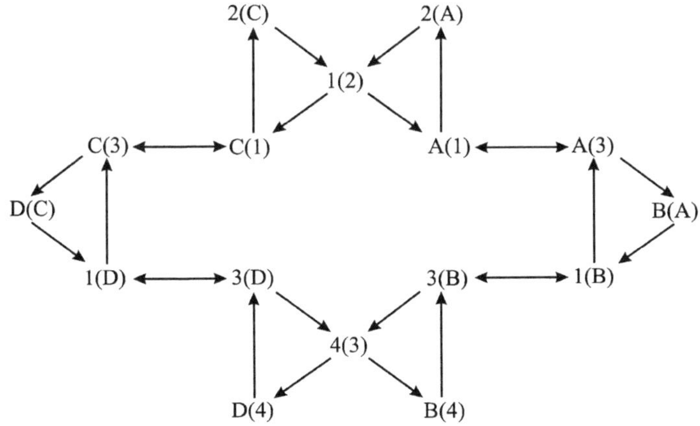

Fig. 11.37 Minor cycle Tuckerman diagram for a bronze even edge flexagon with a skew main position link, triangular intermediate positions flaps

11.2.6.3 A Partial Overlap Bronze Even Edge Flexagon with a Flat Main Position Link

A dual marked net for a partial overlap bronze even edge flexagon with a flat main position link is shown in Fig. 11.38. This was derived by using a silver even edge flexagon with a flat main position link (Section 11.2.5.2) as a precursor, and stretching it in common subsidiary main position 3(1) so that the silver triangles became bronze triangles. The leaves are identical, but do not always overlap exactly, so it is a second outcome stretch flexagon (Section 10.1.2). Other derivations are possible, but it cannot be derived by partially stellating a dodecagon flexagon (Section 10.2.7). It could also be called a partial overlap flexagon, or a scalene triangle even edge flexagon. It is the slit diamond flexagon described by Mitchell (2002), a reference to the shape of its net.

As assembled, the flexagon is in common subsidiary main position 3(1). There are two sectors, and the torsion per sector is 4. The dynamic properties are a combination of those of a partial overlap silver even edge flexagon (Section 10.2.7) and a silver even edge flexagon with a flat main position link (Section 11.2.5.2). The two 3-cycles shown in the Tuckerman diagram can be traversed by using the fourfold pinch flex. The Tuckerman diagram is the same as that for the tetrahexaflexagon (Fig. 11.3). Common subsidiary main position 3(1) and subsidiary main positions 2(3) and 1(4) are, in appearance, flat irregular

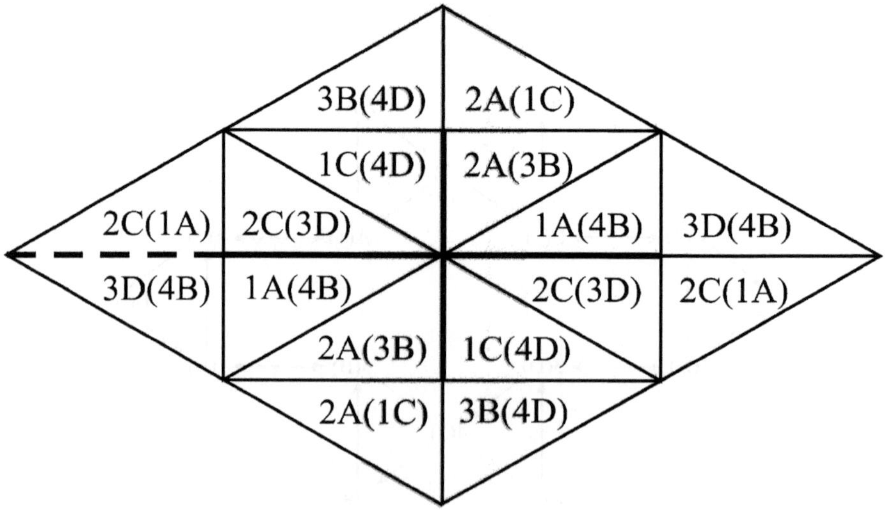

Fig. 11.38 Dual marked net for a partial overlap bronze even edge flexagon with a flat main position link. One copy needed. Fold together pairs of leaves numbered 2 and 4

even edge rings of eight bronze triangles (Fig. 10.6b), and leaves overlap exactly. Subsidiary main positions 1(2) and 4(3) are skew irregular even edge rings of eight bronze triangles, and some leaves do not overlap exactly (Fig. 10.42a). Intermediate positions may be either a bronze triangle edge quadruple (Fig. 10.42b), or have the appearance shown in Fig. 10.43. In Fig. 10.43a leaves have been bent to clarify the structure, and in Fig. 10.43b the intermediate position has been laid flat.

The two minor 4-cycles, shown in the minor cycle Tuckerman diagram can be traversed by using the twofold pinch flex. The minor cycle Tuckerman diagram is the same as that for minor cycles of a silver even edge flexagon with a flat main position link (Fig. 11.28). Face letters have been chosen so that all the leaves in each face either have the same number or the same letter. In the two minor 4-cycles, leaves overlap exactly in all main and intermediate positions. Minor main positions B(A) and D(C) have the same appearance as common subsidiary main position 1(3) (Fig. 10.6b) but the pat structure is different. The other minor main positions are combinations of a flat regular even edge ring of four 30°–60°–90° (bronze) triangles and two edge pairs of 30°–60°–90° (bronze) triangles (Fig. 11.39a). Minor intermediate positions have two different types of rectangular outline (Fig. 11.39b and c). Other flexes are possible, including pocket flexes.

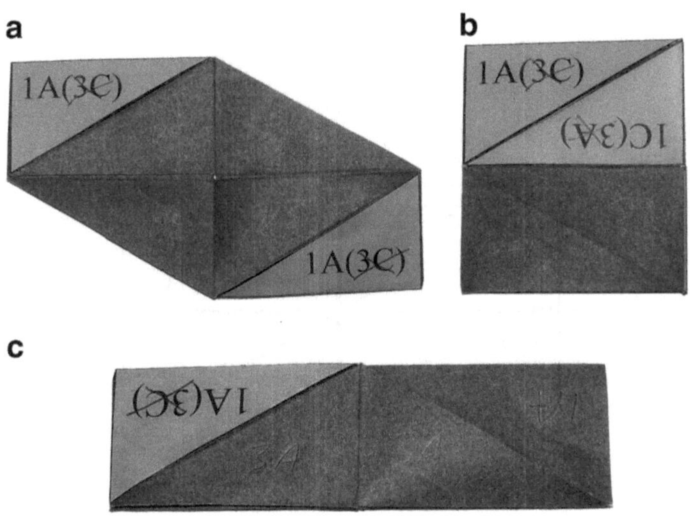

Fig. 11.39 A partial overlap bronze even edge flexagon with a flat main position link. (**a**) Minor main position, a combination of a flat regular even edge ring of four 30°–60°–90° (bronze) triangles and two edge pairs of 30°–60°–90° (bronze) triangles. (**b**) Wide rectangle minor intermediate position. (**c**) Narrow rectangle minor intermediate position

11.2.6.4 An Uncut Bronze Even Edge Flexagon with a Flat Main Position Link

A dual marked net for an uncut stretch silver even edge flexagon with a flat main position link is shown in Fig. 11.40. This was derived by using an uncut silver even edge flexagon with a flat main position link (Section 11.2.5.3) as a precursor, and stretching it in common subsidiary main position 3(1) so that the silver triangles became bronze triangles. The leaves are identical, but do not always overlap exactly, so it is a second outcome stretch flexagon (Section 10.1.2). Other derivations are possible, but it cannot be derived by partially stellating a dodecagon flexagon (Section 10.2.7). It could also be called a partial overlap flexagon, or a scalene triangle even edge flexagon. It is an uncut flexagon because it can be assembled without cutting and rejoining the net. The assembly sequence is similar to that for an uncut silver even edge flexagon with a flat main position link (Fig. 11.31). It can also be assembled as the enantiomorph, but a different set of face markings is needed (cf. Section 11.2.3.1).

As assembled, the flexagon is in common subsidiary main position 3(1). There are two sectors, and the torsion is zero. Apart from differences in pat structure, some of the dynamic properties are similar to those of a stretch silver even edge flexagon with a flat main position link (previous section). A difference is that pocket flexes are not possible. The two 3-cycles shown in the Tuckerman diagram can be traversed by using the fourfold pinch flex. The Tuckerman diagram is the

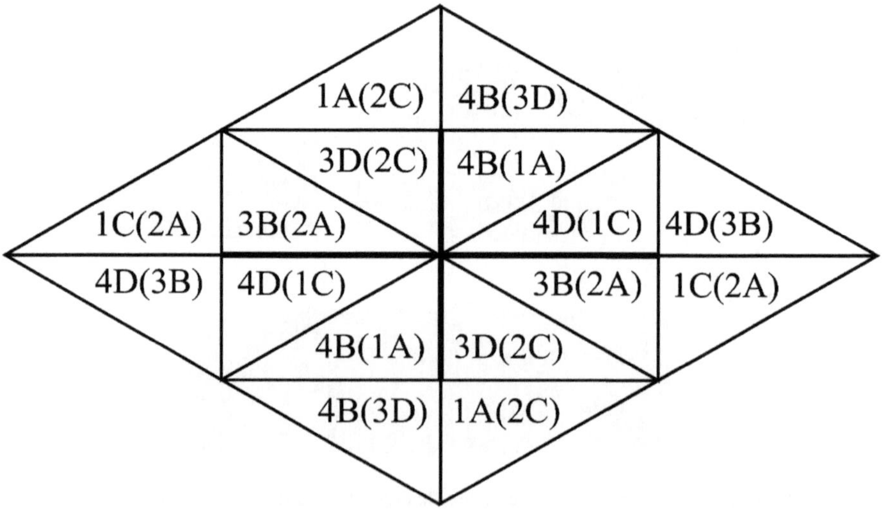

Fig. 11.40 Dual marked net for an uncut bronze even edge flexagon with a flat main position link. One copy needed. On one side of the flexagon, turn over the pair of leaves numbered 1(4) twice so that pairs of leaves numbered 4 are folded together. Then do the same on the other side of the flexagon. See Fig. 11.31

same as that for the tetrahexaflexagon (Fig. 11.3). Common subsidiary main position 3(1) and subsidiary main positions 2(3) and 1(4) are, in appearance, flat irregular even edge rings of eight bronze triangles (Fig. 10.6b), and leaves overlap exactly. Subsidiary main positions 1(2) and 4(3) are skew irregular even edge rings of eight bronze triangles, and some leaves do not overlap exactly (Fig. 10.42a). Intermediate positions may be either a bronze triangle edge quadruple (Fig. 10.42b), or have the appearance shown in Fig. 10.43. In Fig. 10.43a, leaves have been bent to clarify the structure, and in Fig. 10.43b the intermediate position has been laid flat.

The two minor 4-cycles, shown in the minor cycle Tuckerman diagram, can be traversed by using the twofold pinch flex. The minor Tuckerman diagram is the same as that for minor cycles of a silver even edge flexagon with a flat main position link (Fig. 11.28). Face letters have been chosen so that all the leaves in each face either have the same number or the same letter. In the two minor 4-cycles, leaves overlap exactly in all main and intermediate positions. Minor main positions B(A) and D(C) have the same appearance (Fig. 10.6b) as common subsidiary main position 1(3) but the pat structure is different. The other minor main are combinations of a flat regular even edge ring of four 30°–60°–90° (bronze) triangles and two edge pairs of 30°–60°–90° (bronze) triangles (Fig. 11.39a). Minor intermediate positions have two different types of rectangular outline (Fig. 11.39b and c).

11.3 Linked Point Flexagons

11.3.1 Methods of Linking

Methods of linking two point flexagons are analogous to those used for even edge flexagons (Sections 11.1, 11.2.1 and 11.2.2). In a main position link, paper models of the two point flexagons to be linked are flexed into main positions in which one of the pats is a single leaf. The two single leaves are removed, and the remaining two pats are assembled into a single, linked point flexagon. The torsion of a linked point flexagon is the algebraic sum of the torsions of the precursor point flexagons.

Distinctions between types of main position link between point flexagons are made on the basis of the relative positions of the hinges that are joined when the linked flexagon is assembled. An adjacent link is a link where point hinges that appear to be adjacent are connected. This corresponds to a flat main position link (Section 11.2.1). Only adjacent links are possible between triangle point flexagons. In a diagonal link the hinges appear to be diagonally opposite, and this corresponds to a box position link in a square flexagon (Section 11.2.3.2). Both adjacent links and diagonal links are possible between square point flexagons.

When two point flexagons of the same type but opposite hands are linked, the resulting flexagons has zero torsion. The additional degree of freedom in a point hinge (Section 1.2) means that it is always possible to arrange the net for a linked point flexagon with zero torsion as a flat ring, but is not always possible to arrange the flat ring so that leaves do not overlap. Hence, all linked point flexagons with zero torsion are uncut flexagons that can be assembled without cutting and rejoining a hinge. Some simple examples of linked point flexagons are given below to illustrate the range of possibilities.

Point flexagons can also be linked by end links. To construct an end link between paper models of two point flexagons flex each to an intermediate position in which the top and bottom leaves are connected by a point hinge. Disconnect these point hinges and reconnect the hinge halves to form two new point hinges linking the precursor flexagons. Linking using end links always leads to point flexagons that can be constructed in other ways. For example, end linking two fundamental triangle point flexagons 1⟨3, 3⟩ (Fig. 5.2c) of the same hand results in the augmented fundamental triangle point flexagon 1⟨3, 3, 6⟩ (Fig. 5.15).

11.3.2 Linked Triangle Point Flexagons

There are two triangle point flexagons that can be formed by linking two fundamental triangle point flexagons 1⟨3, 3⟩ (Section 5.3.2) using adjacent links. The net for one of these, a triangle point flexagon with an adjacent link, is shown in Fig. 11.41. This was derived by linking two fundamental triangle point flexagons 1⟨3, 3⟩ (Fig. 5.2c) of the same hand. As assembled, the flexagon is in intermediate position 1. It is interleaved and the torsion is 2. The intermediate position map (Fig. 11.42) can be traversed by using the simple flex. The dynamic properties are analogous to those of the tetrahexaflexagon (Section 11.2.2).The Tuckerman diagram (Fig. 11.3) and face numbering sequence (Fig. 11.1) are the same. Common main position 3(1) is. in appearance, an equilateral triangle vertex pair linked by one point hinge (Fig. 1.8b). The other main positions are equilateral triangle vertex pairs linked by pairs of point hinges (Fig. 1.8a). Intermediate positions are single equilateral triangles.

The net for the other, an uncut triangle point flexagon with an adjacent link is shown in Fig. 11.43. This was derived by linking two fundamental triangle point flexagons 1⟨3, 3⟩ (Fig. 5.2c) of opposite hands. The torsion is zero. It is an uncut flexagon because it can be assembled without cutting and rejoining the net. The dynamic properties are more complex than those of a triangle point flexagon with an adjacent link. There is no triangle even edge flexagon equivalent.

As assembled, the flexagon is in intermediate position 1. The intermediate position map can be traversed by using the simple flex. The intermediate position map is the same as that for the first order fundamental square even edge flexagons *S*⟨4, 4⟩ (Fig. 4.8). Main position 2(4) is an equilateral triangle vertex pair linked by one point hinge (Fig. 1.8b). The other main positions are equilateral triangle vertex pairs linked by pairs of point hinges (Fig. 1.8a). Intermediate positions are single equilateral triangles.

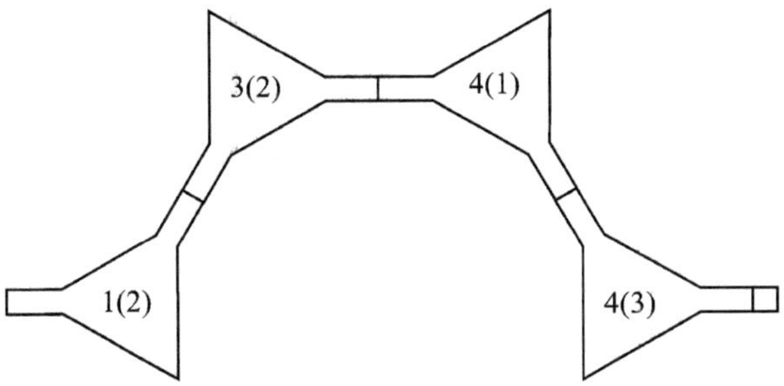

Fig. 11.41 Net for a triangle point flexagon with an adjacent link. One copy needed

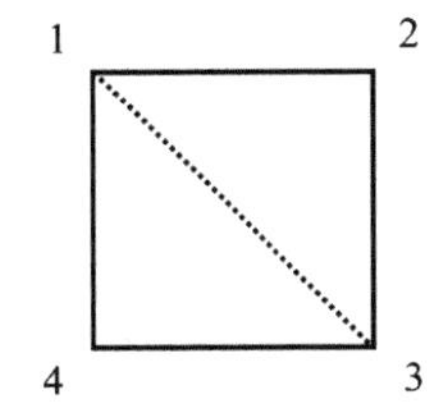

Fig. 11.42 Intermediate position map for a triangle point flexagon an adjacent link

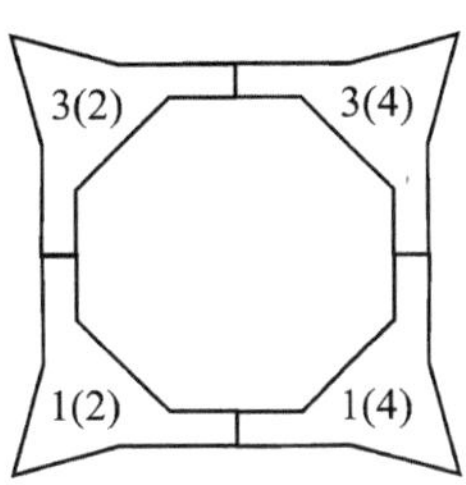

Fig. 11.43 Net for an uncut triangle point flexagon with an adjacent link. One copy needed

11.3.3 Linked Square Point Flexagons

Square point flexagons can be linked by adjacent links or by diagonal links. Enumeration of possible types is difficult. There are seven known linked square point flexagons consisting of two linked degenerate square point flexagons (Section 8.3.2). Four of these are described below.

The net for a square point flexagon with an adjacent link is shown in Fig. 11.44. This was derived by linking two degenerate non interleaved square point flexagons (Fig. 8.35) of the same hand. As assembled, the flexagon is in intermediate position 1. It is interleaved and the torsion is 2. The intermediate position map (Fig. 11.45) can be traversed by using the simple flex. The dynamic properties are analogous to those of a square even edge flexagon with two incomplete cycles and a flat main

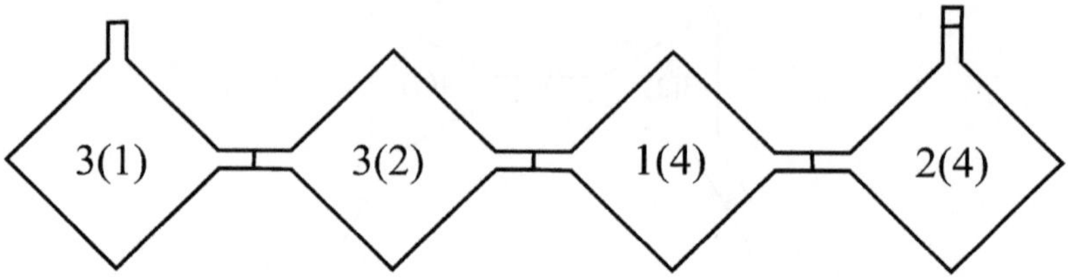

Fig. 11.44 Net for a square point flexagon with an adjacent link. One copy needed

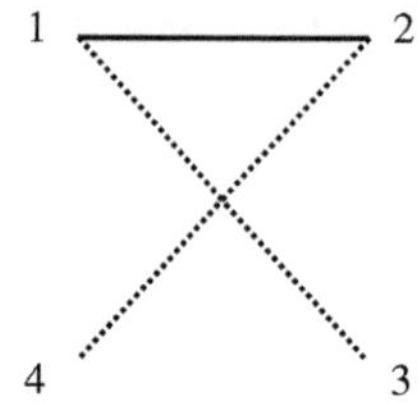

Fig. 11.45 Intermediate position map for a square point flexagon with an adjacent link

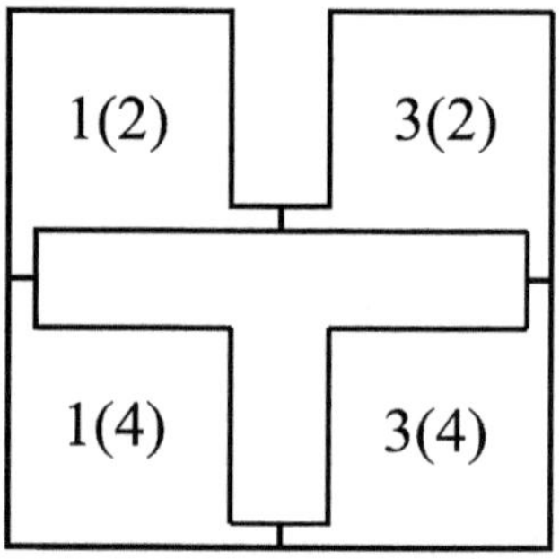

Fig. 11.46 Net for an uncut square point flexagon with an adjacent link. One copy needed

position link (Section 11.2.3.1). The Tuckerman diagram (Fig. 11.11) and face numbering sequence (Fig. 11.10) are the same. Main positions are, in appearance, square vertex pairs linked by pairs of point hinges (Fig. 3.23). Intermediate positions are single squares.

The net for an uncut square point flexagon with an adjacent link is shown in Fig. 11.46. This was derived by linking two degenerate non interleaved square point flexagons (Fig. 8.35) of opposite hands. The torsion is zero. It is an *uncut flexagon* that can be assembled without cutting and rejoining the net. The dynamic properties are more complex than those of a square point flexagon with an adjacent link. There is no square even edge flexagon equivalent. As assembled, the flexagon is in intermediate position 1. The intermediate position map (Fig. 11.47) can be traversed by using the simple flex. Main positions are square vertex pairs linked by pairs of point hinges (Fig. 3.23). Intermediate positions are single squares.

There are two ways of linking two degenerate non interleaved square point flexagons (Fig. 8.35), of the same hand, with diagonal links. One way leads to the

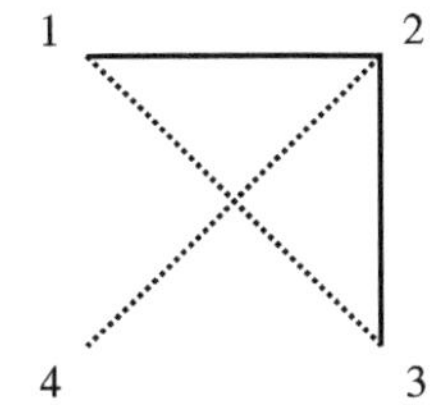

Fig. 11.47 Intermediate position map for an uncut square point flexagon with an adjacent link

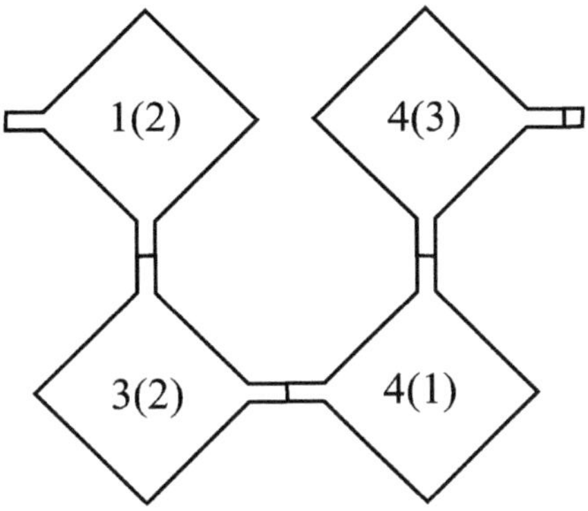

Fig. 11.48 Net for a square point flexagon with a diagonal link. One copy needed

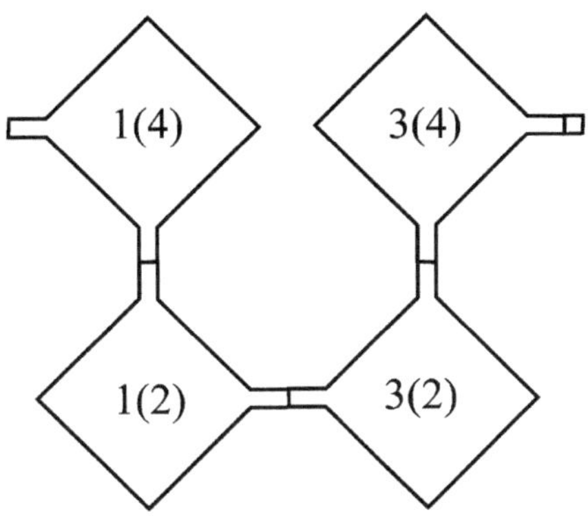

Fig. 11.49 Net for an uncut square point flexagon with a diagonal link. One copy needed. Join the ends of the net as an untwisted band before assembly

irregular cycle interleaved square point flexagon (Fig. 7.17). The other way leads to a square point flexagon with a diagonal link whose net is shown in Fig. 11.48. As assembled the flexagon is in intermediate position 1. It is interleaved and the torsion is 2. The intermediate position map can be traversed by using the simple flex. The intermediate position map is the same as that for the irregular cycle interleaved square point flexagon and also that for the fundamental square point flexagon 1⟨4, 4⟩ (Fig. 5.5). Main positions are square vertex pairs linked by pairs of point hinges (Fig. 3.23). Intermediate positions are single squares.

The net for an uncut square point flexagon with a diagonal link is shown in Fig. 11.49. This was derived by linking two degenerate non interleaved square point flexagons (Fig. 8.33) of opposite hands. The ends of the net can be joined as an untwisted band and the flexagon laid flat, but two leaves of the band overlap. The torsion is zero. It is an uncut flexagon that, after joining the ends of the net, can then be assembled without cutting and rejoining the net. There is no square even edge flexagon equivalent. As assembled, the flexagon is in intermediate position 1. The intermediate position map can be traversed by using the simple flex. The intermediate position map is the same as that for a triangle point flexagon with an adjacent link (Fig. 11.42). The dynamic properties are more complex than those of a square point flexagon with a diagonal link. Main position 1(3) is a square vertex pair linked by one point hinge. The other main positions are square vertex pairs linked by pairs of point hinges (Fig. 3.23). Intermediate positions are single squares.

11.4 Conjoined Point Flexagons

11.4.1 General Properties

Conjoined point flexagons differ from linked point flexagons in that no leaves are removed when two precursor point flexagons are conjoined. Instead, one leaf in each of the two flexagons is conjoined into a single conjoined leaf. In paper models this can be done by flexing the point flexagons so that leaves to be conjoined are at the ends of intermediate positions and gluing them together. Point flexagons to be conjoined do not have to be of the same type, but they must have the same type of leaf.

In general, point flexagons are simple bands, that is the band is a single loop of polygons. Conjoined point flexagons are never simple bands. There are always two or more separate loops, with a conjoined leaf at the connection between two loops. The conjoined leaf is sometimes a loose flap attached by a pair of point hinges. Whether such loose flaps are acceptable. is a matter of taste. The torsion of a conjoined point flexagon is the algebraic sum of the torsions of the precursor point flexagons. A double point flexagon (Section 8.4.1) is a figure-of-eight band with two loops. It can be regarded as a conjoined flexagon, and is an example of a flexagon that belongs to more than one family of flexagons.

The number of relative orientations in which two leaves that are regular convex polygons can be conjoined, is equal to the number of edges on the polygons. Different relative orientations, in general, result in conjoined point flexagons with different dynamic properties. Different relative orientations can lead to the conjoined leaf having different numbers of hinged vertices. In a double conjoin the conjoined leaf has two hinged vertices, in a triple conjoin three hinged vertices, and in a quadruple conjoin four hinged vertices.

Enumeration of the number of distinct ways in which two fundamental point flexagons of the same type and either the same hand or opposite hand can be conjoined is difficult. All the leaves in a fundamental point flexagon have the same status. Hence, with relative orientation taken into account, the same result is obtained whichever leaf is conjoined. The simple examples given below illustrate the range of possibilities. In other point flexagons, not all the leaves have the same status, and there are more possibilities.

11.4.2 Conjoined Triangle Point Flexagons

There are six distinct ways in which two fundamental triangle point flexagons 1⟨3, 3⟩ (Fig. 5.2c) can be conjoined. Two of them have a double conjoin and four a triple conjoin. The net for a conjoined triangle point flexagon with a double conjoin is shown in Fig. 11.50. This was derived by conjoining two fundamental triangle point flexagons of the same hand. The flexagon is a figure-of-eight band and the conjoined leaf is a loose flap. The torsion is 2. As assembled, the flexagon is in intermediate position 1. The intermediate position map (Fig. 11.51) can be traversed by using the simple flex. Main positions are, in appearance, equilateral triangle vertex pairs linked by pairs of point hinges (Fig. 1.8a). Intermediate positions are single equilateral triangles.

The net for a conjoined triangle point flexagon with a triple conjoin is shown in Fig. 11.52. This was derived by conjoining two fundamental triangle point flexagons of the same hand. The flexagon is a figure-of-eight band. As assembled, it is in intermediate position 1. The torsion is 2. The dynamic properties differ from those of a conjoined triangle point flexagon with a double conjoin. The intermediate position map (Fig. 11.53) can be traversed by using the simple flex. Main positions are equilateral triangle vertex pairs, linked by pairs of point hinges (Fig. 1.8a). Intermediate positions are single equilateral triangles.

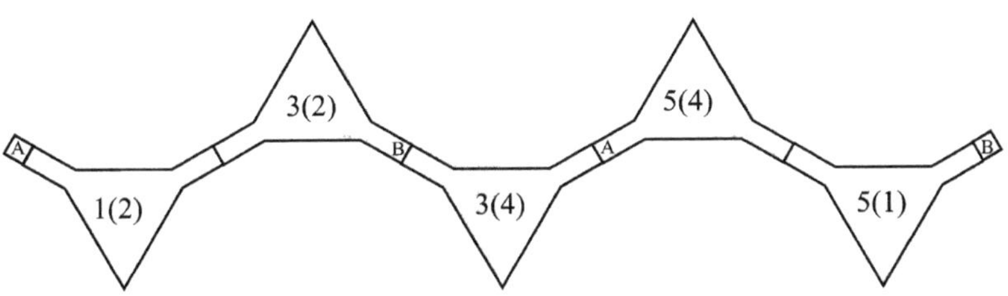

Fig. 11.50 Net for a conjoined triangle point flexagon with a double conjoin. One copy needed. Fold until leaves numbered 1 are visible. Join at A-A and B-B

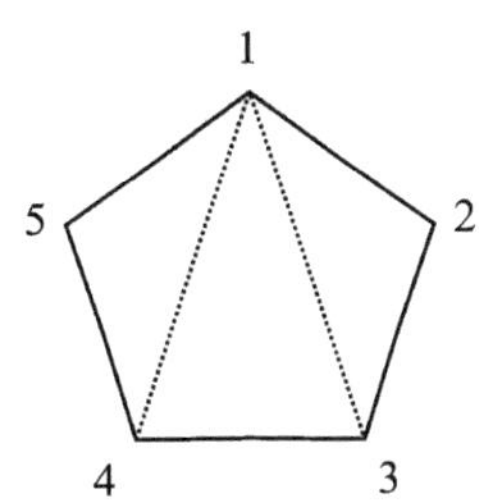

Fig. 11.51 Intermediate position map for a conjoined triangle point flexagon with a double conjoin

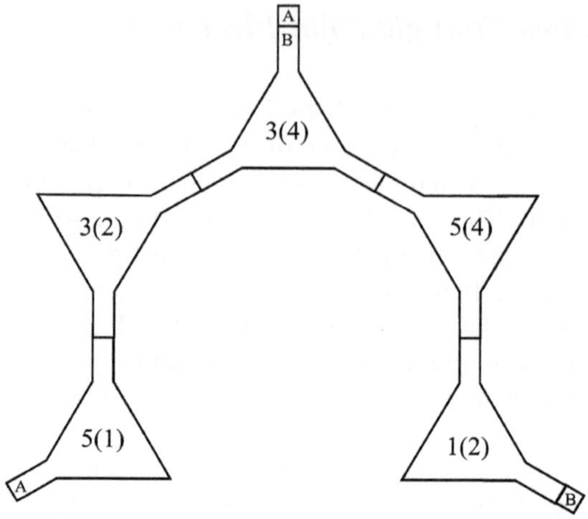

Fig. 11.52 Net for a conjoined triangle point flexagon with a triple conjoin. One copy needed. Fold until leaves numbered 1 are visible. Join at A-A and B-B

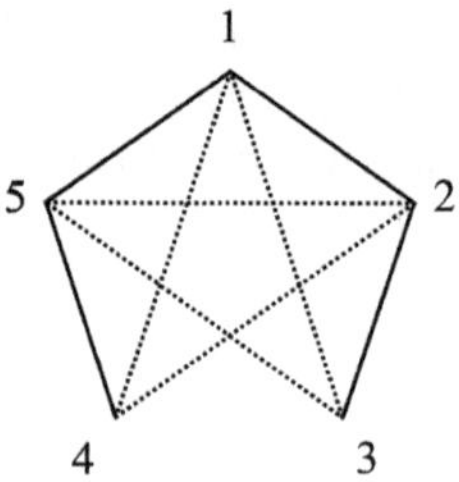

Fig. 11.53 Intermediate position map for a conjoined triangle point flexagon with a triple conjoin

11.4.3 A Conjoined Pentagon Point Flexagon

The net for a conjoined pentagon point flexagon with a quadruple conjoin is shown in Fig. 11.54. This was derived by conjoining two fundamental pentagon point flexagons 1⟨5, 5⟩ (Fig. 5.6) of opposite hands. The flexagon is a figure-of-eight band. The torsion is zero. As assembled, the flexagon is in intermediate position 1. The intermediate position map (Fig. 11.55) can be traversed by using the simple flex. Main positions are, in appearance, regular pentagon vertex pairs linked by pairs of point hinges. Intermediate positions are single regular pentagons.

11.5 Bundled Odd Edge Flexagons

11.5.1 General Properties

There are an odd number of pats in a main position of a an odd edge flexagon, and each pat is a sector (Section 4.3.1). Linking of even edge flexagons requires the

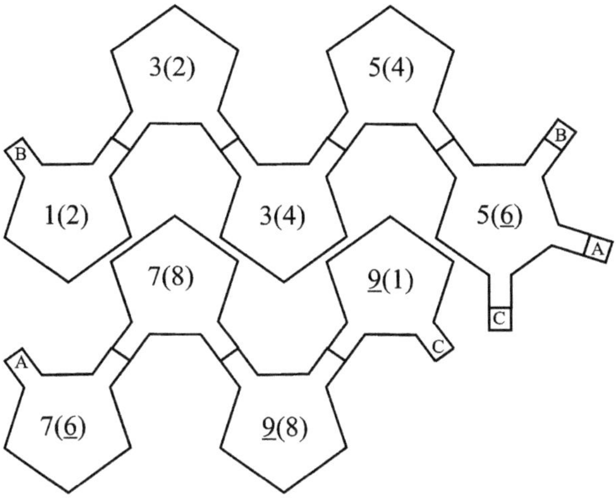

Fig. 11.54 Net for a conjoined pentagon point flexagon with a quadruple conjoin. One copy needed. Join the two parts of the net at A-A. Fold until leaves numbered 1 are visible. Join at B-B and C-C

Fig. 11.55 Intermediate position map for a conjoined pentagon point flexagon with a quadruple conjoin

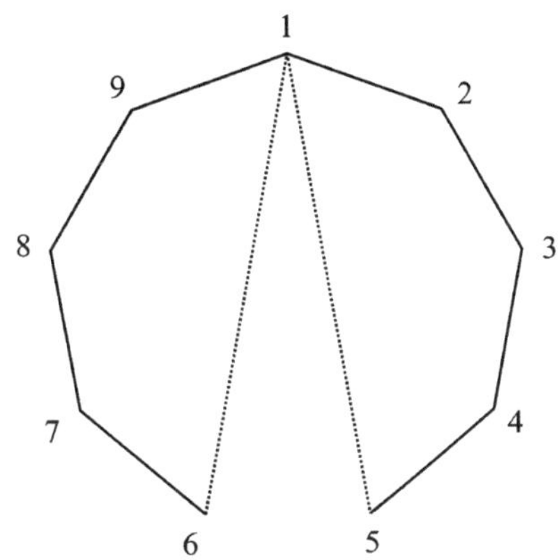

presence, in a main position, of alternate pats that are single leaves (Section 11.3.1). This is impossible in main positions of odd edge flexagons, so linking is also impossible. In a bundled odd edge flexagon, one or more leaves in a sector of an odd edge flexagon is replaced by a pair of leaves from another edge flexagon. The characteristic flex for an odd edge flexagon is the pocket flex (Section 4.3.1). This is a local flex because some pats are left unchanged. Hence each sector does not have to be treated in the same way. Bundling can be extended indefinitely to create very complicated flexagons. Sherman (2007b) describes bundled odd edge flexagons, but not under that name. Two examples are given below. In these, all the sectors in a main position have been bundled in the same way.

11.5.2 A Five Sector Bundled Triangle Odd Edge Flexagon

The net for a five sector bundled triangle odd edge flexagon is shown in Fig. 11.56. This was derived by using the second order fundamental triangle odd edge flexagon $5\langle 3, 3\rangle_2$ (Fig. 4.40) as a precursor, and replacing all the leaves by folded pairs of leaves. If pairs of leaves numbered 5 and 6, and 7 and 8, are glued together the precursor net is recovered. The dynamic properties are an extension of those of the precursor flexagon (Section 4.3.2).

As assembled, the flexagon is either in principal main position 1(2) or in principal main position 2(1). These are, in appearance, slant regular odd edge rings of five equilateral triangles (Fig. 1.4). In the principal main position codes, the number outside the brackets indicates the face number visible on the outside of the ring. It is not possible to flex between the two forms of a principal main position. The torsion per sector in a principal main position is 3.

The flexagon can be flexed from a principal main position to a first minor main position by using a pocket flex. A first minor main position is, in appearance, a combination of a regular slant odd edge ring of three equilateral triangles and an equilateral triangle edge pair (Fig. 4.41c). The pocket flex can be carried out in five different ways, so there are five first minor main positions that all have the same appearance. A second pocket flex leads to a second minor main position. This is a combination of a slant regular odd edge ring of three equilateral triangles and an equilateral triangle edge triple with two equilateral triangles in common (Fig. 2.3b). This can be done in only one way, so there are five second minor main positions.

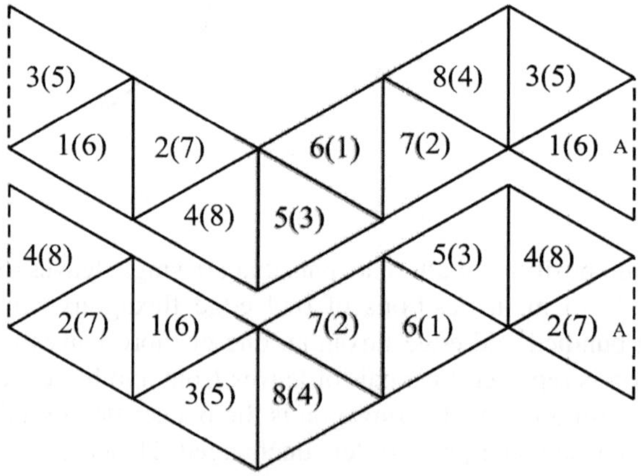

Fig. 11.56 Net for a bundled five sector triangle odd edge flexagon. One copy needed. Join the two parts of the net at A-A. Fold leaves together in the order 8 onto 7, 6 onto 5, and 4 onto 3

11.5.3 The Seven and Six Flexagon

The seven and six flexagon is a seven sector bundled triangle odd edge flexagon with six leaves in each sector. Its net is shown in Fig. 11.57. This was derived by using the net for the second order fundamental triangle odd edge flexagon $7\langle 3, 3\rangle_2$ (Fig. 4.42) as a precursor, and repeatedly replacing leaves by pairs of leaves so that, in a principal main position, there are six leaves in each sector. Despite the large number of leaves the flexagon is easy to assemble. The dynamic properties are an extension of those of the precursor flexagon (Section 4.3.2). They are very complicated and difficult to enumerate. The torsion per sector in a principal main position is 5.

As assembled, the flexagon is in principal main position 1(2). This is, in appearance, a skew regular odd edge ring of seven equilateral triangles (Fig. 4.43), and cannot be arranged with rotational symmetry or laid flat. Various minor main positions can be reached by using 1, 2 or 3 pocket flexes. These include a combination of a slant regular odd edge ring of three equilateral triangles and an equilateral triangle edge triple (Fig. 2.3a), a combination of a slant regular odd edge ring of three equilateral triangles and an equilateral triangle edge quadruple (Fig. 4.44a), and a combination of a flat regular even edge ring of six equilateral triangles and a slant regular odd edge ring of three equilateral triangles (Fig. 4.44b). There are some positions that have the same appearance, but have different pat structures. Asymmetric threefold pinch flexes, as used for the precursor flexagon, are theoretically possible, but they are impractical because the flexagon very easily becomes muddled.

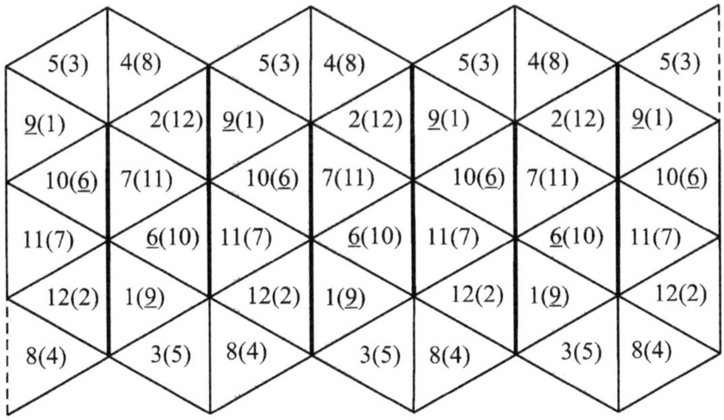

Fig. 11.57 Net for the seven and six flexagon. One copy needed. Fold leaves together in the order 12 onto 11, 10 onto 9, 8 onto 7, 6 onto 5, and 4 onto 3

11.6 Slipagons

11.6.1 General Properties

A slipagon consists of two or more interlinked even edge flexagons of the same type. All the main positions of the individual even edge flexagons can be traversed by using appropriate flexing sequences. The constituent flexagons cannot be separated by flexing because the even edge flexagons are interlinked during assembly so that they become linked bands. The torsion of a slipagon is the algebraic sum of the torsions of the individual even edge flexagons.

In their original form (Engel 1993) a slipagon consisted of two or more trihexaflexagons (Section 4.2.3) of the same hand linked together. The name appears to have been coined because the constituent flexagons sometimes have to be slipped relative to each other in a slip flex order to reach some main positions. Slipagons can be constructed from other types of edge flexagon. Two slipagons are described below. Both consist of two interlinked even edge flexagons.

11.6.2 A Trihexaflexagon Slipagon

The nets for a trihexaflexagon slipagon consisting of two trihexaflexagons 3⟨3, 3⟩ (Fig. 4.14) of the same hand are shown in Fig. 11.58. The two trihexaflexagons are distinguished by using face numbers for one of them, and face letters for the other. To assemble the slipagon, first assemble the numbered net, with pairs of leaves numbered 3 folded together. Then, with face 1 uppermost and the lettered net printed side downwards, pass the lettered net through a slot, oriented as in Fig. 11.59a. Next, fold pairs of leaves lettered C together and join the ends. This is the only distinct way in which two trihexaflexagons can be interlinked to form a trihexaflexagon slipagon. There are five other ways of assembling the slipagon but they are not distinct. The structure is the same, but different combinations of numbers and letters appear during flexing.

As assembled, the slipagon is, in appearance, a block of ten equilateral triangles (Fig. 11.59b), with five leaves numbered 1 and five leaves lettered A visible on one face, and leaves numbered 2 and lettered B visible on the other face. This is principal main position 2B(1A). A principal main position has two forms, with the same appearance. A slip flex, in which the two parts of the slipagon are slipped relative to each other, is used to traverse between them. Figure 11.59c shows the slipagon part way through a slip flex.

Each of the trihexaflexagons can be flexed separately by using the threefold pinch flex. A combination of threefold pinch flexes and slip flexes is used to traverse the principal 3-cycle shown in the Tuckerman diagram (Fig. 11.60). Starting from principal main position 2B(1A), to flex to principal main position 3A(2C), first flex the numbered trihexaflexagon to its principal main position 3(2) to reach

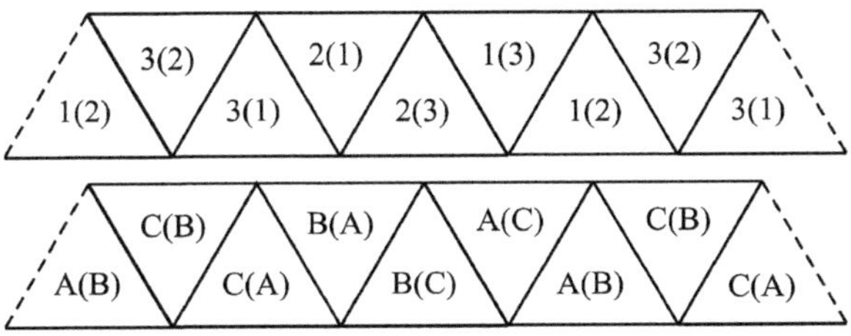

Fig. 11.58 Nets for a trihexaflexagon slipagon. One copy needed. See text for assembly instructions

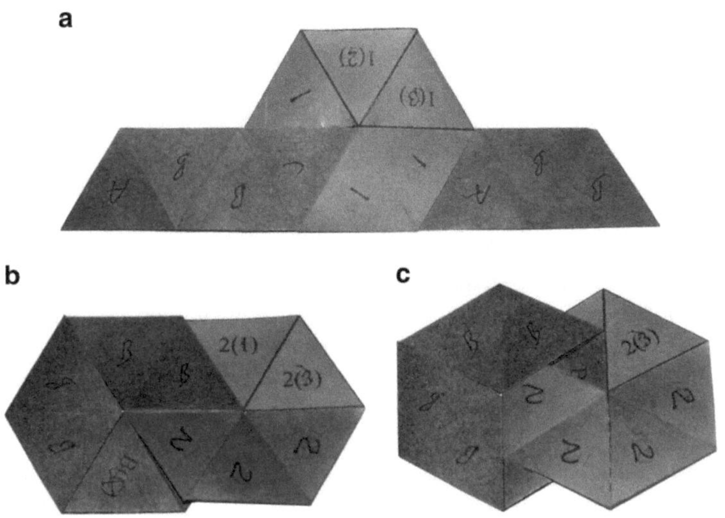

Fig. 11.59 A trihexaflexagon slipagon. (**a**) Partly assembled. (**b**) Principal main position. (**c**) Part way through a slip flex

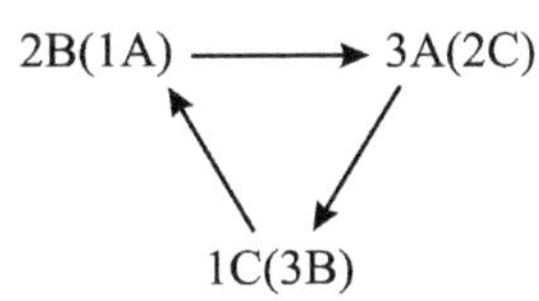

Fig. 11.60 Tuckerman diagram for the 3-cycle of a trihexaflexagon slipagon

an intermediate position. This is a flat regular even edge ring of six equilateral triangles (Fig. 1.1a). A preliminary slip flex may be needed. Each face of an intermediate position has five leaves of one of the trihexaflexagons visible, and one leaf of the other. Next, flex the lettered trihexaflexagon to its principal main position C(A) to reach principal main position 3A(2C). Slip flex to the other form of principal main position 3A(2C), and so on round the principal 3-cycle.

11.6.3 A Partial Overlap Silver Even Edge Slipagon

The nets for a partial overlap silver even edge slipagon consisting of two partial overlap silver even edge flexagons (Fig. 10.15) of the same hand, are shown in Fig. 11.61. The two partial overlap silver even edge flexagons are distinguished by using face numbers for one of them, and face letters for the other. To assemble the slipagon, first assemble the numbered net, with pairs of leaves numbered 3 and 4 folded together. Then, with face 1 uppermost and the lettered net printed side upwards, pass the lettered net through a slot, oriented as in Fig. 11.62a. Next, fold together pairs of leaves lettered C and D, and join the ends. This is the only distinct way in which two partial overlap silver even edge flexagons can be interlinked to form a partial overlap silver even edge slipagon. There is one other way of assembling the slipagon, but it is not distinct. The structure is the same, but different combinations of numbers and letters appear during flexing.

In a principal main position of the slipagon, both the constituent flexagons are in principal main positions. As assembled, these are principal main positions 1(2) and A(B). In position codes, partly visible leaves are ignored. The slipagon will be flat and will probably have a square outline (Fig. 11.62b). It can be manoeuvred into several different shapes by using slip flexes. These include an irregular hexagon (Fig. 11.62c) and an L-shape (Fig. 11.62d). The irregular hexagon is principal main position 1A(2B) with leaves marked 1 and A visible on one face and leaves marked 2 and B on the other. The constituent flexagons can be turned round so a principal main position has four different forms. These all have the same appearance, including face markings, so are not regarded as distinct. If one of the constituent flexagons is turned over lengthways, then leaves marked 1 and B will be visible on one face and leaves marked 2 and A. This is not a principal main position and the flexes described below will not work.

Each partial overlap silver even edge flexagon can be flexed independently by using the twofold pinch flex. To flex the numbered partial overlap silver even edge flexagon from its principal main position 1(2) to its principal main position 2(3), start by making sure that principal main position 1(2) is in a hexagonal position (Fig. 11.62c). Then fold principal main position 1(2) in two to reach its intermediate position 2. The slipagon then has the appearance shown in Fig. 11.62e. Next, unfold the numbered partial overlap silver even edge flexagon into its principal

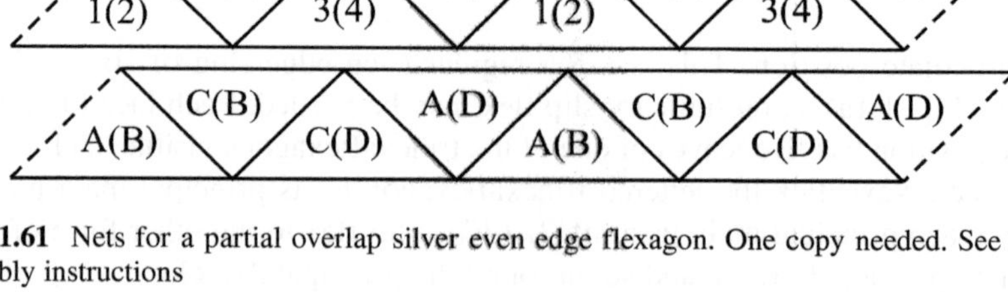

Fig. 11.61 Nets for a partial overlap silver even edge flexagon. One copy needed. See text for assembly instructions

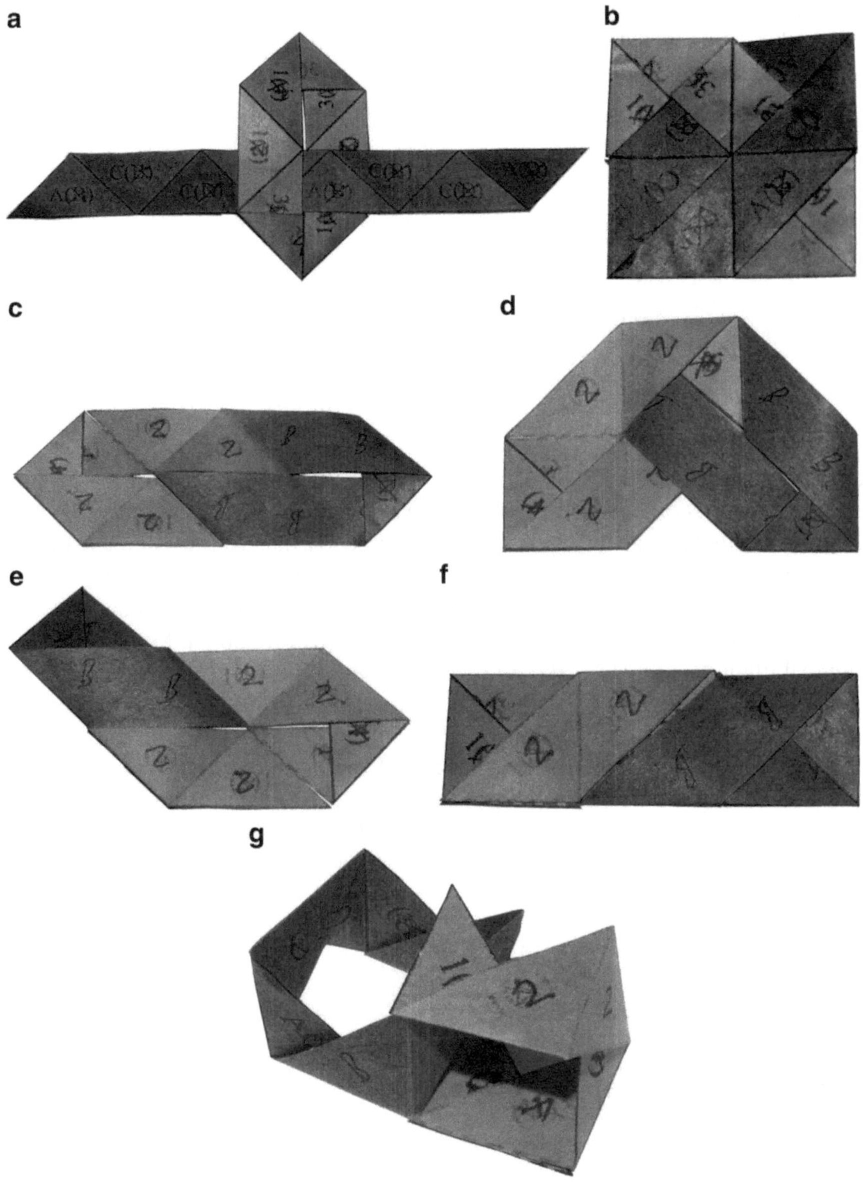

Fig. 11.62 A partial overlap silver even edge slipagon. (**a**) Partly assembled. (**b**) Square principal main position. (**c**) Hexagonal principal main position. (**d**) L-shaped principal main position. (**e**) Irregular intermediate position. (**f**) Rectangular intermediate position. (**g**) Double box position

main position 2(3). The slipagon is then in an L-shaped principal main position (Fig. 11.62d). Slip flex this into a hexagonal principal main position, and so on round the principal 4-cycle shown in the Tuckerman diagram for the numbered

Fig. 11.63 Tuckerman diagram for the principal 4-cycle of a partial overlap silver even edge slipagon, lettered net

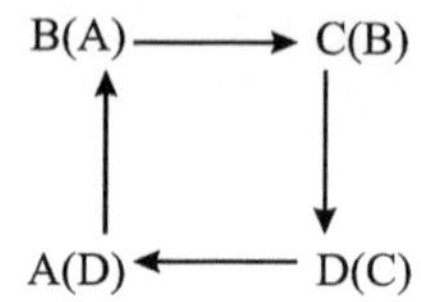

Fig. 11.64 Intermediate position map for a partial overlap silver even edge slipagon

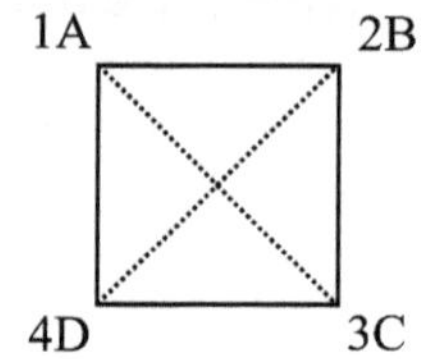

partial overlap silver even edge flexagon. This is the same as that for the principal 4-cycle of the first order fundamental square even edge flexagons $S\langle 4, 4\rangle$ (Fig. 4.17). A similar sequence can be used for the lettered partial overlap silver even edge flexagon, the corresponding Tuckerman diagram is shown in Fig. 11.63.

The principal 4-cycle of each of the two partial overlap silver even edge flexagons can be traversed simultaneously by using a double twofold pinch flex. Starting from principal main position 1A(2B), fold the constituent flexagons in two to reach their intermediate positions 2 and B. This is intermediate position 2B of the slipagon, and has a rectangular outline (Fig. 11.62f). Then unfold the slipagon to reach principal main position 2B(3C), and so on round the principal 4-cycle shown by the solid lines in the intermediate position map (Fig. 11.64). From intermediate positions, the slipagon can be opened into double box positions shown by the dotted lines in the intermediate position map. These show clearly that the slipagon consists of two linked bands (Fig. 11.62g).

11.7 Coupled Point Flexagons

11.7.1 General Properties

A coupled point flexagon consists of two or more point flexagons, with the same type of leaf, interleaved with each other during assembly. Leaves overlap exactly, and a coupled point flexagon can be flexed as if it were a single point flexagon. Unlike slipagons (Section 11.6.1), the constituent point flexagons do not have to be slipped relative to each other in order to reach certain positions. In other words slip flexes are not needed. Interleaf flexes (Section 7.3.2) are sometimes possible, and this means that transformations between flexagons are also possible (Section 7.3.1). The torsion is the algebraic sum of the torsions of the constituent point flexagons.

Possible coupled point flexagons, made from two fundamental point flexagons of the same type, can be enumerated by enumerating possible overlapping polygon twins that have the same vertices as an appropriate regular polygon (cf. Section 7.1). Each polygon has the same number of vertices as the leaves of the fundamental point flexagons. Thus, for two fundamental triangle point flexagons 1⟨3, 3⟩ (Section 5.3.2) the triangle twins have the same vertices as a regular hexagon. The two possible overlapping triangle twins are shown in Fig. 11.65, and are associated polygon twins for corresponding coupled point flexagons. The polygons must overlap for the point flexagons to be interleaved and hence be a coupled point flexagon. If they are separate then they are associated polygons for separate flexagons. Arbitrary type letters are used in the figure and for corresponding coupled point flexagons.

The figure shows that there are two solutions. The two fundamental triangle point flexagons being coupled can be either be of the same hand or of opposite hand. Hence the two associated triangle twins correspond to a total of four types of coupled triangle point flexagons. Furthermore, the two fundamental triangle point flexagons being coupled can have three different relative orientations, making a total of 12 distinct types. Associated polygon twins can be used to derive face numbering sequences for coupled point flexagons. The method used is the same as that for interleaved fundamental point flexagons (Section 5.4.1) with due attention to detail. Some properties of triangle coupled point flexagons, types A and B, are given in Table. 11.3 (cf. Table. 7.6). The notation 1/2 in a face numbering sequence means that the upper face of a leaf is numbered 1 and the lower face is numbered 2. The two typical separate numbering sequences shown for each type are for two fundamental triangle point flexagon of the same hand. For flexagons of opposite hands, invert one of the sequences.

For two fundamental square point flexagons 1⟨4, 4⟩ (Section 5.3.3), the convex quadrilateral twins have the same vertices as a regular octagon. The possible overlapping convex quadrilateral twins are shown in Fig. 11.66. Arbitrary type letters

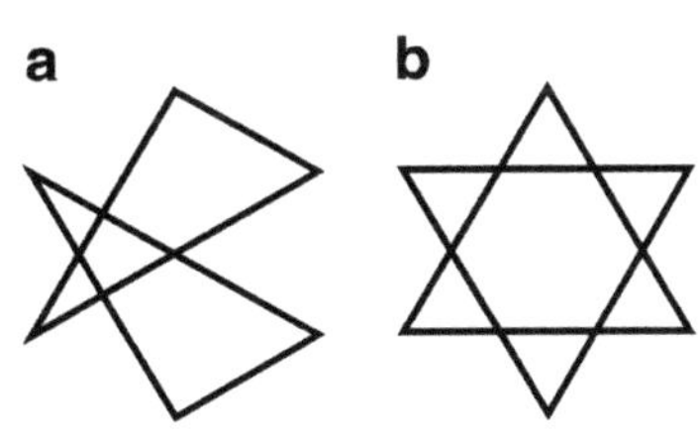

Fig. 11.65 Overlapping triangle twins with the same vertices as a regular hexagon. (**a**) Type A. (**b**) Type B

Table 11.3 Properties of coupled triangle point flexagons, types A and B

Type	Rotational symmetry of associated triangle twins	Distinct face numbering sequences	Typical face numbering sequences
A	Onefold	6	1/2, 3/2, 5/6 and 4/3, 4/5, 1/6
B	Sixfold	1	1/2, 4/3, 5/6 and 3/2, 4/5, 1/6

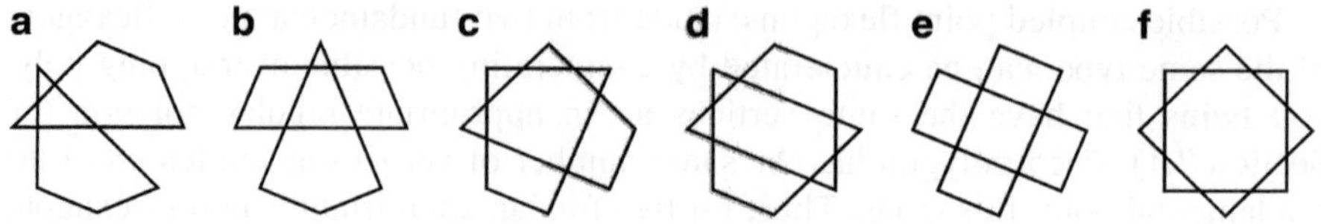

Fig. 11.66 Overlapping convex quadrilateral twins with the same vertices as a regular octagon. (**a**) Type A. (**b**) Type B. (**c**) Type C. (**d**) Type D. (**e**) Type E. (**f**) Type F

Table 11.4 Properties of coupled square point flexagons, types A–F

Type	Rotational symmetry of associated triangle twins	Distinct face numbering sequences	Typical face numbering sequences
A	1	8	1/2, 3/2, 3/4, 8/7 and 4/5, 6/5, 6/7, 1/8
B	1	8	1/2, 5/4, 5/6, 6/7 and 2/3, 4/3, 7/8, 1/8
C	1	8	1/2, 4/3, 4/5, 8/7 and 2/3. 6/5, 6/7, 1/8
D	2	4	1/2, 3/2, 4/5, 8/7 and 3/4, 6/5, 6/7, 1/8
E	4	2	1/2, 3/2, 5/6, 7/6 and 3/4, 5/4, 7/8, 1/8
F	8	1	1/2, 4/3, 5/6, 8/7 and 2/3, 5/4, 6/7, 1/8

are used in the figure and for corresponding coupled point flexagons. There are six solutions, types A–F. The two fundamental square point flexagons being coupled can be either be of the same hand or of opposite hand, and there are four possible relative orientations, making a total of 48 distinct types of coupled square point flexagons. Some properties of coupled square point flexagons, types A–F, are given in Table. 11.4. The two typical separate numbering sequences shown for each type are for two fundamental triangle point flexagon of the same hand. For flexagons of opposite hands, invert one of the sequences.

If the point flexagons being coupled are not fundamental point flexagons, or are not of the same type, then there are numerous possibilities, and enumeration of possible distinct types and distinct face numbering sequences is very difficult.

Intermediate positions of coupled point flexagons have overlapping hinges at some vertices. There is more than one possible method of arranging overlapping hinges and hence more than one method of assembly. In paper models, with paper strips used as approximations to point hinges, coupled point flexagons have a neater appearance, and are easier to handle, if overlapping hinges are nested during assembly, but this is not always possible. When nesting is possible, there is only one method of assembly in which all the hinges are nested, and this is regarded as the correct method (cf. Section 5.6.1).

The examples given below illustrate some possibilities.

11.7.2 A Coupled Triangle Point Flexagon

The net for a coupled triangle point flexagon type B is shown in Fig. 11.67. This was derived by coupling two fundamental triangle point flexagons 1⟨3, 3⟩ (Fig. 5.2c) of the same hand. The corresponding overlapping triangle twins are shown in Fig. 11.65b. The torsion is 2. As assembled, the flexagon is in intermediate position 1. The intermediate position map (Fig. 11.68) can be traversed by using the simple flex. Ensure that leaves remain superimposed during flexing. They can easily slip out of position. Main positions are, in appearance, equilateral triangle vertex pairs linked by pairs of point hinges (Fig. 1.8a). Intermediate positions are single equilateral triangles.

11.7.2.1 An Interleaved Conjoined Triangle Point Flexagon

The point flexagons that make up a coupled point flexagon can be conjoined in the same way that separate point flexagons are conjoined (Section 11.4.1). The result is an interleaved conjoined point flexagon. The net for an interleaved conjoined

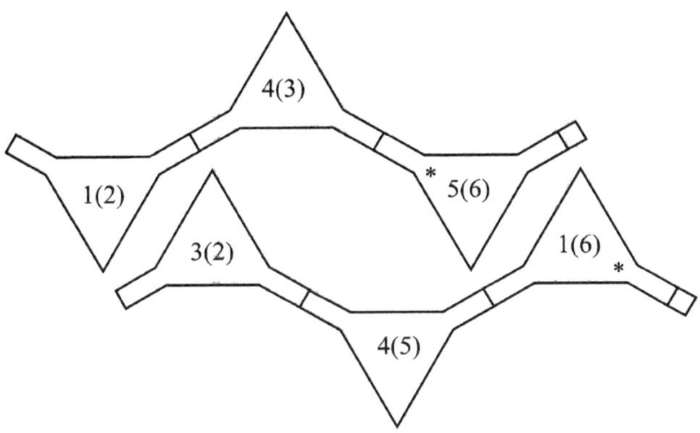

Fig. 11.67 Net for a coupled triangle point flexagon type B. Place together leaves numbered 6 with the asterisks adjacent, fold together like numbered pairs of leaves in the order 5, 4, 3 and 2, nesting hinges, and join the ends

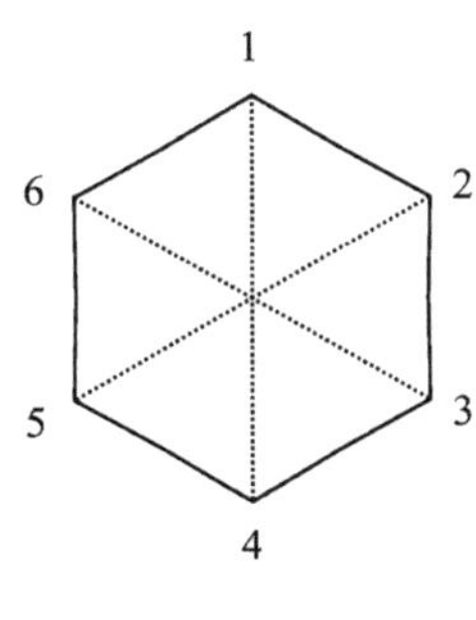

Fig. 11.68 Intermediate position map for a coupled triangle point flexagon type B

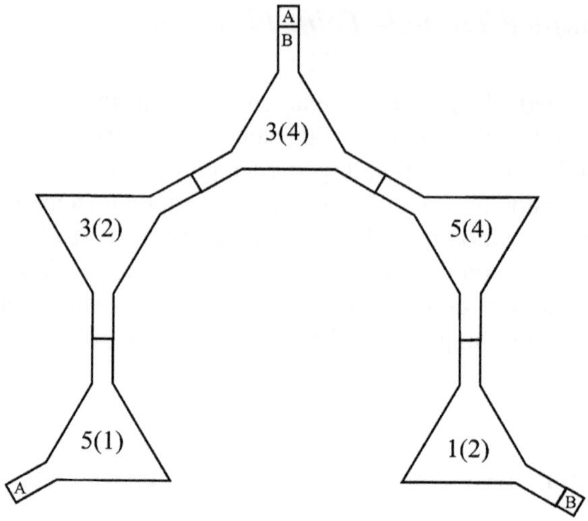

Fig. 11.69 Net for an interleaved conjoined triangle point flexagon with a triple conjoin. One copy needed. Fold together like numbered pairs of leaves in the order 3, 4, 5 and 2. Join at A-A and B-B

Fig. 11.70 Intermediate position map an interleaved conjoined triangle point flexagon with a triple conjoin

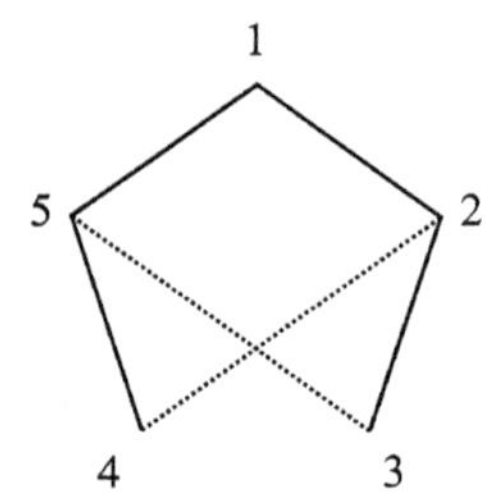

triangle point flexagon with a triple conjoin is shown in Fig. 11.69. This was derived by conjoining the two fundamental triangle point flexagons of the same hand that make up a coupled triangle point flexagon type B (Fig. 11.67). The torsion is 2. The flexagon is a figure-of-eight band. It is an interleaved version of a conjoined triangle point flexagon with a triple conjoin (Section 11.4.2). As assembled, the flexagon is in intermediate position 1. The intermediate position map (Fig. 11.70) can be traversed by using the simple flex. Main positions are, in appearance, equilateral triangle vertex pairs linked by pairs of point hinges (Fig. 1.8a). Intermediate positions are single equilateral triangles.

11.7.3 A Coupled Square Point Flexagon

A dual marked net for a coupled square point flexagon type F is shown in Fig. 11.71. The net was derived by coupling two fundamental square point flexagons 1⟨4, 4⟩ (Fig. 5.4b) of the same hand. The corresponding overlapping square couple is shown

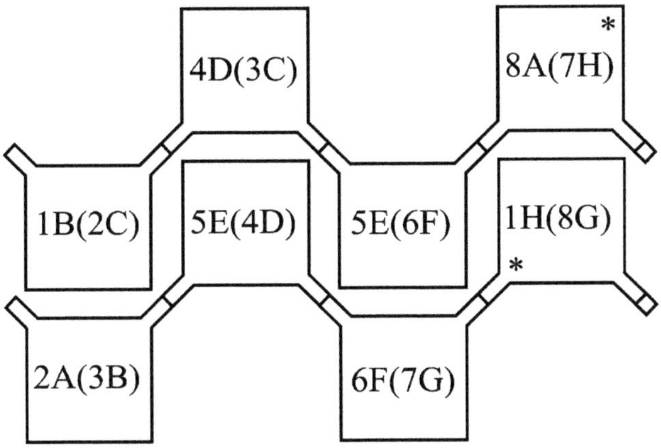

Fig. 11.71 Dual marked net for a coupled square point flexagon. One copy needed. Place together leaves numbered 8 with the asterisks adjacent, fold together like numbered pairs of leaves in the order 7, 6, 5, 4, 3 and 2, nesting hinges, and join the ends

Fig. 11.72 Intermediate position map for a coupled square point flexagon type D

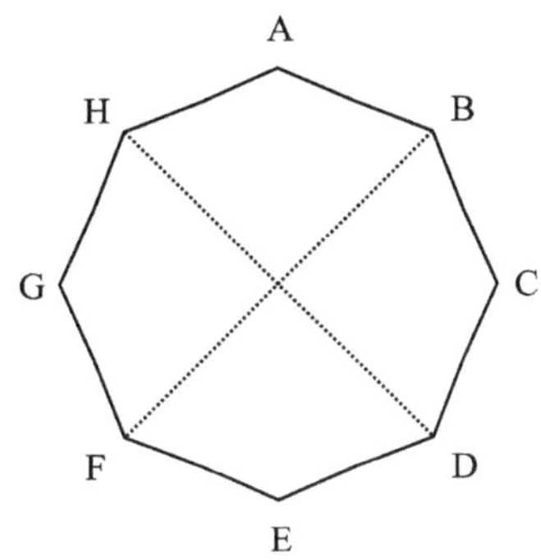

in Fig. 11.66f. The torsion is 4. Transformations between flexagons are possible. As assembled, the flexagon is in intermediate position 1. The intermediate position map can be traversed by using the simple flex. This is the same as that for the augmented interleaved fundamental square point flexagon 1⟨4, 4, 8/3⟩ (Fig. 5.25). Ensure that leaves remain superimposed during flexing. They can easily slip out of position. Main positions are, in appearance, square vertex pairs (Fig. 3.23) linked by pairs of point hinges. Intermediate positions are single squares.

A double interleaf flex at any intermediate position transforms the flexagon into type D (Fig. 11.66d). The letters on the dual marked net are for a double interleaf flex at intermediate position 1, which then becomes intermediate position A. The corresponding intermediate position map (Fig. 11.72) can be traversed by using the simple flex. A single interleaf flex at any intermediate position transforms the flexagon into type C (Fig. 11.66c). Face markings for this type are not shown on the dual marked net.

References

Belenky Y (2009) Theory of hexaflexagons. Flexagon Lovers Group posting 554. http://tech.groups.yahoo.com/group/Flexagon_Lovers/. Accessed 3 February 2009

Conrad AS, Hartline DK (1962) *Flexagons*. RIAS Technical Report 62-11. Baltimore, MD: RIAS

Cromwell PR (1997) *Polyhedra*. Cambridge University Press, Cambridge

Engel DA (1993) Hexaflexagons + HFG = slipagon. Journal of Recreational Mathematics 25(3):161–166

Gardner M (1965) *Mathematical Puzzles and Diversions*. Penguin, Harmondsworth

Gardner M (2008) *Hexaflexagons, Probability Paradoxes, and the Tower of Hanoi*. Cambridge University Press, Cambridge

Highland Games (2008). www.halfpast.demon.co.uk. Accessed 10 December 2008

Kosters M (1999) A theory of hexaflexagons. Nieuw Archief Wisk 17:349–362

McLean B (2008) Dodecaflexagon videos. Flexagon Lovers Group posting 497. http://tech.groups.yahoo.com/group/Flexagon_Lovers/. Accessed 30 October 2008

McLean TB (1979) V-flexing the hexahexaflexagon. American Mathematical Monthly 86:457–466

Mitchell D (1999) *The Magic of Flexagons: Manipulative Paper Puzzles to Cut Out and Make*. Tarquin, Stradbrooke, Diss

Mitchell D (2002) Silverflexagons 2. The slit-square silverflexagons and related forms. Water Trade. Distributed by the British Origami Society

Moseley R (2008) Flexagon portal. www.flexagon.net. Accessed 25 April 2008

Pook LP (2003) Flexagons Inside Out. Cambridge University Press, Cambridge

Pook LP (2008a) Silver flexagons. Flexagon Lovers Group posting 444. http://tech.groups.yahoo.com/group/Flexagon_Lovers/. Accessed 7 February 2008

Pook LP (2008b) Regular star flexagons. Flexagon Lovers Group posting 440. http://tech.groups.yahoo.com/group/Flexagon_Lovers/. Accessed 3 February 2008.

Schwartz A (2008) Flexagon discovery: the shape shifting 12-gon. http://www.eighthsquare.com/12-gon.html. Accessed 22 April 2008

Sherman S (2007a) Point flexagons. http://loki3.com/flex/point-flexagon.html. Accessed 17 April 2007

Sherman S (2007b) Triangle flexagon bestiary. http://loki3.com/flex/triangles.html. Accessed 3 June 2007

Sherman S (2008) Hexaflexagon video. Flexagon Lovers Group posting 504. http://tech.groups.yahoo.com/group/Flexagon_Lovers/. Accessed 14 November 2008

Chapter 12
Miscellaneous Flexagons

12.1 Introduction

The miscellaneous flexagons described in this final chapter include some flexagons and related structures that do not fit conveniently into the classification schemes used in earlier chapters. Examples are included.

Impossible flexagons are structures that look like flexagons, and have a flexagon like pat structure, but they are not flexagons because they cannot be flexed. Three sector odd edge flexagons and three sector skeletal flexagons are examples of impossible flexagons.

Degenerate odd edge flexagons are derived by replacing one or more of the pats in a principal main position of a second order fundamental odd edge flexagon by single leaves. Pats in a principal main position consist of folded pairs of leaves. This is the reverse of bundling, where leaves are replaced by folded pairs of leaves (Section 11.5.1). It is not possible to delete complete faces from a second order fundamental odd edge flexagon.

Alternating odd edge flexagons are irregular in that minor main positions reached by one pocket flex from a principal main position have two types of pat structure. They differ from second order fundamental odd edge flexagons, which are regular in the sense that all the minor main positions reached by one pocket flex from a principal main position have the same pat structure.

Flapagons are related to flexagons. Like some flexagons, they can be flexed to display pairs of faces in cyclic order, but the structure is different. A fundamental flapagon is a skew regular edge ring of convex polygons, whereas a flexagon is a band of hinged convex polygons (Section 1.1). A characteristic feature of flapagons is that loose flaps are present on faces of main positions. Flapagons are flexed by using flap flexes in which loose flaps are turned over. The cycles that can be traversed are called flap cycles. Some types of fundamental flapagon can be linked together to form linked flapagons. Linking of an appropriate flapagon to an appropriate even edge flexagon to form a flapagon–flexagon hybrid is also possible.

Multiplex edge flexagons are derived from solitary edge flexagons by increasing the number of sectors by a factor that is a whole number. Multiplex edge flexagons always have at least some main positions which are, in appearance, flat regular

L. Pook, *Serious Fun with Flexagons,* Solid Mechanics and Its Applications 164,
DOI 10.1007/978-90-481-2503-6_12, © Springer Science+Business Media B.V. 2009

edge rings of convex polygons. A characteristic feature is that loose flaps are present on at least some faces of flat main positions. There is always at least one flap cycle that can be traversed by using a flap flex. This is sometimes an incomplete cycle. At least one other type of flex is always possible. In general, at least one transformation between flexagons is possible. There are numerous possibilities, including some spectacular examples. In general, the dynamic properties of multiplex flexagons are complicated, and they are difficult to handle.

Hooke's joint flexagons have leaves that are hinged together by Hooke's joints, as used in motor vehicle drivelines (Dunkerley 1910). A Hooke's joint has two degrees of freedom but, as applied to hinged polygons, these differ from those of a point hinge (Section 1.2). Hence, the dynamic properties of Hooke's joint flexagons differ significantly from those of point flexagons.

12.2 Three Sector Odd Flexagons

12.2.1 General Properties

There are some structures that look like flexagons, and have a flexagon like pat structure, but which are impossible flexagons because they cannot be flexed. Three sector odd edge flexagons and three sector skeletal flexagons are examples of impossible flexagons.

In a second order fundamental odd edge flexagon made from regular convex polygons (Section 4.3.1), and also in a fundamental isosceles triangle odd edge flexagon (Section 10.2.5), each pat in a principal main position is a sector. Examples of these flexagons described earlier all have at least five sectors. If made from rigid leaves, they are mechanisms and they can be flexed using the pocket flex, which is their characteristic flex.

Three sector second order fundamental odd edge flexagons and three sector fundamental isosceles triangle odd edge flexagons are rigid if made from rigid leaves. That is, they are mechanisms with no degrees of freedom. They cannot be flexed, even if leaf bending is allowed, so they are impossible flexagons. For methods of determining whether a mechanism is rigid see Demaine and O'Rourke (2007).

There is a three sector skeletal flexagon corresponding to every three sector odd edge flexagon (Section 5.2.1). Three sector skeletal flexagons look as if flexing should be possible, but it is not, so they are also impossible flexagons.

The examples of impossible flexagons given below are all solitary flexagons.

12.2.2 The Three Sector Fundamental Dodecagon Odd Edge Flexagon

Second order fundamental odd edge flexagons (Section 4.3.1) are made from second order fundamental edge nets (Section 3.3). The three sector second order fundamental dodecagon odd edge flexagon $3\langle 12, 12/5\rangle_2$ has flat main positions. This is unusual for a three sector fundamental odd edge flexagon. The net for $3\langle 12, 12/5\rangle_2$ is shown in Fig. 12.1. The torsion per sector is 1. The sector diagram is shown in Fig. 12.2. As assembled, the flexagon is in principal main position 1(2). This is, in appearance, a flat regular odd edge ring of three regular dodecagons (Fig. 12.3). It is an impossible flexagon that cannot be flexed, but the other face numbers can be seen by bending the leaves.

12.2.3 A Three Sector Isosceles Triangle Odd Edge Flexagon

The net for a three sector 30°–30°–120° isosceles triangle odd edge flexagon is shown in Fig. 12.4. In the terminology of Section 10.2.5 it can be called a triflexagon where 'tri' refers to the triangular outline of a main position. The torsion per sector is 1. The sector diagram is shown in Fig. 12.5. As assembled, the flexagon is in principal main position 1(2). This is, in appearance, a flat regular odd edge ring of three 30°–30°–120° isosceles triangles (Fig. 10.2). The pat structure is similar to that of a pentaflexagon (Section 10.2.5). It is an impossible flexagon that cannot be flexed, but the other face numbers can be seen by bending the leaves.

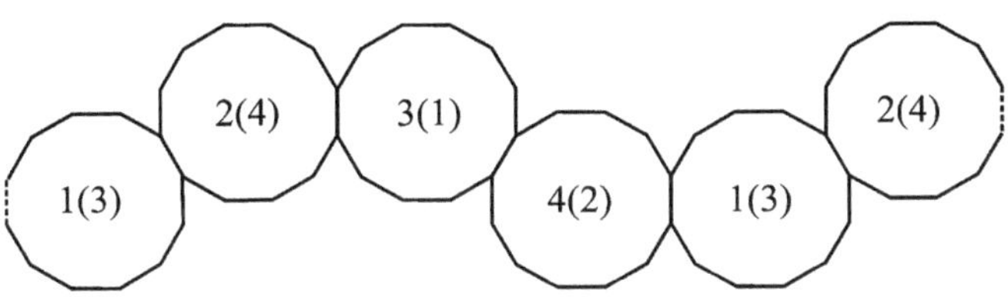

Fig. 12.1 Net for the three sector second order fundamental dodecagon odd flexagon $3\langle 12, 12/5\rangle_2$. One copy needed. Fold each leaf numbered 3 onto a leaf numbered 4

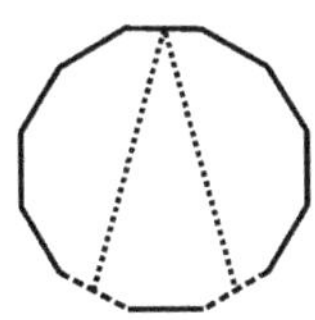

Fig. 12.2 Sector diagram for the three sector second order fundamental dodecagon odd flexagon $3\langle 12, 12/5\rangle_2$

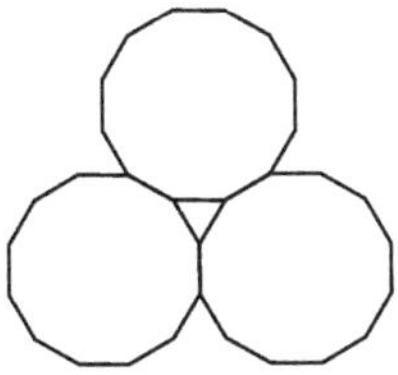

Fig. 12.3 Flat regular odd edge ring of three regular dodecagons

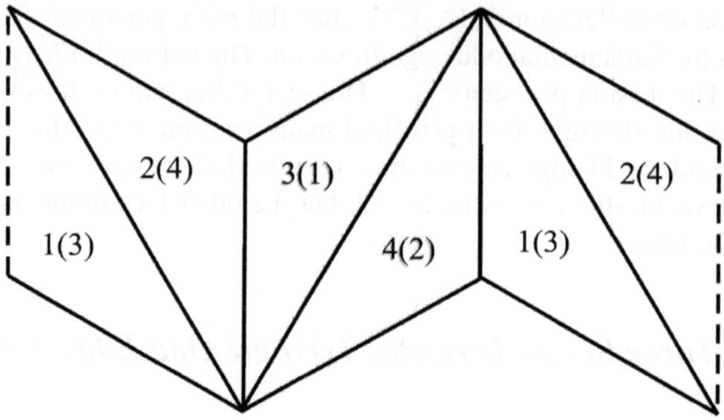

Fig. 12.4 Net for a three sector 30°–30°–120° isosceles triangle odd edge flexagon. One copy needed. Fold leaves numbered 3 onto leaves numbered 4

Fig. 12.5 Sector diagram for a three sector 30°–30°–120° isosceles triangle odd edge flexagon

12.2.4 A Three Sector Hexagon Odd Skeletal Flexagon

The net for a three sector hexagon odd skeletal flexagon is shown in Fig. 12.6. This was derived by using the net for the second order fundamental triangle odd edge flexagon $5\langle 3, 3\rangle_2$ (Fig. 4.40) as a precursor, reducing the number of sectors from five to three, truncating the equilateral triangles to regular hexagons and then inscribing hexagons with vertices at midpoints of the edges of the original hexagons. Thus, the edge hinges along edges of the original hexagons became point hinges at vertices of the inscribed hexagons (Section 5.2.1). Point hinges are impossible in a paper model, but short paper strips provide a workable approximation (Section 3.4). The torsion per sector is 1. The sector diagram is shown in Fig. 12.7.

As assembled, the flexagon is in principal main position 1(2). This is, in appearance, a regular odd vertex ring of three regular hexagons (Fig. 12.8a). The flexagon can be flexed into what appears to be principal main position 3(4) by applying a single interleaf flex to each pat in turn. However, the paper strips used to approximate point hinges become twisted (Fig. 12.8b) whereas they are untwisted in Fig. 12.8a. A close up of a twisted paper strip is shown in Fig. 12.9. The reason for the twisting is that principal main positions 1(2) and 3(4) have the same torsion but of opposite sign (Section 4.3.1).

The twisting that is possible in a paper strip used as an approximation to a point hinge is due to an additional degree of freedom that is not present in an ideal point hinge (Section 1.2). This additional degree of freedom means that the torsion of the flexagon is indeterminate. Whether the use of this additional degree of freedom is allowable is a matter of taste. If it is not allowable then the flexagon is an impossible flexagon that

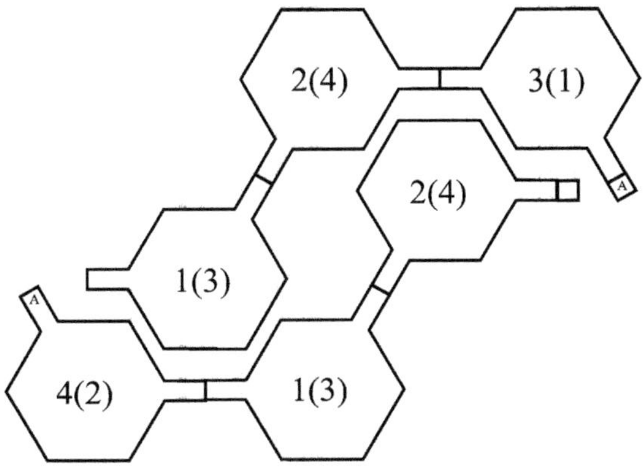

Fig. 12.6 Net for a three sector hexagon odd skeletal flexagon. One copy needed. Join the two parts of the net at A-A. Fold each leaf numbered 3 onto a leaf numbered 4

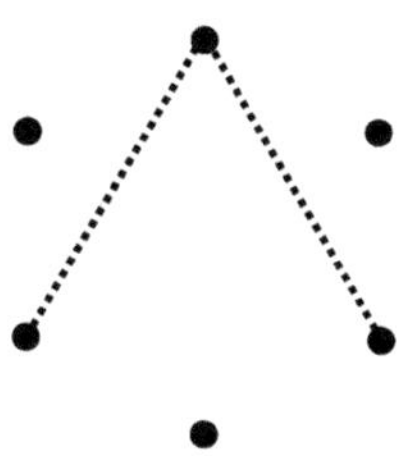

Fig. 12.7 Sector diagram for a three sector hexagon odd skeletal flexagon

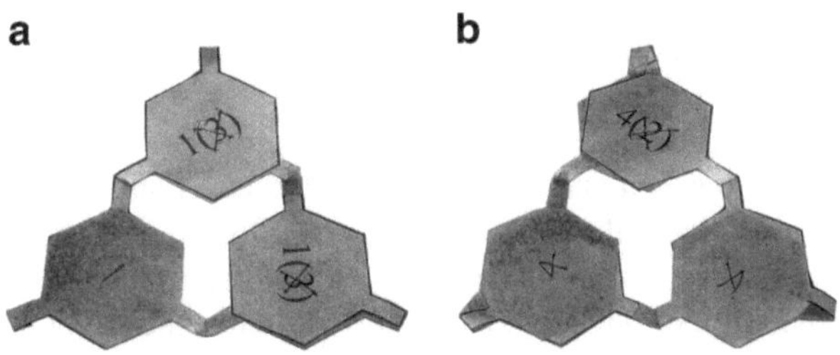

Fig. 12.8 A flexagon as a regular odd vertex ring of three regular hexagons. Point hinges approximated by paper strips. (**a**) Untwisted paper strips. (**b**) Twisted paper strips

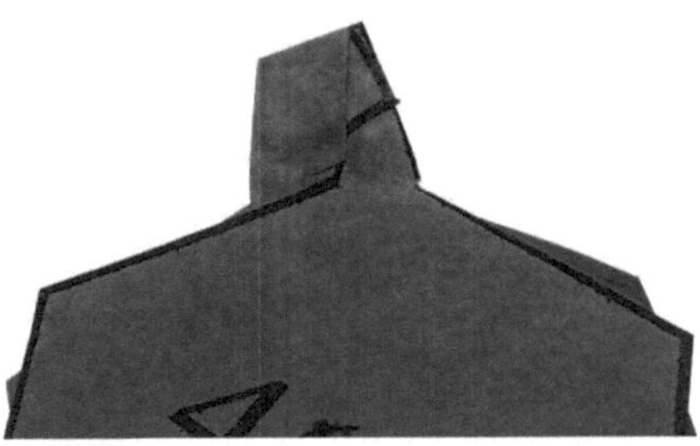

Fig. 12.9 Close up of a twisted paper strip

cannot be flexed. In Section 7.3.2 it is assumed, implicitly, that the use of the additional degree of freedom is not permissible when interleaf flexing point flexagons.

12.3 Degenerate Odd Edge Flexagons

12.3.1 General Properties

Each pat in a principal main position of a second order fundamental odd edge flexagon consists of a folded pair of leaves, and each pat is a sector (Section 4.3.1). Degenerate even edge flexagons are derived by deleting faces from precursor flexagons (Section 8.1). However, it is not possible to delete complete faces from a second order fundamental odd edge flexagon.

Degenerate odd edge flexagons can be derived by replacing one or more of the pats in a principal main position of a second order fundamental odd edge flexagon by single leaves. This procedure deletes parts of two faces. It is the reverse of bundling, where single leaves are replaced by folded pairs of leaves (Section 11.5.1). The characteristic flex for a second order fundamental odd edge flexagon is the pocket flex. This is also the characteristic flex for a degenerate odd edge flexagon. An example is given below.

12.3.2 A Degenerate Square Odd Edge Flexagon

A dual marked net for a degenerate square odd edge flexagon is shown in Fig. 12.10. This was derived by using the net for the five sector second order fundamental square odd edge flexagon $5\langle 4, 4\rangle_2$ (Fig. 4.48) as a precursor, increasing the number of sectors from five to seven and deleting parts of faces 3 and 4. As assembled, the flexagon is either in principal main position 1(2) or in principal main position 2(1). These are, in appearance, box edge rings of seven squares (Fig.12.11a). In the main position codes the number outside the brackets indicates the face number visible on the outside of the ring. In a principal main position the flexagon can be arranged with sevenfold rotational symmetry, but it cannot be laid flat.

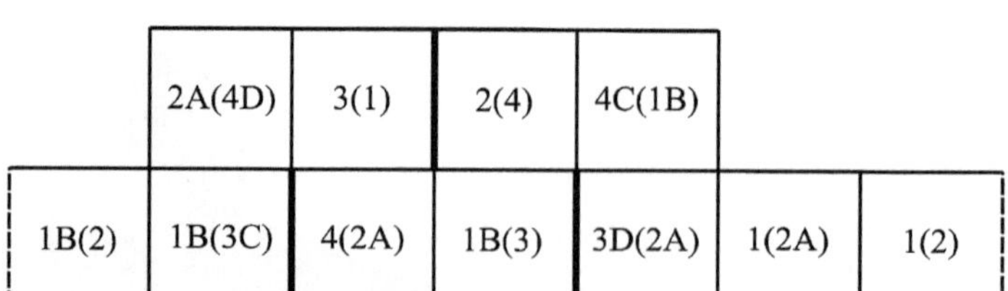

Fig. 12.10 Dual marked net for a degenerate square odd edge flexagon. One copy needed. Starting from one end of the net fold each leaf numbered 3 onto a leaf numbered 4

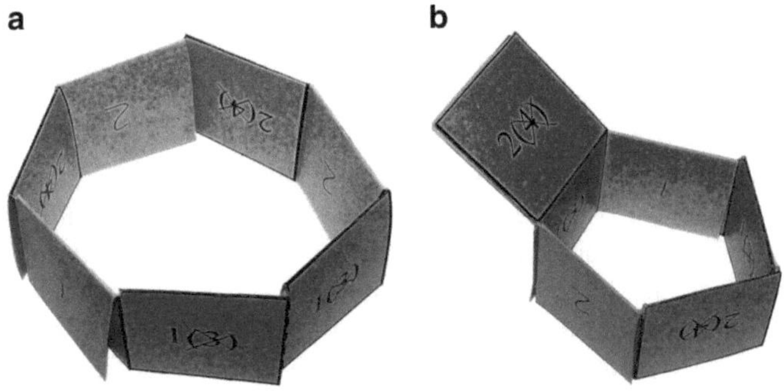

Fig. 12.11 A flexagon as the following. (**a**) A box edge ring of seven squares. (**b**) A combination of a box edge ring of five squares and a square edge pair

The flexagon can be flexed to a minor main positions by using a pocket flex. This works only in two places on a principal main position. Start by pinching two pats together and adjust so that a flap consisting of two pats can be unfolded to reach an intermediate position. This is a combination of a box edge ring of five squares and a regular even edge ring of four squares. Then fold pairs of leaves together, using the hinges at right angles to those used to reach the intermediate position, to reach a minor main position. This is a combination of a box edge ring of five squares and a square edge pair (Fig. 12.11b). The pocket flex can be carried out in two different ways so there are two minor main positions.

The principal main position can be turned inside out, so it is possible to flex between its two forms. Starting from principal main position 1(2), pinch together two pairs of pats so that leaves marked A are folded together. Adjust the flexagon so that a flap of three leaves can be unfolded to reach an intermediate position. This is an irregular odd edge ring of seven squares (Fig. 2.4). Next, hold the two leaves marked D and turn them outwards to reach another intermediate position. Then fold the flexagon in two lengthways to reach principal main position 2(1). If starting from principal main position 2(1), pinch together two pairs of pats so that leaves marked B are folded together, and turn the two leaves marked C outwards. Turning the principal main position inside out makes another two minor main positions available, making a total of four.

12.4 Alternating Odd Edge Flexagons

12.4.1 General Properties

Second order fundamental odd edge flexagons are made from second order fundamental edge nets (Section 3.3). Each pat in a principal main position consists of a folded pair of leaves, each pat is a sector, and all the pats have the same structure

(Section 4.3.1). The characteristic flex is the pocket flex. Second order fundamental odd edge flexagons are regular in that, starting from a principal main position, only one type of pocket flex is possible, and these lead to only one type of minor main position. Sector diagrams always have a plane of reflection symmetry.

Similarly, each pat in a principal main position of an alternating odd edge flexagon consists of a folded pair of leaves, and each pat is a sector, and all the pats have the same structure. They are not made from second order fundamental edge nets, and sector diagrams do not have reflectional symmetry. The characteristic flex is the pocket flex. However, starting from a principal main position, two types of pocket flex are possible. Minor main positions reached by one pocket flex from a principal main position all have the same appearance, but there are two types of pat structure.

Alternating odd edge flexagons are solitary flexagons. They are twisted bands, and the torsion per sector is 1. Since there is an odd number of sectors, the total torsion is equal to the number of sectors, and is always odd, so the flexagons are Möbius bands. If a standard face numbering sequence (Section 4.1.1) is used, then alternating odd edge flexagons can be assembled either in principal main position 1(2) or principal main position 3(4). The torsion of these principal main positions is of opposite sign, so it is never possible to flex between them.

Possible distinct types of alternating odd edge flexagons can be enumerated by counting possible sector diagrams that do not have reflectional symmetry. There are none for equilateral triangles, one distinct type for squares, and two distinct types for regular pentagons. The sector diagram for square alternating odd edge flexagons is shown in Fig. 12.12. Sector diagrams can be used to derive nets for alternating odd edge flexagons by using the method described in Section 7.2.2 for even edge flexagons, with due attention to detail.

12.4.2 A Square Alternating Odd Edge Flexagon

The net for the five sector square alternating odd edge flexagon is shown in Fig. 12.13. As assembled, the flexagon is in principal main position 1(2). This is, in appearance, a skew regular odd edge ring of five squares (Fig. 2.8). Two types of pocket flex can be used to flex to minor main positions. In the first type, pinch two pats together, with leaves numbered 1 left visible, and adjust the flexagon so that a flap consisting of two pats can be unfolded to reach a first intermediate position (Fig. 12.14a). This is a combination of a slant regular odd edge ring of three squares and a flat regular even edge ring of four squares. Then fold pairs of leaves together, using the hinges

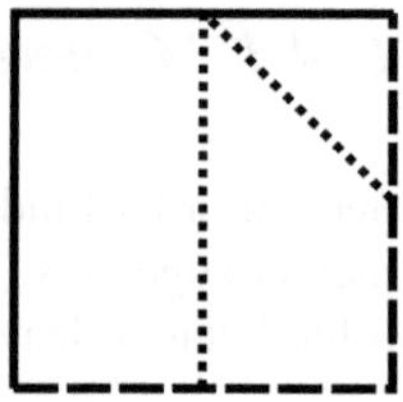

Fig. 12.12 Sector diagram for square alternating odd edge flexagons

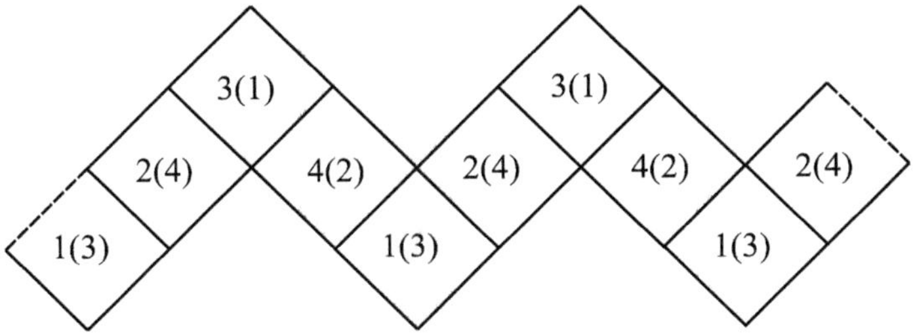

Fig. 12.13 Net for the five sector square alternating odd edge flexagon. One copy needed. Starting from one end of the net, fold each leaf numbered 3 onto a leaf numbered 4

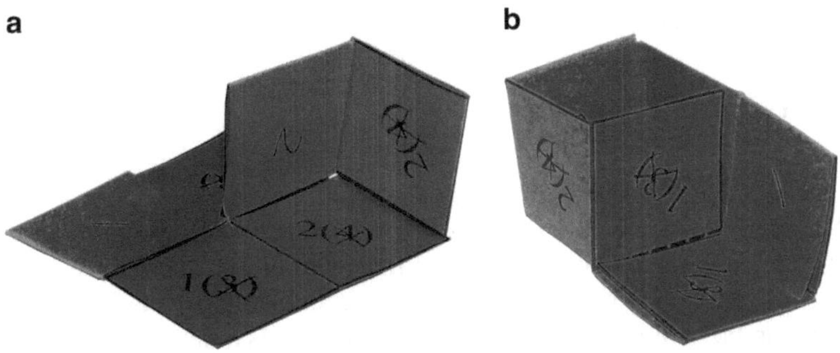

Fig. 12.14 The five sector square alternating odd edge flexagon. (**a**) First intermediate position, a combination of a slant regular odd edge ring of three squares and a flat regular even edge ring of four squares. (**b**) Second intermediate position, a combination of a slant regular odd edge ring of three squares and a box edge ring of four squares

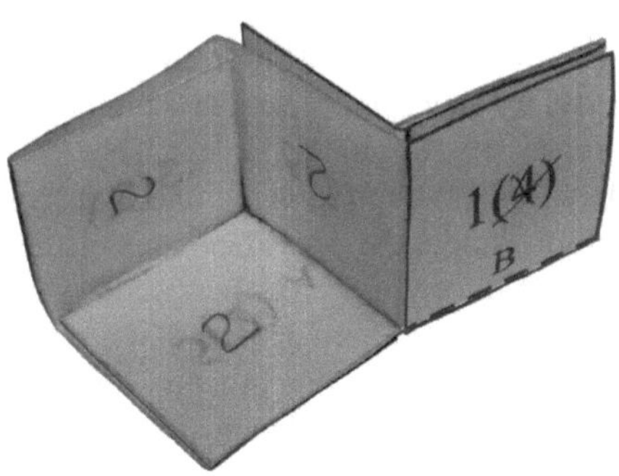

Fig. 12.15 A flexagon as a combination of a slant regular odd edge ring of three squares and a square edge pair with one square in common

at right angles to those used to reach the first intermediate position, to reach a minor main position. This is a combination of a slant regular odd edge ring of three squares and a square edge pair (Fig. 12.15). This pocket flex can be carried out in five different ways. No further pocket flexes are possible.

In the second type of pocket flex, pinch two pats together, with leaves numbered 2 left visible, and adjust the flexagon so that adjacent pats can be opened into an open ended box to reach a second intermediate position (Fig. 12.14b). This is a combination of a slant regular odd edge ring of three squares and a box edge ring of four squares. Then close the box in the opposite direction, in the second part of a box flex, to reach a minor main position. This minor main position has the same appearance as that for the first type of pocket flex (Fig. 12.15), except that it is the enantiomorph. The pat structure is also different to that obtained by using the first type of pocket flex. The second type of pocket flex can be carried out in five different ways so there a total of ten different minor main positions.

12.5 Flapagons

12.5.1 General Properties

Flapagons are related to flexagons in that they can be flexed to display pairs of faces in cyclic order. A fundamental flapagon is a skew regular edge ring of convex polygons. Its torsion is zero. Main positions of flapagons are, in appearance, flat regular edge rings of convex polygons. A characteristic feature of flapagons is that loose flaps are present on faces of main positions. A null flapagon is a flat polygon edge ring. A fundamental duplex flapagon contains twice as many leaves as the corresponding null flapagon, and main positions are always even edge rings. Similarly, a fundamental triplex flapagon has three times as many leaves as the corresponding null flapagon. Its main positions can be either even edge rings or odd edge rings. Fundamental flapagons are solitary flapagons, and correspond to solitary flexagons (Section 4.2.1).

In a duplex flapagon, pairs of adjacent pats are sectors. In a triplex flapagon each pat is a sector. Flexagon figures are not appropriate for the characterisation of flapagons. They are flexed by using flap flexes in which loose flaps are turned over. The flap flex is the characteristic flex for flapagons. The cycles that can be traversed are called flap cycles. In a fundamental flapagon one flap cycle can be traversed. For fundamental duplex and fundamental triplex flapagons the number of main positions in a flap cycle is twice the number of sectors in a main position. Main positions of fundamental duplex flapagons have alternate pats that are single leaves, so flat main position links (Section 11.2.1) between fundamental duplex flapagons are possible. Hence linked flapagons are possible. It is also possible to link a fundamental duplex flapagon to an appropriate even edge flexagon to create a flapagon-flexagon hybrid. The examples below illustrate some of the possibilities.

12.5.2 The Fundamental Square Duplex Flapagon

The net for the fundamental square duplex flapagon is shown Fig. 12.16. This is a skew regular even edge ring of eight squares. As assembled, the flapagon is in main position 1(2). This is, in appearance, a flat regular even edge ring of four squares (Fig. 1.1b).

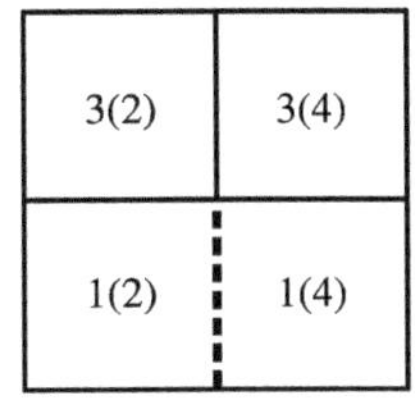

Fig. 12.16 Net for the fundamental square duplex flapagon. Two copies needed

The flap 4-cycle shown in the Tuckerman diagram can be traversed by using a flap flex. The Tuckerman diagram is the same as that for the principal 4-cycle of the first order fundamental square even edge flexagons *S*⟨4, 4⟩ (Fig. 4.17). Starting from main position 1(2), turn over the two flaps on which leaves numbered 1 are visible to reach main position 3(2), turn the flapagon over and turn over the two flaps on which leaves numbered 2 are visible, and so on round the flap 4-cycle. If only one flap is turned over, this is a local flex because some pats are left unchanged (Section 4.3.1) and face numbers become mixed up.

12.5.3 A Square Flapagon–Flexagon Hybrid

The net for a square flapagon–flexagon hybrid is shown in Fig. 12.17. This was derived by linking the fundamental square duplex flapagon (Fig. 12.16) and the first order fundamental square even edge flexagon 2⟨4, 4⟩ (Figs. 1.2 and 4.16a). As assembled, the hybrid is in common main position 2(1). The dynamic properties of the precursor flexagons appear in the 4-cycles shown in the Tuckerman diagram. The Tuckerman diagram is the same as that for the Janus flexagon (Fig. 11.15a). Cycle A is a flap 4-cycle and can be traversed by using a flap flex. Cycle B is a principal 4-cycle and can be traversed by using the twofold pinch flex. Main positions are, in appearance, flat regular even edge rings of four squares (Fig. 1.1b). Intermediate positions in the principal 4-cycle are square edge pairs (Fig. 1.15).

12.5.4 The Fundamental Isosceles Triangle Triplex Flapagon

The net for the fundamental 30°–30°–120° isosceles triangle triplex flapagon is shown in Fig. 12.18. This is a skew regular odd edge ring of nine 30°–30°–120° isosceles triangles. As assembled, the flapagon is in main position 1(2). This is, in appearance, a flat regular even edge ring of six 30°–30°–120° isosceles triangles (Fig. 10.2). The flap 6-cycle shown in the Tuckerman diagram (Fig. 12.19) can be traversed by using a flap flex. Starting from main position 1(2), turn over three flaps on which leaves numbered 1 are visible to reach main position 3(2). Because of interference between flaps. some backtracking is needed to get all three turned over. Then turn the flapagon over and turn over the three flaps on which leaves numbered 2 are visible, and so on round the flap 6-cycle. If only one flap is turned over, this a local

Fig. 12.17 Net for a square flapagon–flexagon hybrid. Two copies needed. Fold together pairs of leaves in the order 3, 4, 6 and 5

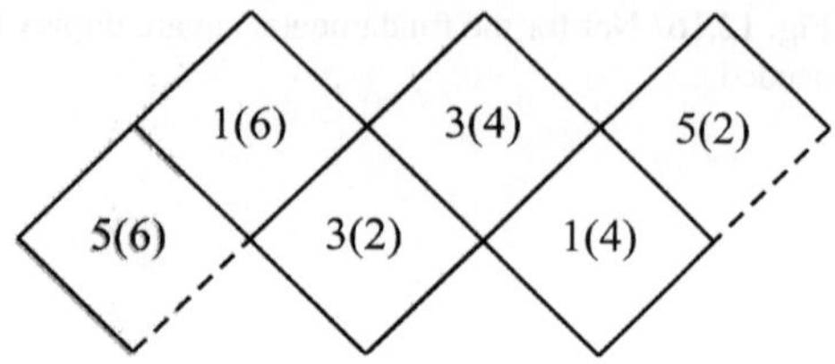

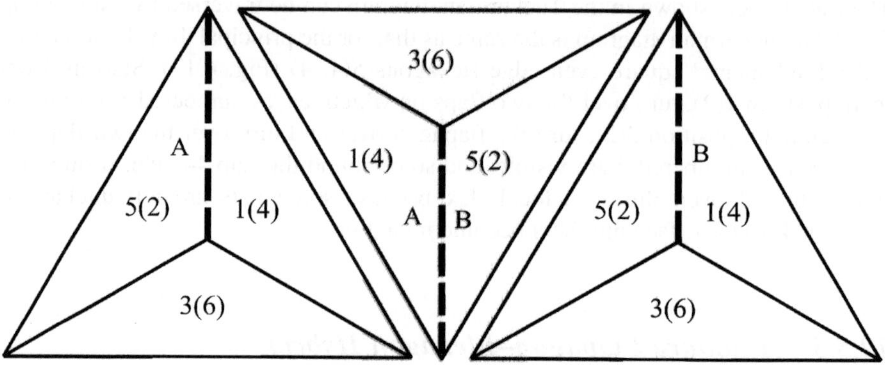

Fig. 12.18 Net for the fundamental 30°–30°–120° isosceles triangle triplex flapagon. One copy needed. Join the three parts of the net at A-A and B-B. Fold a leaf numbered 3 onto a leaf numbered 5 and in the same pat fold the leaf numbered 4 onto the leaf numbered 6, then repeat with the other two pats

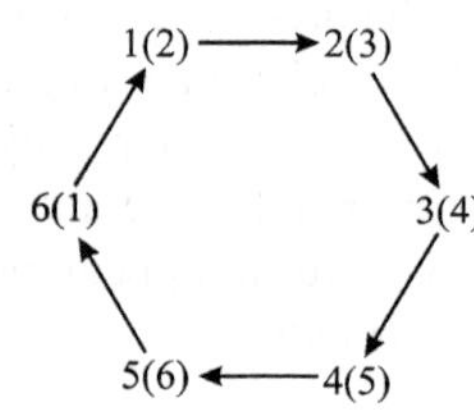

Fig. 12.19 Tuckerman diagram for the 6-cycle of the fundamental 30°–30°–120° isosceles triangle triplex flapagon

flex because one pat is left unchanged, and face numbers become mixed up. If two flaps are turned over, all three pats are changed and face numbers are mixed up.

12.6 Multiplex Edge Flexagons

12.6.1 General Properties

In general a multiplex edge flexagon can be derived by using a solitary edge flexagon as a precursor and increasing the number of sectors by a factor that is a whole number. Hence, the resulting multiplex flexagon is also a solitary flexagon, and there are always at least some main positions which are, in appearance, flat regular edge rings of convex polygons. Impossible flexagons (Section 12.2.1) can sometimes be used as precursors.

A characteristic feature of multiplex edge flexagons is that loose flaps are present on at least some faces of flat main positions. The characteristic flex is the flap flex in which loose flaps on a face are turned over. There is always at least one flap cycle, which may be an incomplete cycle, that can be traversed by using a flap flex. At least one other type of flex is always possible. In general, at least one transformation between flexagons is possible. There are numerous possibilities, including some spectacular examples. In general, the dynamic properties of multiplex flexagons are complicated, and they are difficult to handle. The examples given below are reasonably easy to handle.

To construct a duplex edge flexagon, select a solitary even edge flexagon that has at least one flat main position, and double the number of sectors. Odd edge flexagons cannot be used as precursors. What was a flat main position in the precursor flexagon becomes a skew main position in the duplex edge flexagon. Re-arrange the skew main position so that it has the same appearance and symmetries as those of the precursor's flat main position. This re-arrangement always results in the appearance of loose flaps. Number the faces of the duplex edge flexagon so that pairs of faces appear in cyclic order. Similarly, to construct a triplex edge flexagon, select either a solitary even edge flexagon or a solitary odd edge flexagon that has at least one flat main position, and triple the number of sectors. Then proceed as for duplex edge flexagons.

12.6.2 A Square Duplex Edge Flexagon

The net for a square duplex edge flexagon is shown in Fig. 12.20. This is a form of the octopus flexagon. Some other forms are described in Sections 4.2.4 and 9.3.3. Transformations between flexagons with these other forms are possible. The net was derived by using the two sector version of the first order fundamental square even edge flexagon $2\langle 4, 4\rangle$ (Fig. 4.16a) as a precursor, increasing the number of sectors from two to four and renumbering. The torsion per sector is 4. The flexagon can be flexed by using the twofold pinch flex, and also by a flap flex. The latter is the characteristic flex.

As assembled, the flexagon is in principal main position 1(2). The principal 8-cycle (solid lines) and subsidiary 8-cycle (dotted lines) shown in the intermediate position map (Fig. 12.21) can be traversed by using the twofold pinch flex. The presence of loose flaps means that care is needed to ensure that the flexagon is opened into the desired main position from an intermediate position. Main positions are, in appearance, flat regular even edge rings of four squares (Fig. 1.1b). Intermediate positions are square edge pairs (Fig. 1.15). In principal main positions, alternate pats contain one leaf and seven leaves, so flat main position links (Section 11.2.1) are possible. In subsidiary main positions alternate pats contain three leaves and five leaves.

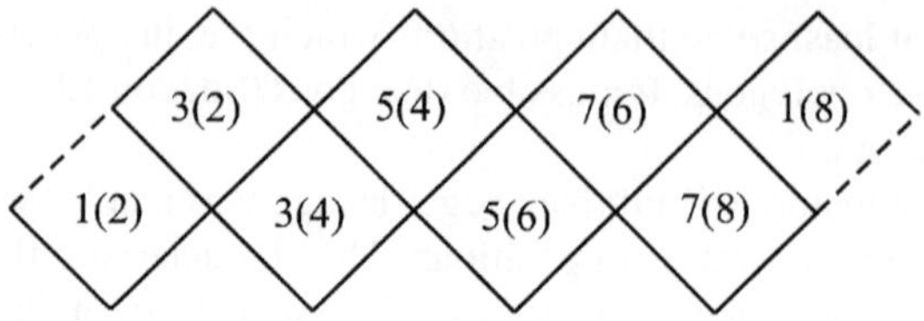

Fig. 12.20 Net for a square duplex edge flexagon. Two copies needed

Fig. 12.21 Intermediate position map for a square duplex edge flexagon

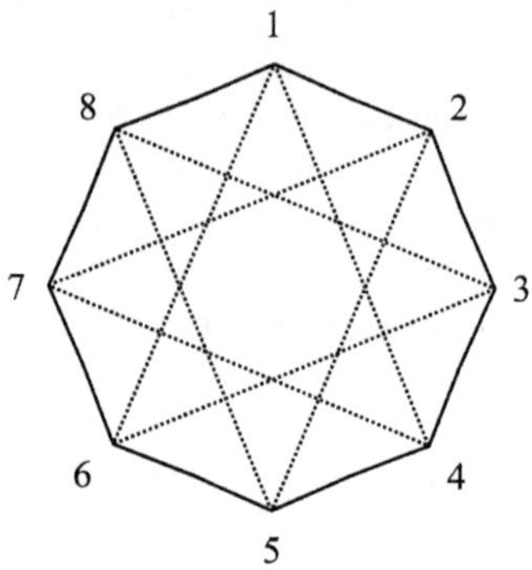

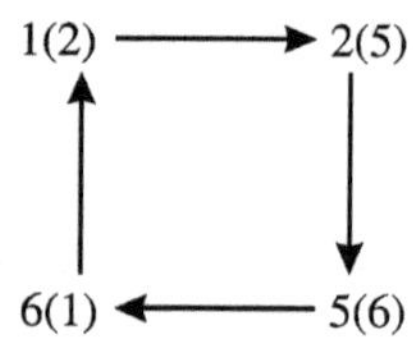

Fig. 12.22 Tuckerman diagram for a flap 4-cycle of a square duplex edge flexagon

Four different flap 4-cycles can be traversed by using the flap flex. These are represented by the four rectangles that can be discerned in the intermediate position map. In a flap 4-cycle, principal main positions and subsidiary main positions appear alternately. One of the flap 4-cycles is shown in a Tuckerman diagram (Fig. 12.22). To traverse this flap 4-cycle, starting from principal main position 1(2), turn over the two flaps on face 1 to reach subsidiary main position 2(5). Then turn over the two flaps on face 2 to reach principal main position 5(6), and so on round the flap 4-cycle.

To transform to the four sector version of the first order fundamental square even edge flexagon 4⟨4, 4⟩ (Section 4.2.4), start at a principle main position. Fold up the two flaps on a face so that they are at right angles to the other pats, then fold the flexagon in two to reach a square edge quadruple. This is an intermediate position of 4⟨4, 4⟩. To transform to the fundamental irregular ring eight square even edge flexagon type A (Section 9.3.3), start from a subsidiary main position. Unfold two pairs of pats from a face to reach a flat irregular even edge ring of eight squares (Fig. 9.1a). This is a main position of type A.

12.6.3 *The Thrice Threefold Flexagon*

The thrice threefold flexagon is a triplex edge flexagon. Its special name was chosen because of the flexagon's structure in some positions, and also because of the bewildering complexity of its dynamic properties (Pook 2007). The thrice threefold flexagon can be flexed to numerous positions, and could also be called a 30°–30°–120° isosceles triangle odd edge flexagon or a trapezoidal flexagon. All four names are used below, as appropriate. Its complexity is comparable with the octopus flexagon (Sections 4.2.4, 9.3.3 and 12.6.2), and also with a bronze even edge flexagon with a skew main position link (Section 11.2.6.2). Despite its complexity the flexagon is stable and easy to handle, but mistakes are easily made. It is therefore advisable to make one of the hinges easy to disconnect and reconnect in order that the flexagon can be unfolded and re-assembled in an initial position. Some flexing sequences are complicated, and cannot be described concisely. The torsion is 9. Transformations between flexagons are possible.

Four different dual marked versions of the net for the thrice threefold flexagon are shown in Figs. 12.23 to 12.26. These illustrate some of the numerous flexing possibilities. Face numbers are the same in all four nets. Face letters differ and have been chosen to facilitate both assembly in various positions, and subsequent flexing to further positions. The net was derived by using the net for a 30°–30°–120° isosceles triangle odd edge flexagon (Fig. 12.4) as a precursor, retaining the face numbers and tripling the number of sectors. An alternative derivation is to use the net for the second order fundamental triangle odd edge flexagon $9\langle 3, 3\rangle_2$ (Fig. 4.45) as a precursor, retaining the face numbers and replacing the equilateral triangles with 30°–30°–90° isosceles triangles.

12.6.3.1 Flexing as a Triplex Edge Flexagon

As assembled using the dual marked net for triangle and hexagon positions (Fig. 12.23), the thrice threefold flexagon can be regarded as a triplex edge flexagon. It is in main position A(B), with leaves lettered A visible on one face, and leaves lettered B on the other. This is a triangle position which is, in appearance, a flat regular odd edge ring of three 30°–30°–120° isosceles triangles (Fig. 10.2). Ignoring face markings, the flexagon has threefold rotational symmetry. The flap 6-cycle shown in the Tuckerman diagram (Fig. 12.27) can be traversed by using a flap flex. This is the characteristic flex for the flexagon. Starting from main position A(B), turn over three flaps on which leaves lettered A are visible to reach main position C(B). Because of interference between flaps, some backtracking is needed to get all three flaps turned over. Then turn the flexagon over and turn over the three flaps on which leaves lettered B are visible, and so on round the flap 6-cycle. If only one flap is turned over this a local flex because one pat is left unchanged, and face letters become mixed up. If two flaps are turned over all three pats are changed and face letters are mixed up.

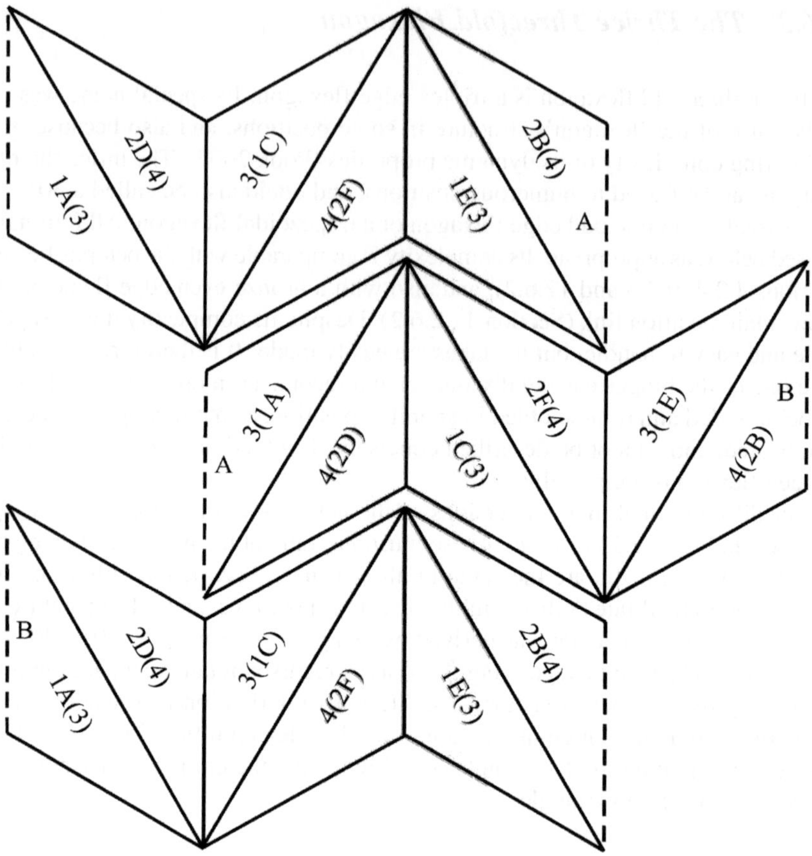

Fig. 12.23 Dual marked net (triangle and hexagon positions) for the thrice threefold flexagon. One copy needed. Join the three parts of the net at A-A and B-B. Fold each leaf numbered 3 onto a leaf numbered 4, and temporarily secure with a paper clip. Next, starting at one end, fold a leaf lettered C onto a leaf lettered E and in the same pat fold the leaf lettered D onto the leaf lettered F, then repeat with the other two pats, working along the flexagon

If one of the flaps on a face of a triangle position is turned over, then a pocket flex leads to a kite position, which is a combination of a flat regular odd edge ring of three 30°–30°–90° isosceles triangles and a 30°–30°–90° isosceles triangle edge pair (Fig. 12.28a). Figure 12.28b shows the flexagon part way through a pocket flex. A pocket flex can be done in 36 different ways. There are two distinct ways in which two pocket flexes can be carried out, and three distinct ways in which three pocket flexes can be carried out. One of these leads to a hexagon position, which is a combination of a flat regular odd edge ring of three 30°–30°–90° isosceles triangles and three 30°–30°–90° isosceles triangle edge pairs (Fig. 12.29). There are 12 different hexagon positions.

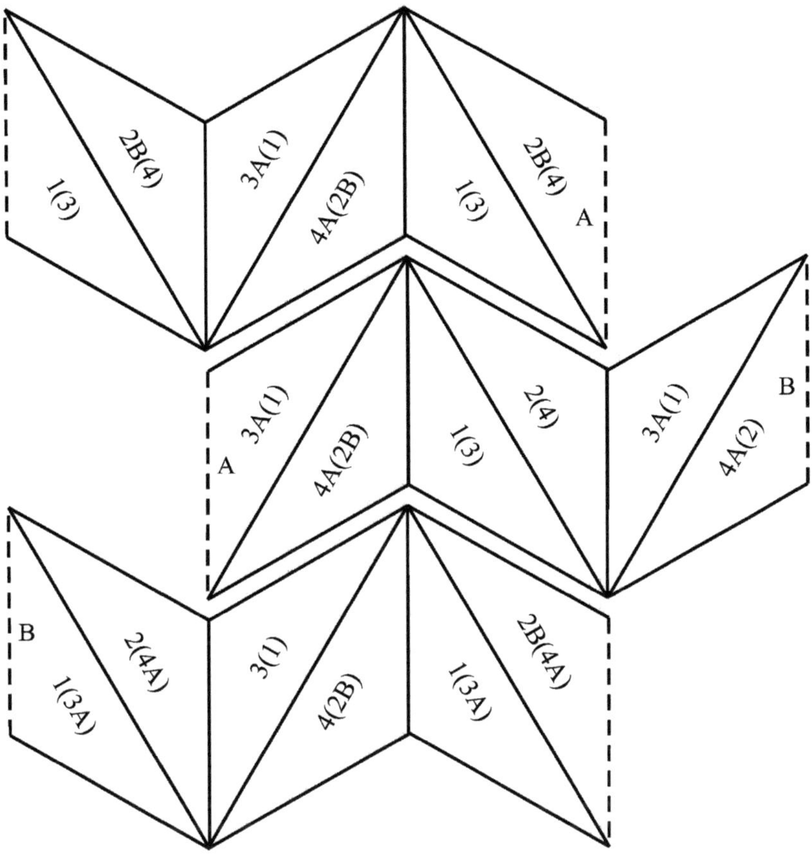

Fig. 12.24 Dual marked net (irregular pyramid and flat positions) for the thrice threefold flexagon. One copy needed. Join the three parts of the net at A-A and B-B. Fold pairs of leaves together in the order A and B. Ensure that leaves numbered 3 and 4 are on the inside of the pyramid

12.6.3.2 Transformations and Flexing Using a Threefold Flex

As assembled using the dual marked net for triangle and hexagon positions (Fig. 12.23) the thrice threefold flexagon can be regarded as a triplex edge flexagon in main position A(B), which is a triangle position (previous section). The triplex edge flexagon can be transformed into a pyramid position (Fig. 12.30a). This is an open triangular pyramid (Fig. 12.30a) which has threefold rotational symmetry. The three triangles on the right hand side of the photograph make up one of the faces of the pyramid. The pyramid is actually an irregular odd edge ring of nine 30°–30°–90° isosceles triangles: there is no connection between the two triangles on the right hand side of the photograph. Leaves on the outside of the pyramid all have the same face number, but face numbers inside the pyramid are mixed up. To transform into a pyramid position, fold up the three loose flaps on one of the faces

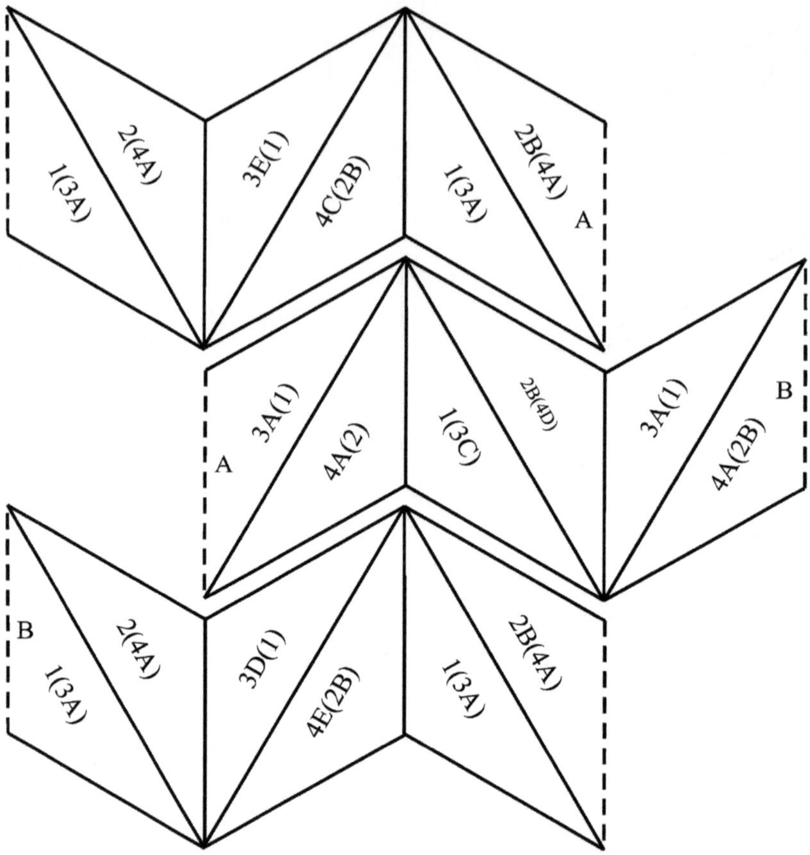

Fig. 12.25 Dual marked net (pyramid and three flap intermediate positions) for the thrice threefold flexagon. One copy needed. Join the three parts of the net at A-A and B-B. Fold pairs of leaves together in the order A and B. Ensure that leaves numbered 3 and 4 are on the inside of the pyramid

of a main position of the triplex flexagon to reach a threefold intermediate position (Fig. 12.30a). This has threefold rotational symmetry, the ends of the flaps coincide, and some leaves are bent. Then open the flexagon into a pyramid position from underneath. There are two different pyramid positions.

Main position A(B) of the triplex edge flexagon can be transformed into principal main position 1(2) of a 30°–30°–120° isosceles triangle odd edge flexagon, with leaves numbered 1 visible on one face, and leaves numbered 2 on the other. To do this, fold up the loose flaps on face A of main position A(B) to reach a threefold intermediate position (Fig. 12.30b). All nine leaves numbered 1 can then be seen. Then arrange the flexagon so that all the leaves of faces 1 and 2 are visible, thus completing a transformation between flexagons. The 30°–30°–120° isosceles triangle odd edge flexagon cannot be arranged with rotational symmetry with all the leaves numbered 1 and 2 visible. The 30°–30°–120° isosceles triangle odd edge flexagon can be returned to a pyramid position (Fig. 12.30a) by using a threefold pinch flex.

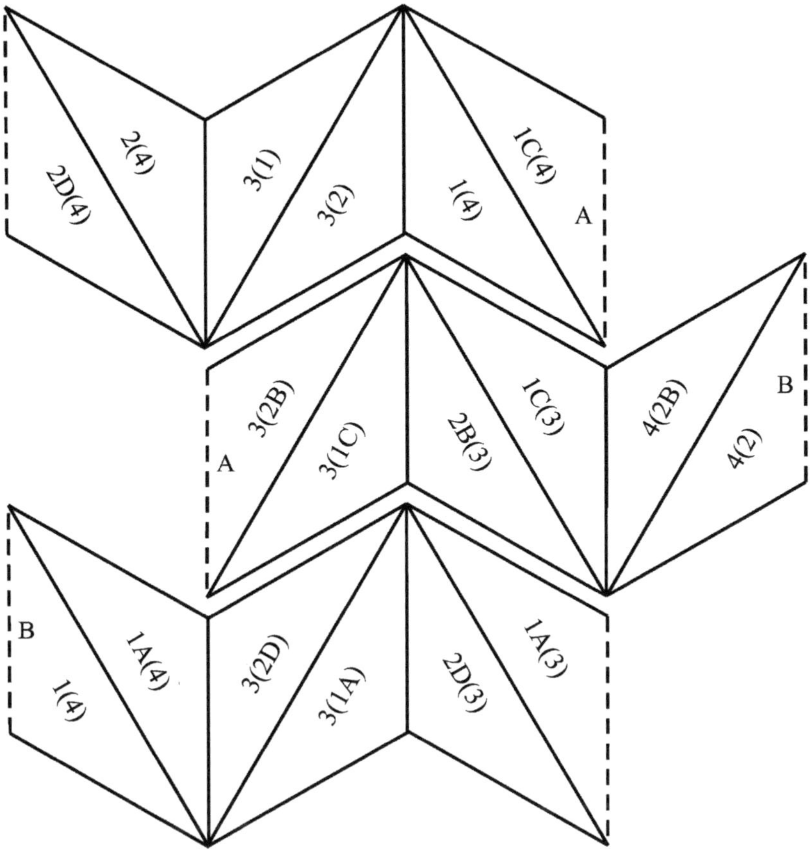

Fig. 12.26 Dual marked net (trapezium, rhombus and triangle positions) for the thrice threefold flexagon. One copy needed. Join the three parts of the net at A-A and B-B. Fold pairs of leaves together in the order A and B. Ensure that leaves numbered 3 and 4 are on the inside of the pyramid

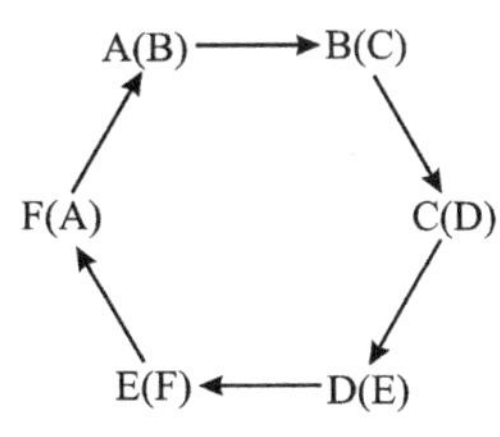

Fig. 12.27 Tuckerman diagram for the flap 6-cycle of the thrice threefold flexagon

To do this, pinch together three pairs of pats, separated by single pats, to reach a threefold intermediate position (Fig. 12.30b). Then open the flexagon into a pyramid position from underneath. There are six different intermediate positions which can be reached from main position 1(2) and hence six different pyramid positions. Reversing the flex returns a pyramid position to main position 1(2).

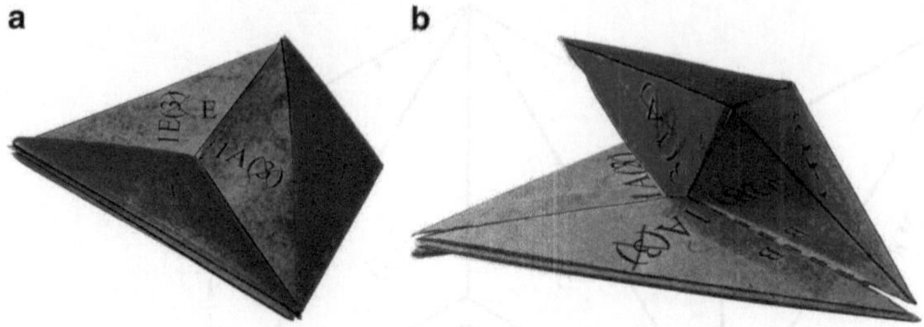

Fig. 12.28 The thrice threefold flexagon. (**a**) Kite position, a combination of a flat regular odd edge ring of three 30°–30°–90° isosceles triangles and a 30°–30°–90° isosceles triangle edge pair. (**b**) Part way through a pocket flex

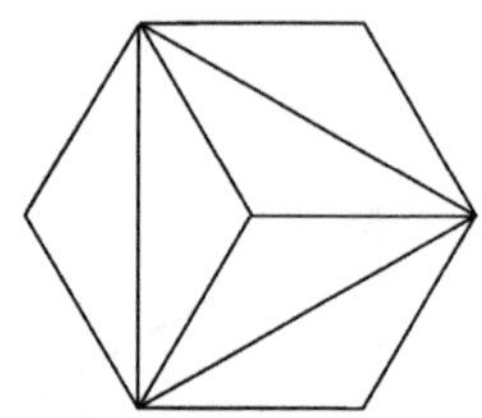

Fig. 12.29 A combination of a flat regular odd edge ring of three 30°–30°–90° isosceles triangles and three 30°–30°–90° isosceles triangle edge pairs

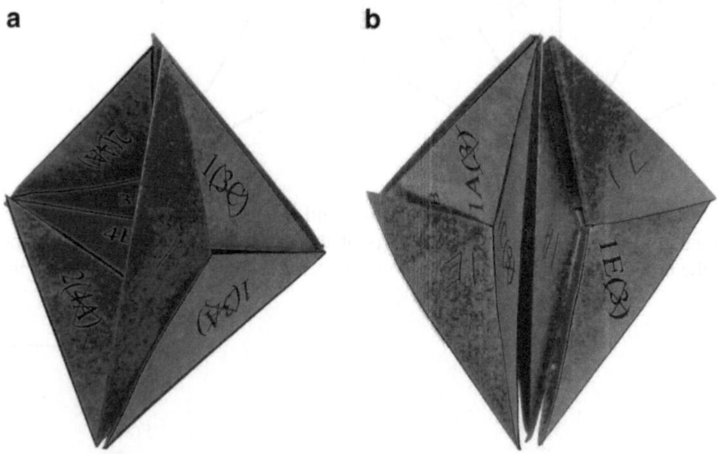

Fig. 12.30 The thrice threefold flexagon. (**a**) Pyramid position, an irregular odd edge ring of nine 30°–30°–90° isosceles triangles. (**b**) Threefold intermediate position

12.6.3.3 Transformation and Flexing Using an Asymmetric Fourfold Pinch Flex

As assembled using the dual marked net for triangle and hexagon positions (Fig. 12.23), the thrice threefold flexagon can be regarded as a triplex edge flexagon in main position A(B). This is a triangle position (Section 12.6.3.1) which is, in appearance, a flat regular odd edge ring of three 30°–30°–120° isosceles triangles (Fig. 10.2). The triplex edge flexagon can be flexed into irregular four sided pyramids by using an asymmetric fourfold pinch flex. Start from a main position of the triplex edge flexagon and transform it into a main position of a 30°–30°–120° isosceles triangle odd edge flexagon (previous section). Then pinch six of the pats together in three pairs, leaving the remaining three pats in a flat triangle, to reach an asymmetric fourfold intermediate position (Fig. 12.31a). This can be done in 18 different ways. To complete the asymmetric fourfold pinch flex open the intermediate position into an irregular four sided open pyramid with a flap attached (Fig. 12.31b). Apart from the flap, and ignoring pat structure, this irregular pyramid has fourfold rotational symmetry. Next, close the flap against the pyramid to reach a slightly different irregular four sided open pyramid (Fig. 12.31c). Ignoring pat structure, this irregular pyramid has twofold reflectional symmetry. Either form of the irregular pyramid can be flexed into four distinct irregular flat positions, one of which is shown in Fig. 12.31d. These irregular flat positions can be opened into various asymmetric positions, but it does not appear to be possible to traverse any cycles which include an asymmetric fourfold pinch flex.

The asymmetric fourfold pinch flex is difficult. It is much easier to assemble the flexagon directly into an irregular four sided open pyramid by using the dual marked net for irregular pyramid and flat positions (Fig. 12.24).

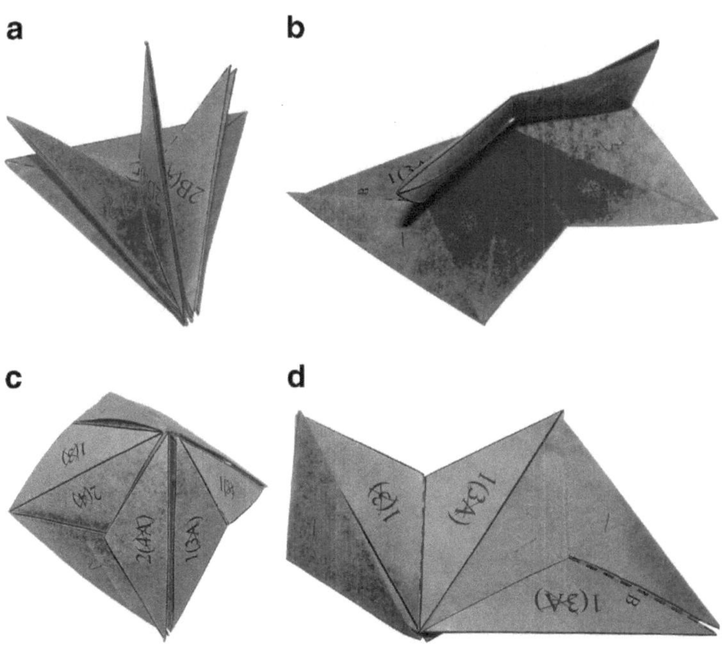

Fig. 12.31 The thrice threefold flexagon. (**a**) Asymmetric fourfold intermediate position. (**b**) Four sided pyramid, flap open. (**c**) Four sided pyramid, flap closed. (**d**) An irregular flat position

12.6.3.4 Traversing 3-Cycles by Using Threefold Flexes

There are two different pyramid positions that can be reached by flexing from main position 1(2) of a 30°–30°–120° isosceles triangle odd edge flexagon (Section 12.6.3.2). Pyramid position P(1) has leaves numbered 1 on the outside of the pyramid and pyramid position P(2) leaves numbered 2. A flap flex can be used to traverse between the two pyramid positions. Hence, a 3-cycle can be traversed, as shown in the Tuckerman diagram (Fig. 12.32), where P(1) and P(2) indicate the pyramid positions.

The dual marked net for pyramid and three flap intermediate positions (Fig. 12.25) demonstrates the flap flex. As assembled, the flexagon is in pyramid position P(1). To flex to pyramid position P(2), fold together pairs of leaves inside P(1) that have the same letter, to reach a three flap intermediate position (Fig. 12.33). This has threefold rotational symmetry, and all the leaves numbered 1 and 2 are visible. Then open the intermediate position into pyramid position P(2) by a reverse of the first part of the flap flex.

12.6.3.5 A Trapezoidal Flexagon

As assembled using the dual marked net for trapezium, rhombus and triangle positions (Fig. 12.26), the thrice threefold flexagon can be regarded as main position 1(2) of a trapezoidal flexagon. This is a trapezium position (Fig. 12.34a). The pat structure of a trapezium position has twofold rotational symmetry about a line in the plane of the flexagon. This is surprising for a flexagon that has an odd number of leaves in its net. A trapezium position has a distinctive appearance and structure, and is unusual in that it cannot be described as a polygon ring or a combination (Section 2.1.2). A transformation between flexagons is possible by using a twist flex from a triangle position of a triplex edge flexagon (Section 12.6.3.1), but this is difficult. To carry out the twist flex hold the middle pair of leaves in a pat, and pull and twist the

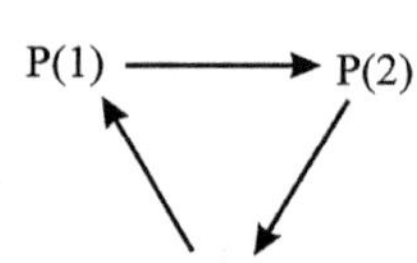

Fig. 12.32 Tuckerman diagram for a 3-cycle of the thrice threefold flexagon

Fig. 12.33 Three flap intermediate position of the thrice threefold flexagon

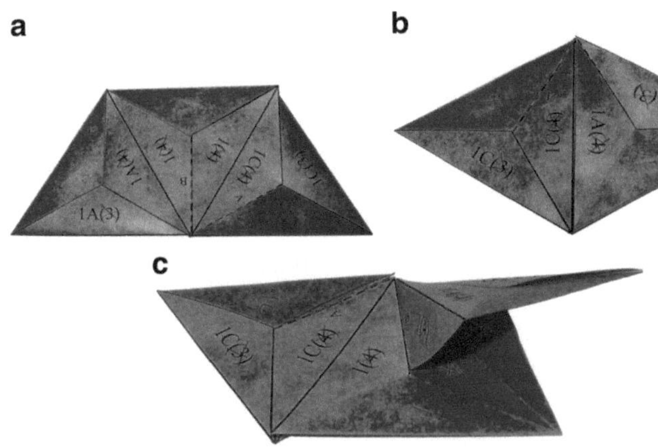

Fig. 12.34 The thrice threefold flexagon. (**a**) Trapezium position. (**b**) Part way through a pocket flex. (**c**) Rhombus position

opposite vertex. This can be done in nine different ways, so here are nine different trapezium positions, all with the same code.

Starting from a trapezium position, the flexagon can be flexed into a rhombus position (Fig. 12.34b) by using a pocket flex. The flexagon part way through a pocket flex is shown in Fig. 12.34c. This can be done in four different ways. A second pocket flex leads to a triangle position, which is a flat regular odd edge ring of three 30°–30°–90° isosceles triangles (Fig. 10.2). Triangle positions are identified by face letters. There are two of these, triangle positions A(B) and C(D). These triangle positions have a different pat structure to those of the triangle positions of a triplex edge flexagon (Section 12.6.3.1).

12.7 A Hooke's Joint Flexagon

A Hooke's joint, as used in motor vehicle drivelines (Dunkerley 1910, Pook 2003) is shown schematically in Fig. 12.35 in a form that can be used to construct a paper model. The two strips are hinged to the central square so a Hooke's joint has two degrees of freedom but, as applied to hinged polygons, these differ from those of point hinges (Section 1.2). Two polygons connected by a point hinge (Fig. 1.10) have one degree of freedom if the polygons are confined to a plane. There is only one position in which two polygons connected by a Hooke's joint are in the same plane and, if the polygons are confined to this plane, there are no degrees of freedom. Hence, the dynamic properties of Hooke's joint flexagons differ significantly from those of point flexagons (Section 5.1).

The net for a truncated square Hooke's joint flexagon is shown in Fig. 12.36. This is the same as one of the hybrid flexahedrons described by Engel (1969), except that the leaves are a different shape. The squares have been truncated to irregular hexagons in order to accommodate the Hooke's joints, which are necessarily of finite size.

To assemble the flexagon, first transfer the numbers in brackets on the upper face of a leaf to the same positions on the lower face, and delete them from the upper face. Then overlap the leaves, as shown in Fig. 12.37, and insert the small squares as shown. Tape over the four diagonals to form hinges between the leaves and small squares. Turn the flexagon over, keeping everything in position, and tape the remaining four diagonals. As assembled, the flexagon is in main position 1, as shown by the numbers at the centre of the flexagon. The torsion is 1. To reach main position 2 rotate the leaf numbered 1 and 2 through 180° about the two Hooke's joints connecting it to adjacent leaves. This simple flex is similar to that used for point flexagons (Section 5.3.2), but not identical. Figure 12.38 shows the flexagon part way through a simple flex. The 8-cycle shown in the main position sequence diagram (Fig. 12.39) can be traversed by repeating the simple flex.

And thick and fast they came at last,
And more, and more, and more.

"Lewis Carroll (Charles Dodgson), *Through the Looking Glass*."

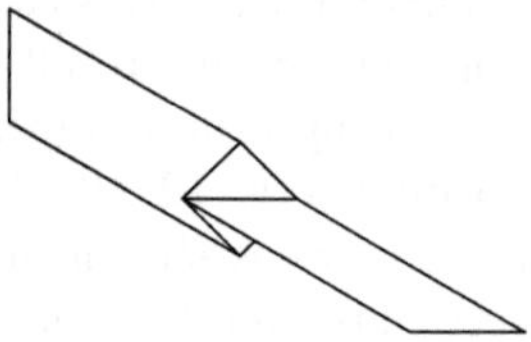

Fig. 12.35 Paper model of a Hooke's joint

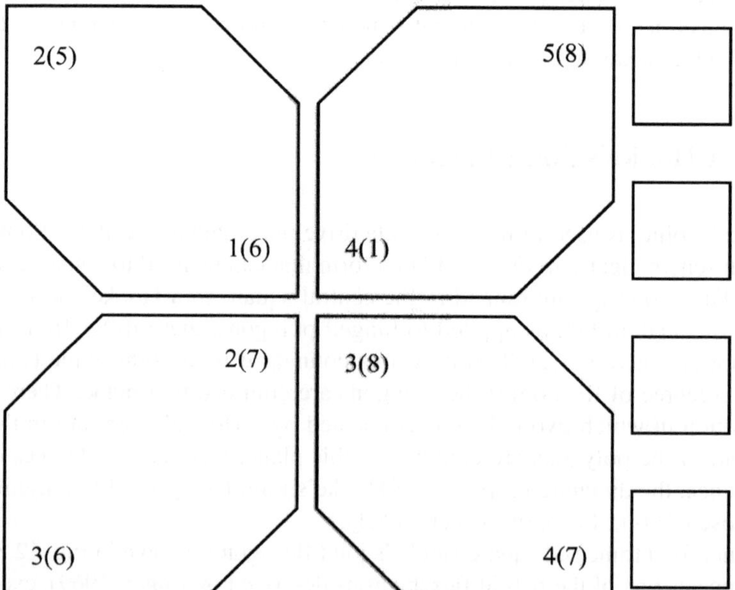

Fig. 12.36 Net for a truncated square Hooke's joint flexagon. One copy needed. See text for assembly instructions

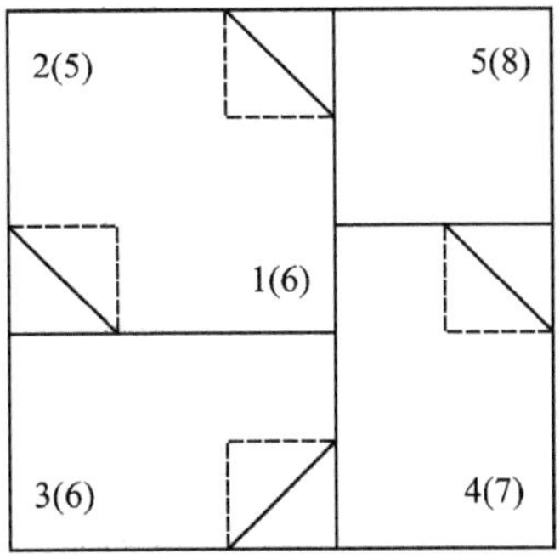

Fig. 12.37 Main position of a truncated square Hooke's joint flexagon

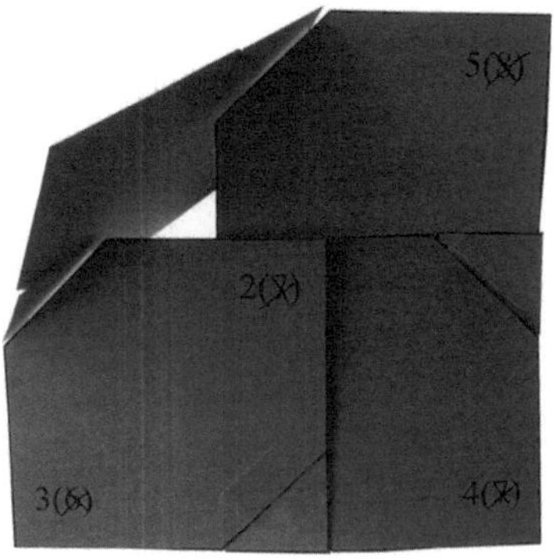

Fig. 12.38 A truncated square Hooke's joint flexagon part way through a simple flex

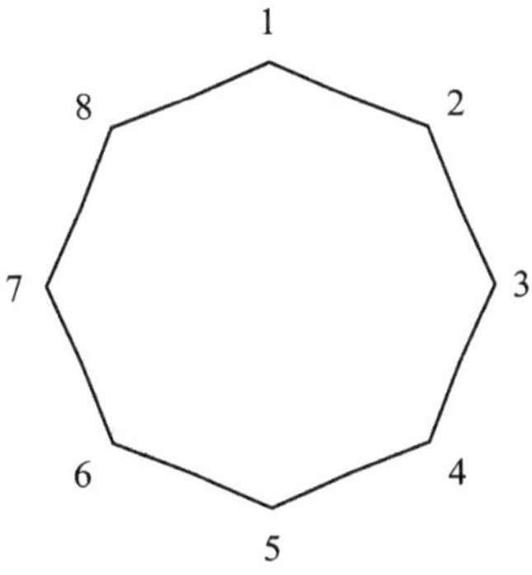

Fig. 12.39 Main position sequence diagram for a truncated square Hooke's joint flexagon

References

Demaine EJ, O'Rourke J (2007) Geometric Folding Algorithms: Linkages, Origami, Polyhedra. Cambridge University Press, Cambridge

Dunkerley S (1910) Mechanisms, 3rd edn. Longmans Green, London

Engel D (1969) Hybrid flexahedrons. Journal of Recreational Mathematics 2:35–41

Pook LP (2003) Flexagons Inside Out. Cambridge University Press, Cambridge

Pook, LP (2007) The Thrice Three-Fold Flexagon. Flexagon Lovers Group posting 396. http://tech.groups.yahoo.com/group/Flexagon_Lovers/.Accessed 5 October 2007.

References

Index

A

Adjacent link, 277
Alternating odd edge flexagons, 299, 306
Angle deficit, 3
Antibox edge ring, 25
Antibox positions, 25
Apex angle, 203
Associated double polygons, 148, 166
Associated polygons, 33, 46, 82, 88, 91, 104, 119, 122
Associated polygon twins, 292
Associated semi double polygons, 148
Asymmetric 2-fold pinch flex, 181
Asymmetric fourfold pinch flex, 72, 319
Asymmetric pinch flexes, 69
Asymmetric threefold pinch flex, 71
Augmented fundamental point flexagon, 77, 91
Augmented fundamental unagon, 91
Augmented interleaved fundamental point flexagons, 77, 95
Augmented interleaved fundamental unagon, 96–97
Augmented point flexagon, 91, 137

B

Band, 1
Band flex, 61, 124
Base angles, 203
Basic compound edge flexagon, 170
Basic even edge flexagon, 151
Basic point flexagon, 163
Book flex, 53
Box edge ring, 18
Box flexes, 53, 184, 259
Box main positions, 18
Box position links, 249, 250, 256
Box positions, 18, 54
Box vertex ring, 30
Bronze even edge flexagons, 202, 217, 219, 265, 271, 274
Bronze even edge rings, 206
Bronze flexagons, 216
Bronze triangle even edge flexagon, 224, 227
Bronze triangles, 28
Bundled odd edge flexagon, 248, 284

C

Cart wheel edge ring, 204
Cart wheel even edge flexagon, 213
Cart wheel odd edge flexagons, 214
Centroid, 20
Characteristic flexes, 6, 46, 68, 78, 82, 103, 121, 129, 137, 142, 175, 201, 247, 304, 306, 308, 311, 313
Circuits, 56
Combination, 17
Common main position, 223
Complex flexagon, 247
Composite number, 20
Compound edge flexagons, 103, 147
Compound edge ring, 8, 23
Conjoined leaves, 248, 282
Conjoined point flexagons, 248, 282, 295
Convex quadrilateral twins, 293
Cord loops, 40
Coupled point flexagon, 249, 292
Crescent hexaflexagon, 253
Curvature, 3, 8, 43, 78
Cycle, 10

D

Decagon compound edge flexagon, 115, 117
Deficient flexagon, 49
Degenerate compound edge flexagons, 169

Degenerate even edge flexagons, 149
Degenerate flexagon, 10, 147
Degenerate interleaved point flexagon, 166
Degenerate irregular polygons, 148
Degenerate irregular ring even edge flexagons, 176
Degenerate odd edge flexagons, 299, 304
Degenerate point flexagons, 165
Degrees of freedom, 7, 57, 300
Degenerate unagon, 151
Deletion of face numbers, 176
Deletion of faces, 147, 176
Diagonal link, 277
Different, 6
Dihedral angle, 6
Distinct, 6
Distorted polygon flexagons, 201
Dodecagon even edge flexagons, 66, 159, 198
Dodecagon odd edge flexagon, 301
Double conjoin, 282
Double 2-fold pinch flex, 184
Double interleaf flex, 131
Double point flexagon, 166, 282
Double twofold pinch flex, 291
Dual, 29
Duplex edge flexagon, 311
Duplex flapagon, 308

E

Edge flexagons, 1, 201
Edge hinges, 1
Edge pair, 18
Edge ring, 17
Edge strip, 18
Enantiomorphic pairs, 6, 121
End links, 248, 278
Enneagon point flexagons, 90
Equiangular polygon, 202
Equivalent flexagon figure, 169
Even edge flexagons, 2, 247, 287
Even edge ring, 2
Even skeletal flexagon, 3
Even vertex ring, 3

F

Face numbering sequences, 44, 141
Faces, 1
Family of flexagons, 6
First order fundamental edge nets, 33, 34
Fivefold pinch flex, 59
Flapagon–flexagon hybrid, 299, 308, 309
Flapagons, 299, 308
Flap cycles, 299, 300, 308
Flap flexes, 211, 273, 299, 308, 320
Flat main position links, 249, 250, 256
Flat main positions, 15
Flexagon diagram, 57
Flexagon figure, 46, 57, 79, 119, 149
Flexagon symbols, 46, 68, 83, 88
Forms, 51, 53
Fourfold pinch flex, 54
Friezes, 1
Full main position codes, 176, 183
Full map, 56
Fundamental bronze edge net, 41
Fundamental bronze even edge flexagon, 217
Fundamental compound edge flexagons, 103, 104
Fundamental compound vertex flexagons, 103
Fundamental edge flexagons, 43
Fundamental edge nets, 33
Fundamental even edge flexagons, 45, 69, 147, 149
Fundamental flapagon, 299, 308
Fundamental irregular ring even edge flexagons, 175
Fundamental odd edge flexagons, 43, 68, 300
Fundamental point flexagons, 77, 82, 282
Fundamental silver edge net, 41
Fundamental skeletal flexagons, 77, 78
Fundamental unagon, 82
Fundamental vertex net, 36

G

20-Gon odd edge flexagons, 74

H

Hand, 249
Heptaflexagon, 215
Heptagonal compound edge flexagons, 115
Heptagon point flexagons, 142
Hexa-dodeca-flexagon, 248, 271
Hexaflexagon, 251
Hexagon compound edge flexagons, 109, 111, 112, 115, 169, 171–173
Hexagonal compound edge flexagon, 113
Hexagon even edge flexagons, 62, 122, 127, 149, 154, 191, 192, 194
Hexagon odd skeletal flexagon, 302
Hexagon point flexagons, 86, 129, 135, 142, 161
Hexahexaflexagon, 247, 252
Hinge angle, 6

Hinged linkage, 7
Hooke's joint, 321
Hooke's joint flexagons, 300, 321
Hybrid flexahedrons, 321

I
Idea leaf, 6
Ideal edge flexagon, 6
Ideal edge hinge, 6
Ideal flexagon, 6
Ideal hinges, 6
Ideal pat, 6
Ideal point flexagon, 6
Ideal point hinge, 8, 29
Impossible flexagons, 299–302
Incomplete cycles, 140, 147, 267
Interleaf flexes, 131, 248
Interleaved even edge flexagons, 248
Interleaved fundamental point flexagons, 77, 87
Interleaved fundamental unagon, 87
Interleaved point flexagons, 87, 89
Intermediate position code, 45
Intermediate position map, 47, 147
Intermediate positions, 13, 16, 45, 47
Invariants, 2, 17
Irregular cycle compound edge flexagons, 121
Irregular cycle even edge flexagons, 119, 121
Irregular cycle even edge unagon, 121
Irregular cycle flexagon, 10, 119
Irregular cycle interleaved point flexagons, 120, 129, 147
Irregular cycle non interleaved point flexagons, 121, 142, 147
Irregular cycle unagon, 142
Irregular edge ring, 8, 26
Irregular hexagon even edge flexagon, 64, 66, 113, 198
Irregular pentagon, 240
Irregular pentagon even edge flexagons, 202, 260
Irregular pentagon even edge ring, 241
Irregular polygon edge flexagon, 201
Irregular polygon edge ring, 27
Irregular polygon edge rings, 27
Irregular polygon even edge flexagon, 201
Irregular polygon even edge flexagons, 201
Irregular polygon point flexagon, 201
Irregular polygons, 27, 119
Irregular quadrilateral, 231
Irregular quadrilateral even edge flexagons, 202
Irregular quadrilateral even edge ring, 232
Irregular ring even edge flexagons, 175, 177
Irregular ring odd edge flexagons, 177
Irregular triangle even edge flexagons, 202
Isosceles triangle, 203
Isosceles triangle edge rings, 204
Isosceles triangle even edge flexagons, 207, 213, 214
Isosceles triangle flexagons, 207
Isosceles triangle odd edge flexagons, 202, 214, 301, 313, 316, 320
Isosceles triangle triplex flapagon, 309

J
Janus flexagon, 257

K
Kaleidocycles, 31
Key pats, 107, 176
Kite position, 314
Kite threefold pinch flex, 223

L
Leaves, 1
Linkage, 7
Linked even edge flexagon, 249
Linked flapagons, 299, 308
Linked point flexagons, 248, 277
Local flex, 69
Locked flexagons, 49
Locked linkage, 49
Loose flaps, 166, 248, 273, 282, 299, 300, 308, 311

M
Main position code, 45, 176
Main position link, 249, 277
Main positions, 2, 45
Main position sequence diagrams, 176
Minitwist flexagon, 196
Minor cycle, 248
Minor main positions, 69
Möbius band, 45
Motion structure, 1
Muddled, 7, 61, 111, 196
Multiple polygon, 16
Multiplex edge flexagons, 299, 310

N
Net, 1
Net symbol, 33, 34

Non interleaved point flexagons, 161
Nonperformable operation, 19, 121
Null flapagon, 308

O
Octagon compound edge flexagon, 113
Octagon even edge flexagons, 64, 157
Octagon point flexagons, 142
Octopus flexagon, 53, 183, 311
Odd edge flexagons, 2, 299–301
Odd edge ring, 2

P
Paper strips, 39
Partial overlap flexagons, 44, 201, 203, 260, 274, 276
Pass through flex, 19
Pats, 2
Pentaflexagon, 214
Pentagon compound edge flexagon, 107
Pentagonal compound edge flexagons, 104, 111, 172
Pentagon even edge flexagons, 58, 122, 126, 149, 187, 190, 262–264
Pentagon point flexagons, 85, 89, 129, 133, 142, 143, 161, 164, 284
Pentagon unagon, 83
Performable operation, 18, 121
Pinch flex, 46
Plane of reflection symmetry, 19
Pocket flex, 69
Point flexagons, 3, 77, 141
Point hinge, 3, 8
Polygon pairs, 16
Polygon rings, 15
Polygon triples, 16
Polygon twins, 292
Precursor flexagon, 10
Principal cycle, 44, 49
Principal main positions, 49, 68

Q
Quadruple, 71
Quadruple conjoin, 282
Quintuple, 60

R
Regular cycle flexagons, 45, 78, 119
Regular edge ring, 2, 20
Rectangle even edge flexagon, 233
Regular even edge rings, 20, 28
Regular even vertex ring, 29
Regular odd edge rings, 21, 28
Regular polygon, 201
Regular polygon edge flexagon, 201
Regular polygon point flexagons, 201
Rhombus even edge flexagon, 231, 235, 236
Rigid origami, 6
Ring even edge flexagon, 50
Ring symbol, 20, 24
Rogue triangles, 273
Rolled, 7
Rotation of leaves, 57
Rotational symmetry, 18

S
Scalene triangle, 203
Scalene triangle edge rings, 206
Scalene triangle even edge flexagons, 216, 217, 265, 276
Scalene triangle even edge rings, 206
Scalene triangle flexagons, 216
Scalene triangle odd edge rings, 206
Schläfli symbol, 33
Second order fundamental edge net, 34
Sector diagrams, 58, 68
Sectors, 43, 175
Sector symbol, 55
Semi double point flexagon, 166
Semi double polygon, 166
Seven and six flexagon, 286
Sextuple, 223
Silver even edge flexagons, 202, 207, 212, 265, 267, 269, 276
Silver flexagons, 207
Silver triangle even edge flexagon, 211
Silver triangles, 28
Simple band, 165
Simple flex, 82, 84, 322
Simplified map, 259
Single interleaf flex, 131
Sixer flexagon, 182
Sixfold pinch flex, 222
Skeletal flexagons, 78, 300
Skew main position links, 249, 250
Skew main positions, 15
Skew ring, 3, 18
Slant flexes, 264
Slant ring, 17
Slant main position links, 249
Slant main positions, 15, 251, 263
Slipagon, 248, 287, 289

Slip flex, 287
Slit diamond flexagon, 274
Slit square flexagon, 209, 265
Snap flex, 19
Solitary flexagons, 43, 46, 247, 308
Splitting, 247
Square compound edge flexagons, 104, 107, 111, 172
Square duplex edge flexagon, 311
Square duplex flapagon, 308
Square edge flexagon, 184
Square even edge flexagons, 1, 53, 54, 121–123, 149, 179, 180, 183, 186, 212, 250, 255, 256, 258
Square even skeletal flexagon, 81
Square-like, 104
Square-like compound edge flexagons, 109
Square odd edge flexagon, 72, 304
Square point flexagons, 84, 93, 100, 129, 131, 141–143, 161, 163, 279, 293, 296
Star edge ring, 204
Star ring even edge flexagons, 207, 214
Stella octangula, 11
Street hexaflexagon, 254
Stretch flexagons, 202, 203, 274, 276
Stretch polygon ring, 203
Subsidiary cycles, 49
Subsidiary main positions, 49

T
Templates, 1
Terminology, 5
Tetrahexaflexagon, 252
Threefold pinch flex, 52
Three sector skeletal flexagons, 299, 300
Thrice threefold flexagon, 313, 315, 318
Topological invariants, 17, 45, 78
Topological structure, 2
Torsion, 45, 88, 166, 247, 249, 277, 282, 287, 292, 302, 308
Transformation between flexagons, 51
Transformation of polygons, 201, 202
Trapezoidal flexagon, 313, 320
Triangle even edge flexagons, 51, 122, 149, 177–179, 265, 274
Triangle even skeletal flexagons, 79
Triangle odd edge flexagons, 69, 284, 286
Triangle point flexagons, 83, 92, 97, 137, 139, 141, 166, 278, 283, 292, 294, 295
Triangle position, 313
Triangle-threefold pinch flex, 223
Triangle twins, 292
Triangle unagon, 83
Triangular compound edge flexagons, 104, 107
Triflexagon, 301
Trihexaflexagon, 1, 51, 247, 249
Triple conjoin, 282
Triplex edge flexagon, 311, 313, 315, 318–319
Triplex flapagon, 308
Truncated flexagons, 44, 198, 203, 212
Truncated trihexaflexagon, 157
Truncation, 44
Tube flex, 259
Tubulating flexagon, 259
Tuckerman diagram, 55
Tuckerman traverse, 223
Twist flex, 108
Twofold pinch flex, 13, 53
Two point flexagons, 277
Type of flexagon, 6

U
Unagons, 82, 91, 97, 121, 142
Unagon symbol, 83
Uncut flexagons, 258, 267, 276, 278, 281

V
Variety of flexagon, 6
Vertex flexagon, 3
Vertex pair, 3
Vertex ring, 29
V-flex, 253

W
Wheel even edge flexagons, 207

Z
Zigzag flex, 181

MIX
Papier aus verantwortungsvollen Quellen
Paper from responsible sources
FSC® C105338

If you have any concerns about our products,
you can contact us on
ProductSafety@springernature.com

In case Publisher is established outside the EU,
the EU authorized representative is:
Springer Nature Customer Service Center GmbH
Europaplatz 3, 69115 Heidelberg, Germany

Printed by Libri Plureos GmbH
in Hamburg, Germany